D1568523

The Road to Galaxy Formation
(Second Edition)

William C. Keel

The Road to Galaxy Formation

Second Edition

Professor William Keel
University of Alabama
Tuscaloosa
Alabama
USA

SPRINGER–PRAXIS BOOKS IN ASTRONOMY AND PLANETARY SCIENCES
SUBJECT *ADVISORY EDITORS*: Philippe Blondel, C.Geol., F.G.S., Ph.D., M.Sc., Senior Scientist, Department of Physics, University of Bath, UK; John Mason, M.Sc., B.Sc., Ph.D.

ISBN 978-3-540-72534-3 Springer Berlin Heidelberg New York

Springer is part of Springer-Science + Business Media (springer.com)

Library of Congress Control Number: 2007928348

Printed in Germany

Cover design: Jim Wilkie
Project management: Originator Publishing Services Ltd, Gt Yarmouth, Norfolk, UK

Printed on acid-free paper

Contents

Preface

A century ago, astronomy concerned itself almost completely with describing the positions and dynamics of the objects that we could see. The standard textbooks of the late nineteenth century carried enormous detail about how to measure stellar positions and calculate the orbits of objects within the Solar System. Newtonian physics had made celestial mechanics the first of the exact sciences, providing tools with which to analyze the motions of planets, their satellites, and even binary stars with impressive accuracy. Theoretical considerations of the origins of these systems—cosmogony—were relegated to the fringe, despite the honored memory of such prominent speculators as Kant and Laplace. Oddly enough, only in the twentieth century did the Universe acquire a history in astronomers' eyes. Understanding that starlight has its origins in nuclear fusion showed that stars must have origins and fates, and ever more detailed views of bodies in the Solar System revealed fascinating differences which must betray histories. Even on the scale of galaxies themselves, putting together the observational pieces has made clear that these systems must change, and therefore at some point they came to be. The situation for astronomers has now reversed, so that most of us are concerned more deeply with how a certain observed situation might have come to be, rather than with simple descriptions of the nature of that situation.

Three of the major problems in contemporary astronomy are the formation of planets, stars, and galaxies. These are not only problems of history, but broadly the same *kind* of history: how diffuse material (gas, often with a mixture of solid grain particles) has been collected into much more compact and collapsed objects. In each case we must deal with the competition between gravity and other opposing forces, but the details play out quite differently in each case. Likewise, on these different scales, the details of the countervailing forces (such as internal thermal pressure, magnetic fields, or motions of subsystems) act in different ways.

Planets form from material which is already orbiting in a disk surrounding a star, and chemical forces ("stickiness") are crucial in early stages of their growth, along with the fundamental distinction between a growing planet's ability to gravitationally bind solid particles and individual atoms of the abundant light elements. In this case, the original material is cool compared with the orbital energies involved, so that its exact temperature does not affect the rate of growth. Planets will heat themselves internally over time, as they accrete smaller bodies at high speeds, and later may be large enough to trap heat from radioactive decay, regardless of how they began.

Stars form from clouds massive enough to make whole clusters. We know a great deal about this process, particularly through advances in infrared and millimeter-wavelength observation. The massive clouds of molecular gas can be traced, and we can see where massive stars ionize and then blow away the surrounding gas. The outflows and jets from less massive newborn stars are clear, as are many examples of disks rich in dust particles which seem fated to become planetary systems. Yet despite our ability to see snapshots of this process in exquisite detail, important pieces of the puzzle still elude us. Is there something fundamental about the efficiency with which a cloud forms stars? Is it necessary that a particular fraction of the gas be expelled rapidly in order for the core of a cloud to collapse into a star? Why are binary stars such a universal outcome? And what is the role of stellar magnetic fields?

Galaxy formation plays out against the cosmological fabric of expanding space-time, so that much of what happens is ultimately conditioned by the initial conditions of the whole Universe. The formation of galaxies is a race between gravity and the cooling of material, pulling denser lumps of gas together, and cosmic expansion, pulling it apart and slowing these two processes. In contrast to the formation of planets and stars, the birth of galaxies must have been dominated by the unfamiliar properties of dark matter, and their formation in turn hosted production of the next level of structure: stars. The initial round of star formation may have been quite different from what we see in the solar neighborhood, since there were no heavy elements to cool the gas as it collapsed. Galaxy formation may also be one end of a whole spectrum of structure formation, which gave rise to clusters, superclusters, and the whole intricate tapestry of the largest structures we know—a process which is visibly continuing to this day.

In attempting to observe all three of these processes, we encounter a well-known astronomical application of the laws of thermodynamics: the most interesting events happen in environments which make them as difficult as possible to see, and they happen so quickly compared with the vastness of cosmic time that we can seldom catch them in the act. In the formation of galaxies, the local Universe shows that much of the action happened long ago in the early Universe. Fortunately, the finite speed of light comes to our rescue—at least in principle—affording us a view, in Alan Dressler's words, of "if not our own past, somebody else's". In observing the distant Universe, we cannot help looking back through cosmic time.

Galaxy formation may mean different things, and happen at different epochs, depending on one's point of view. For a cosmological theorist testing calculations of the behavior of ordinary and dark matter, "galaxy formation" may happen when lumps of material become gravitationally bound and separate from cosmic expansion. For an observer seeking the earliest signatures from infant galaxies, the magic time may be when the first flash of star formation lights up the surrounding gas and begins the intricate cycling of gas and stars. The various clocks that measure galaxy history—dynamical relaxation, chemical enrichment, conversion of gas into stars—may run at different rates in different conditions, so that a galaxy may be young by some measures but well-evolved by others. From any of these viewpoints, we need to establish a trail of evidence from the earliest epochs to the Universe we see today. How did the remarkably uniform gas in the early Universe manage to clump and begin producing galaxies and quasars in less than a billion years? What were the key players among competing physical processes?

Observational limitations make this a challenging problem. We can see the results of galaxy formation, after 13 billion years or so of further development, in some detail, and find interesting patterns which may trace back to this primeval epoch. We can also see some pieces of the evolution of galaxies over the last several billion years, in decreasing detail as we look back. Most of the action in galaxy formation took place in a sort of cosmic "Dark Age" that we still cannot observe directly. Beyond that, we can see conditions in the pregalactic Universe in reasonable detail, as the cosmic microwave background. The task, then, is to narrow down possible events in the "Dark Ages" to accord with what we see before and after, incorporating such disparate elements as ages and chemistry of old stars in the Milky Way and its neighbors, global relationships between dynamics and luminosity for galaxies, and the amplitudes of various scales of fluctuations in the cosmic microwave background. This book sets out the boundary conditions for a picture of galaxy formation, explores some of the approaches to determining what actually happened, and looks ahead to advances we may expect from new instruments and technologies. Recent developments have led to substantial changes in how we view galaxy formation and evolution, especially in the importance of feedback from both star formation and accretion around massive black holes as regulators of galaxy growth.

There is no reason to think that galaxies are fully formed and mature today. It is clear that massive galaxies have grown piecemeal, in smaller and larger bits, over time. We see mergers and acquisitions occurring in our own neighborhood. It may be more accurate to speak of galaxy building than formation as we try to understand their present properties. However, it is equally clear that many of the crucial events shaping galaxies happened in a rather brief epoch early in cosmic history, so that there is still a special period worthy of being called "galaxy formation".

Advances in other areas of technology have led to ready access to information, and most of the technical papers referenced in this book can be retrieved online from the NASA Astrophysics Data Service at:

http://adsabs.harvard.edu/abstract_service.html

Using an older technology—print—a good introduction to the discovery of galaxies is Berendzen, Hart, and Seeley's *Man Discovers the Galaxies*, which illuminates the pathways to our recognition that we live in a vast Universe full of other galaxies.

Acknowledgments

Even a casual scan of the bibliographies and figure captions in this book will make it clear the exploration of the Universe is an endeavor for a community, and not just for an individual. I have benefited from the cooperation of the many colleagues listed as providing data or images or this presentation. Bruce Partridge and the late Don Osterbrock commented on parts of the first edition in manuscript, in each case improving their accuracy and presentation. Deidre Hunter and Mary Guerreri were helpful in tracking down bibliographical items beyond the scope of our library, and Ray White III pointed me to additional helpful resources. Cheryl Schmidt of the Space Telescope Science Institute helped to verify some subtle credit issues for Hubble data and images. Fred Nixon pointed out a mathematical typesetting error in the first edition. My own research into the history of galaxies has been supported by the National Science Foundation, particularly through operation of the National Optical Astronomy Observatories, and especially by the National Aeronautics and Space Administration. Much of this work has been carried out with several long-time collaborators, among whom Rogier Windhorst and Frazer Owen have been especially important in shaping my approach to the field. This book owes its existence (now in two editions) to the polite yet persistent initiative of Clive Horwood, and the simultaneously pleasant and professional work of Neil Shuttlewood.

Figures

* See also color section.

TABLES

Abbreviations and acronyms

2dF	2-degree Field spectroscopic facility (Anglo-Australian Telescope)
2MASS	2-Micron All Sky Survey
3C	Third Cambridge Catalog of radio sources
ACS	Advanced Camera for Surveys
ADM	After Dark Matter
AGN	Active Galactic Nuclei
ALMA	Atacama Large Millimeter Array
AO	Adaptive Optics
ASCA	Advanced Satellite for Cosmology and Astrophysics
BAT	Burst Alert Telescope (onboard Swift)
BATSE	Burst and Transient Source Experiment
BDM	Before Dark Matter
BOOMERANG	Balloon Observations Of Millimetric Extragalactic Radiation ANd Geophysics (satellite)
c.g.s.	Centimeter-gram-second
CAS	Concentration, Asymmetry, and Clumpiness
CASTLES	CfA–Arizona Space Telescope LEns Survey
CBI	Cosmic Background Imager
CCD	Charge-Coupled Device
CDM	Cold Dark Matter
CfA	Harvard–Smithsonian Center for Astrophysics
CGRO	Compton Gamma-Ray Observatory
CMB	Cosmic Microwave Background
CMBR	Cosmic Microwave Background Radiation
CNO	Carbon–Nitrogen–Oxygen cycle
COBE	COsmic Background Explorer
COSMOS	COSMOlogical Evolution Survey

CVZ	Continuous-Viewing Zone
DEEP 2	Deep Extragalactic Evolutionary Probe 2 (project at Keck Observatory)
DEIMOS	Deep Imaging Multi-Object Spectrograph (Keck Observatory)
DMR	Differential Microwave Radiometer
DRAO	Dominion Radio Astronomy Observatory
EELT	European Extremely Large Telescope
ELS	Olin Eggen, Donald Lynden-Bell, and Allan Sandage
ESA	European Space Agency
ESO	European Southern Observatory
FIRAS	Far-InfraRed Absolute Spectrophotometer
FIRST	Far InfraRed Space ObservaTory
FOS	Faint-Object Spectrograph
FUSE	Far-Ultraviolet Spectroscopic Explorer
GALEX	GALaxy Evolution EXplorer
GDDS	Gemini Deep Deep Survey
GEMS	Galaxy Evolution and Morphology from Spectral Energy Distributions
GLAST	Gamma-Ray Large Area Space Telescope
GMOS	Gemini Multi-Object Spectrograph
GMRT	Giant Metre-Wavelength Radio Telescope
GMT	Giant Magellan Telescope
GOODS	Great Observatories Origins Deep Survey
GRB	Gamma-Ray Burst
HD	Hydrogen–deuterium molecule
HDF	Hubble Deep Field
HEASARC	High-Energy Astrophysics Science Archive Research Center
HR	Hertzsprung–Russell
HST	Hubble Space Telescope
HUDF	Hubble Ultra-Deep Field
HUT	Hopkins Ultraviolet Telescope
ICM	IntraCluster Medium
IGM	InterGalactic Medium
IMF	Initial Mass Function
IR	InfraRed
IRAS	InfraRed Astronomical Satellite
ISM	Interstellar medium
ISO	Infrared Space Observatory
IUE	International Ultraviolet Explorer
JWST	James Webb Space Telescope
LAR	Large Adaptive Reflector
LSB	Low Surface Brightness
MAP	Microwave Anisotropy Probe

MAST	Multimission Archive at Space Telescope
MCAO	Multi-Conjugate Adaptive Optics
MERLIN	Multi-Element Radio-Linked Interferometer Network
M/L	Mass-to-Light ratio
NAO	National Astronomical Observatory
NASA	National Aeronautics and Space Administration
NED	NASA Extragalactic Database
NGC	New General Catalog of Nebulae and Clusters of Stars
NICMOS	Near-Infrared Camera and Multi-Object Spectrometer
NOAO	National Optical Astronomy Observatory
NRAO	National Radio Astronomy Observatory
NRO	Nobeyama Radio Observatory
NSF	National Science Foundation
OVV	Optically Violently Variable
PI	Principal Investigator
QSO	QuasiStellar Object
QSRS	QuasiStellar Radio Source (quasar)
r.m.s.	Root-mean-square
ROSAT	RÖntgen SATellite
ROTSE	Robotic Optical Transient Search Experiment
S	Ordinary spiral galaxy
SB	Barred spiral galaxy
SCUBA	Submillimetre Common-User Bolometer Array (at James Clerk Maxwell Telescope)
SDSS	Sloan Digital Sky Survey4
SETI	Search for ExtraTerrestrial Intelligence
SFR	Star Formation Rate
SKA	Square Kilometer Array
SN	Supernova
SSC	Super Star Cluster
SST	Spitzer Space Telescope
STIS	Space Telescope Image Spectrograph
STScI	Space Telescope Science Institute
TMT	Thirty-Meter Telescope
UIT	Ultraviolet Imaging Telescope190
UV	UltraViolet
UVOT	UV/Optical Telescope
VLA	Very Large Array
VLT	Very Large Telescope
VMO	Very Massive Object
VMS	Very Massive Star
WFPC2	Wide-Field Planetary Camera 2
WMAP	Wilkinson Microwave Anisotropy Probe
WWW	World Wide Web

XEUS	X-ray Evolving Universe Spectroscopy (mission)
XMM	X-ray Multimirror Mission (XMM-Newton)
XRT	X-Ray Telescope
XUV	Extended ultraviolet

1

A cosmological cartoon

While many observational programs have ended up (unavoidably) mixing the results of galaxy evolution with effects of cosmology, I will try to separate them here as far as practical. The basic problem in doing so is that we may trace changes in the luminosity or size of galaxies, which can be mimicked by the properties of space itself through departures from the familiar inverse-square law for light propagation. Unless we have other means of tracing the behavior of size, distance, and light in an expanding Universe, some of the effects of galaxy evolution cannot be separated from those produced by the properties of space in the Universe—its cosmology. There are already several excellent treatments of cosmology itself, at all levels, so there is not much to add if we are to focus on galaxy formation. I will sketch here a "standard" cosmology and review its empirical and theoretical underpinnings. These results will be largely taken as a "given" in succeeding chapters, for calculations of sizes, luminosities, and (perhaps most important) the mapping between redshift and cosmic time.

1.1 A STANDARD COSMOLOGY

Cosmology is the effort to comprehend the broad features of the structure and history of the Universe, with some argument among various people about how broad "broad" ought to be. As a properly scientific endeavor (although there are some who will argue that it has yet to achieve this exalted status), cosmology traces back to the time of Einstein and Hubble. The foundations of modern cosmology can be traced plausibly to several key discoveries, theoretical as well as observational. These key ingredients to the cosmological mix include the following.

1.1.1 The cosmological principle

The cosmological principle is usually stated formally as "Viewed on a sufficiently large scale, the properties of the Universe are the same for all observers." This amounts to the strongly philosophical statement that the part of the Universe which we can see is a fair sample, and that the same physical laws apply throughout. In essence, this says that the Universe is in a sense knowable and is playing fair with scientists. In some ways we can check on this principle. Comparison of the spectra of objects far apart on the sky, and at great enough distances, allows us to see whether the laws of physics which enter into formation of lines in galaxy spectra are the same. These include not just the determining constants for emission-line wavelengths, which are the speed of light c, electron mass m_e and charge e, Planck's constant h, and the fine-structure constant α, but indirectly the strength of gravity (traditionally expressed as the Newtonian gravitational constant G). This enters once we can recognize absorption features from stars at high redshift, since the existence of stars requires a particular range of relative strengths of nuclear force to gravity. Outside this range, stars will be unstable or have vastly different lifetimes than we infer from galaxy luminosities and chemistry. By an analogous argument, the presence of fossils of ferns and scale trees in coal mines near my home, whose leaves are similar to those I can see on ferns in my yard today, shows that the Sun has been shining at about the same brightness for about 300 million years. Returning to the cosmological context, wavelength ratios of different ionic species from QSO absorption-line clouds place a limit on any variation in the fine-structure constant, below a few parts in 10^6 since the time corresponding to redshift $z = 3$. Since α incorporates c, e, and h, possible changes in any one of these are likewise limited.

Intellectually, the cosmological principle harks back to Copernicus. One impact of the Copernican revolution (announced in a volume whose title gave the name "revolution" to an overthrow of the existing order) was to put the Earth among the planets, and implicitly make its location more average. This has been called the assumption of mediocrity—the idea that our physical location is probably as average as it can be and still sustain the conditions for our being here to wonder about it. This assumption has become more plausible (as well as statistically likely) as we first learned that we're not at the center of the solar system, then not near the center of our galaxy, and that our galaxy is not central to any particular structure. This all means not that we're on a typical planet, based on the local examples, but we might be circling a pretty typical type G star, we do seem to be in a typical spiral galaxy, and we have every reason to think that our wider environs are very similar to the more distant reaches of space. The exceptions to mediocrity are instances of the anthropic principle—our location and the Universe must be such as to support thinking life forms, in order for the questions to ever be asked at all. There are some who, reasoning from quantum theory, suggest a strong version of this principle, in which the eventual presence of intelligence is a prerequisite for the Universe to come into existence. This remains, naturally, very controversial; for now it suffices to keep in mind that some of our views of the Universe will be conditioned by what is necessary for us to be here wondering at it.

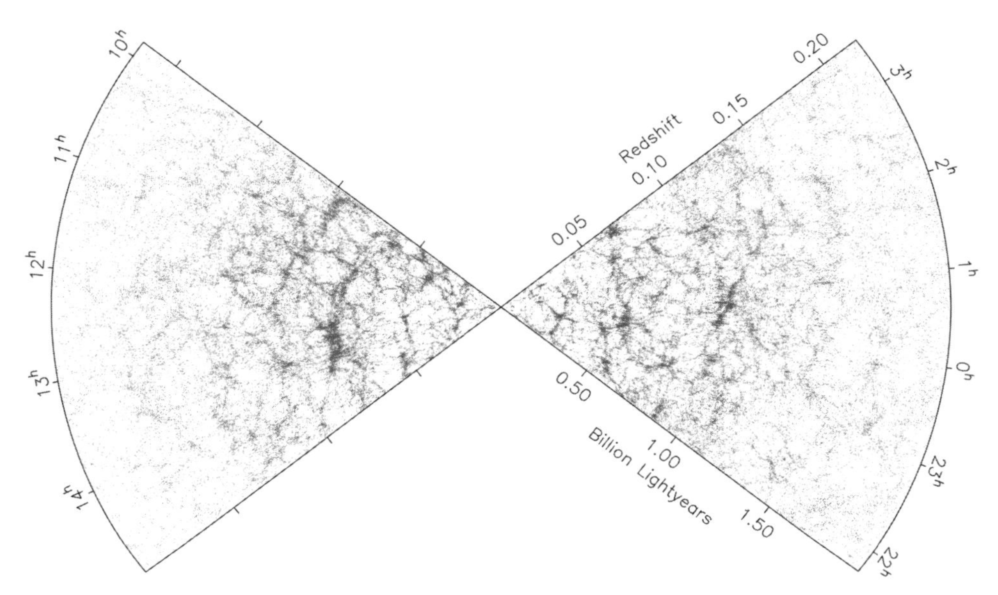

Figure 1.1. The large-scale distribution of galaxies in a slice through space, from the 2dF Galaxy Redshift Survey. This survey, including over 150,000 galaxies, goes deep enough to show an upper limit to the size of galaxy structures. Superclusters and voids occur up to sizes of a few hundred Mpc, but evidently not much larger. This may finally mark the scale on which the cosmological principle can be applied to large-scale structure in the Universe. The thinning out of galaxies near the circular boundaries, farthest from us, represents the sparser sampling of such faint objects in the survey. (Reproduced courtesy of the 2dF Galaxy Redshift Survey team.)

The cosmological principle is supported by the near-uniformity of the cosmic microwave background and by the universality of some physical laws, which can be tested to an interesting extent by observations of distant galaxies and quasars. However, it is possible to imagine situations in which the posited large-scale uniformity is never reached. There have been discussions of fractal structures for the Universe which behave in this way, having no preferred scale and comparable correlation power on all scales including arbitrarily large ones. At this point, though, it seems that surveys of galaxies to distances of billions of light-years have reached the "large enough" needed to smooth out local irregularities and approach cosmic uniformity. The 2dF survey at the Anglo-Australian telescope showed that the trend of larger and larger galaxy groupings as we map larger volumes finally breaks down over sizes of order $\delta z = 0.1$ (about 0.5 Gpc), as shown in Figure 1.1. Expressed somewhat more pedantically, the power in the galaxy–galaxy correlation function is declining inversely with spatial wavenumber k at these scales. This behavior also appears, with better statistics but a somewhat different kind of galaxy sampling, in the million-galaxy mapping by the Sloan Digital Sky Survey (SDSS). On these scales, finally, we have real evidence that the cosmological principle can be applied. The amount of structure in the distribution of galaxies means that in any survey for galaxy counts, for example, there is a "cosmic variance" in excess of Poisson errors which can be reduced for comparison to models only by observations of other samples sufficiently far from the first to avoid significant spatial correlations. This enters into knowing, for example, how to normalize the luminosity function of galaxies, since there is only one locality around us for a zero point, and in doing statistics on the structure in microwave background radiation, since the Universe gives us only a single example to analyze. For some of these global quantities, the best we can do is estimate error bars for models of certain kinds, by creating a large number of numerical realizations of the model and asking what the variation of some observable quantity is if we retrieve it as seen from various locations within the model or for various random seed values used to initialize the calculation.

1.1.2 The Hubble expansion

After demonstrating that a certain class of nebulous objects are indeed independent galaxies, Edwin Hubble conducted a broad reconnaissance of their properties. Among its results was a conjecture, based on a limited set of data, that the redshift in a galaxy's spectrum correlated linearly with its distance, as estimated using a galaxy's apparent size or the brightness of constituent stars and clusters.

Measuring the velocity of a distant object (at least its velocity along the line of sight, remarkable enough over cosmic distances) using shifts in the wavelengths of radiation had been a commonplace for stars for years. The technique employs the well-known Doppler shift, which applies to light and other kinds of radiation as well as to sound waves (sometimes being known as the Doppler–Fizeau shift when speaking specifically of radiation). If we can be sure what spectral features we are seeing in some galaxy spectrum—say, because there are several in an exact and characteristic pattern—we can derive its velocity of approach or recession with respect to our vantage point (as illustrated in Figure 1.2). Because galaxies are mostly space, rather than stars, their light is dim, with surface brightness that may fall below that of the natural airglow in the Earth's atmosphere, so the required observations were very difficult in Hubble's day. In reporting the first redshift measures for the Andromeda galaxy and the bright nuclei of a handful of others, V.M. Slipher listed exposure times on individual photographic plates from 7 to 35 hours, spread over multiple nights, on the Lowell Observatory 60-cm refractor. Even with much larger instruments and more sensitive photographic emulsions, progress was slow until sensitive solid-state detectors and multi-object systems came into use, which could measure hundreds of objects per night. By now, there are well over one million measured galaxy redshifts, most of which were delivered by the Sloan Digital Sky Survey in its initial survey of one-quarter of the celestial sphere (or, in steradians, what the SDSS project has called "π on the sky"). For historical reasons, small redshifts are often quoted in velocity units, as the radial velocity needed to give the observed shift strictly as from the Doppler mechanism due to relative motions. For large values, the redshift is given in dimensionless form as $z = \Delta\lambda/\lambda_0$ (and an

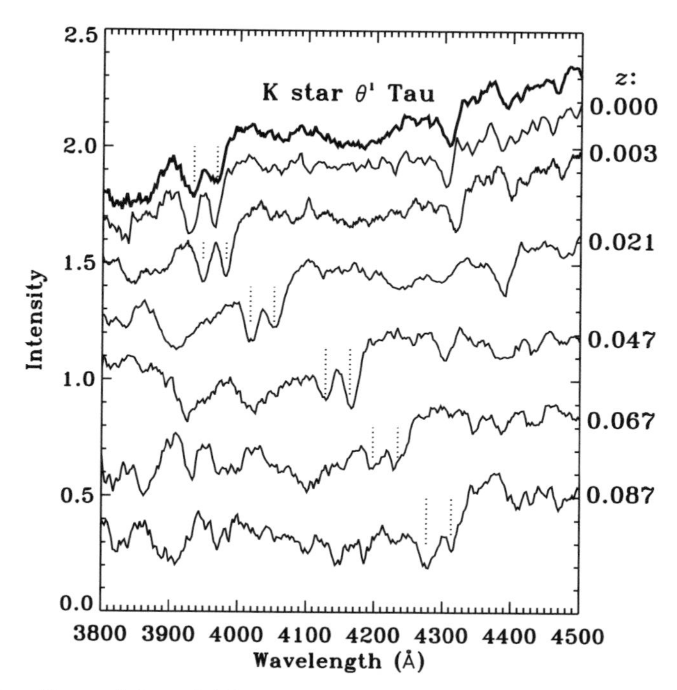

Figure 1.2. The effects of the redshift on spectra of similar elliptical galaxies. A nearby cool giant star is shown for comparison, since these are important in the light of ellipticals and have very similar spectra in this wavelength range. Shown immediately below the star is M32, a companion to the Andromeda galaxy, followed by NGC 4472, the optically brightest galaxy in the Virgo cluster. The other galaxies are radio sources in clusters, with each galaxy labeled by its redshift. The most prominent individual spectral lines are the two H and K lines of calcium, marked with the dashed lines. (These spectra are from observations at Lick Observatory and Kitt Peak National Observatory.)

equivalent formulation for frequency). Here, λ_0 is the laboratory (or emitted-frame) wavelength of each spectral feature, and $\Delta\lambda$ is the difference between observed and emitted wavelengths. Along with a shifting to longer wavelengths, the spectrum is "stretched" by the same ratio $(1+z)$.

Hubble was both bold and lucky in his extrapolation—his data were sparse enough that the correlation he saw could have been due to chance. However, the relation holds quite well with the vast body of subsequent data, so well that some astronomers now try to measure small departures from it as tracers of motions produced by local mass concentrations. The form of the relation he proposed, which has since become known as the Hubble law,[1] is simplicity itself—a galaxy's redshift is

[1] There are two other relationships also known as Hubble laws in different contexts. One of these deals with the distribution of brightness with radius in elliptical galaxies, as described in Chapter 2. The other describes the distribution of reflected starlight in dusty galactic nebulae.

directly proportional to its distance from us. The simplest explanation for such a pattern, and the one which best satisfies such other constraints as the variation in apparent brightness with distance, is that we live in an expanding Universe. A linear expansion of this kind has several interesting (and not necessarily intuitively obvious) properties. We need not be at the center of the expansion to see this—for a linear expansion, any galaxy would observe the same pattern. In the idealized case where no other motions disturb the galaxies, the three-dimensional pattern of galaxies would remain the same over time, changing only in size. This gives rise to the useful concept of *comoving coordinates*, measured in units which share this expansion. We use comoving coordinates to compare the space density of galaxies at various redshifts, so that cosmic expansion is factored out and we can seek genuine changes in the galaxy population. Otherwise, the density of objects will show an increase between our neighborhood and redshift z, by a factor $(1+z)^3$ simply because of cosmic expansion, even if no physical change takes place in the population we observe.

The linear nature of the Hubble pattern, at least as we see it in our neighborhood, suggests a trivial way of estimating galaxy distances. If we can pin down the value of the proportionality constant, namely the expansion rate per unit distance (the *Hubble constant*), we could read off a galaxy's distance from its redshift to within the accuracy allowed by actual scatter in the motion of galaxies. This requires knowing the Hubble constant (sometimes called Hubble ratio or Hubble parameter, to express its possible variations with space or time). Major projects have been carried out over the last five decades to constrain the Hubble constant (H_0, with the subscript 0 denoting here and now), as described in Chapter 2, and with some results shown in Figure 1.3. One prime goal of the Hubble Space Telescope was to determine its value to within 10%, a feat which most (but emphatically not all) workers in the field are satisfied has in fact been achieved. Because it is so common to express redshifts as effective velocities in km/s, and distances in megaparsecs (Mpc), the Hubble constant is generally tabulated in the mixed units km s^{-1} Mpc^{-1}, although from dimensional analysis it is clear that its physical unit is inverse time.

There are other routes to the Hubble constant, independent of the traditional distance ladder involving Cepheid variable stars. Some make use of geometric techniques, such as the time delays between images in gravitational-lensing systems (Chapter 9). Others model the fluctuations in the cosmic microwave background, noting that certain acoustic peaks in the angular correlation shift with the amount of intervening expansion (Chapter 7). These disparate techniques agree well enough to reinforce our confidence that there is not a serious problem with our distance estimates.

The Hubble constant also gives us a unit of time, the Hubble time, defined as $T_H = 1/H_0$. The derivation is simple—this is the length of time objects would have taken, starting all at the same place, to reach their current separations at a constant velocity equal to their cosmological redshift velocities. The Hubble time is closely 20 billion years for $H_0 = 50$, 10 billion years for $H_0 = 100$, and the likely value near 75 gives 13 billion years. This is the age of the expansion if no acceleration or deceleration has occurred, and is often taken as a characteristic age for the Universe.

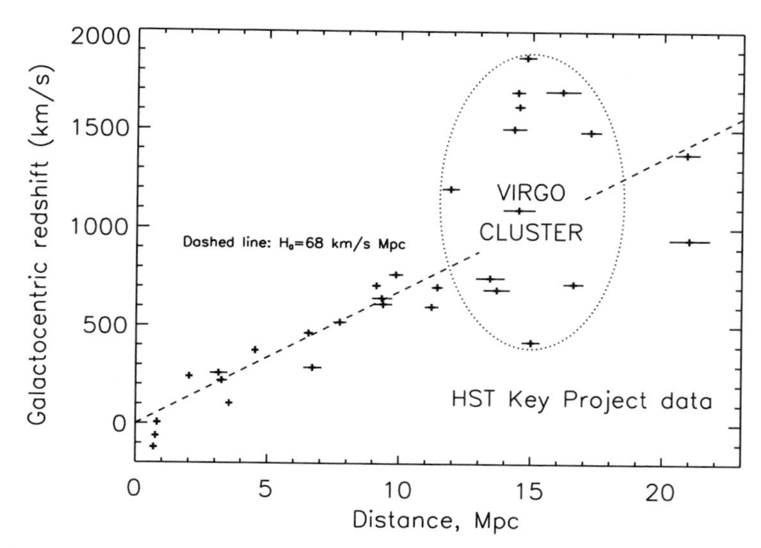

Figure 1.3. The redshift–distance relation, or Hubble law, from the recent data compilation by members of the Hubble Space Telescope Key Project team. All galaxy distances here are derived from Cepheid variables. The Virgo Cluster, where Doppler shifts from internal motions can be comparable with the Hubble redshift, spans roughly the distance and redshift range shown by the ellipse. However, the galaxy between the words at the cluster's center (NGC 7331) is not in Virgo, but is in a much sparser environment seen nearly opposite Virgo, and hence reflects the average cosmic expansion more closely. (These data are from Freedman, W. *et al.*, *Astrophysical Journal*, **553**, 47, 2001.)

The actual age of a cosmological model with critical density and zero cosmological constant is $2/3T_H$; a positive cosmological constant would provide a greater age, the situation favored by recent measurements.

1.2 THE EINSTEIN CONNECTION

The intellectual foundations for modern cosmology are found in Einstein's general theory of relativity. The twin connections between matter density and the geometry of spacetime, and between the geometry and motion of objects, lead to predictions about the behavior of spacetime itself—known as the world model or the world map, depending on the frame of reference chosen. In making this connection, one employs *Weyl's postulate*. Simply put, this is the notion that galaxies (or clusters of galaxies, or some other tracer of interest) have no peculiar motions of their own with respect to the expanding cosmological framework (which is to say that the comoving coordinates of any one of these remain constant). The postulate says that the paths of galaxies through the four-dimensional spacetime of relativity do not intersect and that these paths are orthogonal to the spatial coordinates (which is to say, the galaxies don't move in space).

Box 1.1 The Robertson–Walker metric and the expanding Universe

To see the connection between the relativistic description of spacetime and specific predictions for how distances will behave on cosmic scales, we can begin with the notion of a line element. In ordinary three-dimensional Euclidean geometry, the separation s between two points will be given by

$$s^2 = \Delta x^2 + \Delta y^2 + \Delta z^2$$

where Δ indicates the difference in each of the ordinary Cartesian coordinates x, y, z. While the coordinates of the points may change as we rotate or translate the coordinate systems, s is a constant—an invariant, in the terminology of relativity. Special relativity shows that we should generalize this notion to keep it correct for coordinate systems which are moving with respect to one another, now including a role for the time measured to elapse between two events, so the interval between them spans not just space and time separately but the combination, spacetime:

$$s^2 = \Delta x^2 + \Delta y^2 + \Delta z^2 - c^2 \Delta t^2.$$

Again, s is an invariant, now to any relative motion as well as rotation of the coordinate systems.

The standard cosmological solution comes from a *metric*, a description of the geometry of spacetime. Einstein's general relativity states that the local curvature of spacetime is proportional to the combination of energy and mass density (more properly, momentum density) known as the energy–momentum tensor. This amounts to a more complicated expression for s, in which we start with its differential ds at each point and get the whole path length (e.g., followed by a light ray) upon integration. The coupling between coordinate differences and path length is now more complicated because of this spacetime curvature, meaning that the metric has a separate component linking each of the possible coordinate pairs. If we combine this with the condition that the Universe be homogeneous and isotropic (the cosmological principle), and follow tradition by writing the result in spherical coordinates, the results is the *Robertson–Walker metric*, an expression relating elements ds of length in an expanding Universe:

$$ds^2 = c^2 dt^2 - R^2(t)\left(\frac{dr^2}{1 - kr^2} + r^2\, d\theta^2 + r^2 \sin^2\theta\, d\phi^2\right).$$

Here, R is a scale factor which varies with cosmic time t but not location, a property which makes t, in the words of John Peacock (1999), "suspiciously like a universal time". The angular coordinates θ, ϕ simply represent the direction between the observer and some other object in a polar coordinate frame. The most interesting part of this equation is the radial behavior, obtained by integrating ds along some line of sight. Doing so, for the spatial separation, one obtains

$$R = \frac{1}{c^2}\left(\frac{\ddot{R}}{R} + \frac{2\dot{R} + 2kc^2}{R^2}\right).$$

In both these last expressions, a constant of integration k appears whose value must be determined from the conditions of the problem. This constant has a controlling influence on the behavior of the cosmic expansion that is described by the Robertson–Walker metric.

While it has been the benchmark for descriptions of relativistic cosmology for much of the last century, there is now strong observational evidence that we live in a Universe whose large-scale properties are not well described by the Robertson–Walker metric. For no physically obvious reason, the cosmic expansion appears to be accelerating. Incorporating this behavior requires additional complexity in the expression for the metric, and the mathematical form of this added feature remains highly uncertain.

The geometry described by this relativistic cosmological picture depends on whether the value of k in the Robertson–Walker metric (Box 1.1) is negative, zero, or positive. Since it appears in the denominator for distances as a function of time, its possible values have very different implications. One common convention is to assign values of $k = 0, \pm 1$ only; another, also widely used, is to assign it a value corresponding to the "radius of curvature" of spacetime R such that $k = 1/R^2$. The difference between these forms is taken up in the normalization of the function $R(t)$. This function takes the role of a scale factor which changes with time, giving an expanding Universe. For $k = 0$, any slice of the four-dimensional Universe across time is Euclidean, known as flat spacetime, where the ordinary rules of three-dimensional geometry apply. This is the *Einstein–de Sitter Universe*, for which $R \propto t^{2/3}$ and $T_H = 2/3H$ as noted above. For positive k, such a slice makes a closed volume in four dimensions, so we speak of a closed Universe. Negative k implies a geometry which has its two-dimensional analog as a surface of hyperbolic cross-section and infinite extent, so we speak of an open Universe. If only gravity is operating during the cosmic expansion, the deceleration of the expansion depends solely on the mean mass density ρ. Such a deceleration is quantified through the *deceleration parameter* q, often seen as its local value q_0, which is defined in terms of time derivatives of the scale factor

$$q = \frac{\ddot{R}R}{(\dot{R})^2}.$$

A value of q can be defined whatever influences act on the expansion, but in the particular case of a Universe in which only gravity acts, the mean mass density is directly connected to q through its ratio to the *critical density* $\rho_0 = 3H^2/8\pi G$ where H is the value of the Hubble constant at the time under consideration. Furthermore, $q = \rho/2\rho_0$, making the connection between cosmic mass density and the history of the Universe. In general, the mass density of some particular constituent of the Universe (or the total) is often expressed as the fraction of critical density it contributes: $\Omega = \rho/\rho_0$. For more exotic cosmologies than this sinple case, the relation between q_0 and mass density may be more complicated than the factor of 2 conversion seen above, so that using Ω is more generally applicable.

Looking forward, a closed Universe will someday reach a maximum value of R and collapse thereafter, while a critical-density Universe with $q_0 = 1/2$ will continue to expand but at a rate which asymptotically approaches zero, and an open Universe will expand at a rate asymptotically approaching some value of H > 0. Our particular interest here, though, is in looking backwards. A value of q_0 implies a history of cosmic expansion. The greater q_0, and therefore the cosmic mass density, the more rapidly gravity has been able to decelerate the expansion. This changes the age of the Universe we calculate for a particular Hubble constant today; the larger q_0 is, the younger the Universe must be to match the present conditions.

In an idealized relativistic Universe, the configuration of particles without any peculiar motion of their own will keep the same shape, differing in linear size only by a scale factor $R(t)$ depending on cosmic time. This scale factor is immediately connected to the observable redshift from an object seen at a time such that the scale factor was R_1 by $R_0/R_1 = (1 + z)$ where R_0 is the current scale factor. Clearly, if we are observing multiple epochs and the effects of radiation from one at another, factors of this form will multiply; for example, the scale factor at $z = 1.0$ relative to that at $z = 0.5$ is $(1 + 0.5)/(1 + 1.0) = 0.75$. If we observe a separation x between objects at redshift z, that separation in comoving coordinates today maps to $x(1 + z)$.

It is important to keep in mind that, in relativistic cosmology, the picture is not one of the galaxies flying apart through space, but of the space between them expanding so that the galaxies go along for the ride, plankton floating in the cosmic ocean. The redshift of radiation in this view is not exactly a Doppler shift from motion, but produced as the space through which the radiation passes is stretched over cosmic time. The distinction matters not only conceptually, but because radiation from a distant source is fainter in an expanding Universe than in a static one, since the surface of the sphere which the radiation covers after a given time has expanded while the radiation is in transit.

A crucial concept in observational cosmology, and one that shows us how galaxy evolution and formation might become more than purely theoretical pursuits, is that of lookback time—the time which has elapsed while radiation has been en route to us from some distant object. This can also be described as light-travel time. In the nearby Universe, it will be simply the distance divided by the speed of light—8 minutes from the Sun, 32 minutes from Jupiter at opposition, 4.3 years from the α Centauri system, and so on. For very large distances, we must define more closely just what we mean by distance, since in an expanding Universe, different ways of measuring distance may give quite different results.

In interpreting observations most directly, it is common to employ the luminosity distance D_L and angular-diameter distance D_A. These are the distance measures, for a given cosmological model, which make distant sources satisfy the local and familiar Euclidean relations

$$\text{Flux} = L/4\pi D_L^2$$

and, for angular diameter θ,

$$\theta(\text{radians}) = \text{linear size}/D_A$$

An additional distance measure comes into play for lookback time, the proper distance D_P—the distance integrated along the actual path of a light ray. It has been usual practice to calculate this as a function of the observable redshift z for some assumed expansion history of the Universe. These distance measures are all related via

$$D_L = D_P(1+z)$$

and

$$D_A = D_L/(1+z)^2.$$

For the traditional Robertson–Walker derivation of an expanding Universe, closed-form equations for these and related quantities were derived by Mattig (1958) in a demonstration of considerable algebraic virtuosity. In this "standard" model, Mattig showed that the proper distance, integrated along a light ray to a distance corresponding to redshift z, is given by

$$D_P = \frac{c[q_0 z + (q_0 - 1)[(1 + 2q_0 z)^{0.5} - 1]]}{q_0^2 \mathrm{H}_0(1+z)}.$$

Mattig's expressions simplify in various ways for the special values $q_0 = 0, 1/2, 1$, of which the value $q_0 = 1/2$ seems to be an adequate description of the geometry (though not mass density or deceleration history) of the Universe. Several subtleties come into play when applying these equations to observable quantities, since we frequently measure received flux in energy per unit wavelength. This entails additional factors to make the conversion, since we are now working in a system in which the unit of wavelength is smaller by $(1+z)$ than it was in the reference frame of the emitting object, and each photon has an energy smaller by the same factor when we receive it, strictly because of the redshift.

In an expanding Universe, the redshift–distance relations can have remarkable behavior. The angular-diameter distance reaches a maximum, at a redshift $z = 1.25$ for $q_0 = 0.5$, beyond which more distant objects actually appear larger although much dimmer. A physical interpretation of this is that the objects were so much closer at the time the radiation was emitted that this factor overcomes those of proper distance and cosmic expansion. The proper distance, and hence the lookback time, is a nonlinear function of z. This makes it important that we know something about the cosmology in order to compare galaxies at various redshifts, and hence cosmic times, before making inferences about how they change with time.

1.3 INFLATION IN COSMOLOGY

On the largest scales, the Universe presents several properties that are not explained by the standard elements of a relativistic cosmology (though they do not violate its predictions). One is the *flatness problem*. The overall geometry of spacetime is as close to flat as we can measure—that is, $\Omega = 1$ within our slowly decreasing errors. We know of no physical reason that the expansion and density of the Universe had to be related in this way; in fact, since the critical density of the Universe depends on time through a variation in the expansion rate H, via $\rho_0 = 3\mathrm{H}^2/8\pi G$, we expect a divergence from $\rho = \rho_0$ with time for any value other than ρ_0, so it becomes surprising that the Universe has remained so close to this critical condition (except for the fact that this may be a requirement for our existence to remark on the fact). Any process that would "automatically" yield a flat Universe has to be taken seriously.

A second cosmological puzzle has been *causal connection*. Even under special relativity alone, no stimulus can convey information (such as a change in physical state) faster than the speed of light. This becomes a puzzle when we note that we can observe numerous objects at high redshift which could never have been causally connected to one another in our "standard" cosmology, because the proper distance between them is greater in light-travel time than the age of the Universe at the epoch we observe them. Yet these objects show recognizable spectral features, and these spectral features indicate that the same physical laws and physical constants were operating in each of them. The wavelength ratios of various spectral lines have different dependences on such quantities as the speed of light, Planck's constant, and the electron charge, largely through the fine-structure constant $\alpha = 2\pi e^2/hc$, and seeing the wavelength ratios identical in various objects to high precision tells us that all these constants were already defined at the places and epochs we observe. Any such variations are constrained to be at the level of parts in 10^6 since $z = 3$. In a conventional cosmology, this fact must be inserted "by hand", having no built-in explanation.

Both these puzzles are "solved" (which is to say, they result from a smaller number of basic processes) by the notion of cosmological inflation. First developed by Alan Guth and Andrei Linde, this idea follows from postulates in elementary-particle physics which suggest that high-density matter enters a state known as the false vacuum, a somewhat peculiar name since it shares only a single property with what we normally think of as the vacuum. This single property is that the pressure exerted by a given volume of false vacuum does not decrease with expansion of the "material" in that volume, so matter in this state will enter a state of exponential expansion. When this expansion stops, the original geometry of spacetime will have been so dramatically inflated that it is left closely flat (the frequent analogy is to the surface curvature of an inflating balloon). Furthermore, this expansion is not limited to the speed of light; no information is being transferred, since only the configuration of particles expands without other relative motion, so that special relativity is satisfied. This offers a deeper explanation of why disparate regions of the Universe have identical physics—they were once causally connected for a brief instant, then whipped apart by inflation and are only now entering each other's connected regions once again. The attraction of this double explanatory power has helped drive many cosmologists to the conviction that $\Omega = 1$ to very high precision, beyond the level actually warranted by the accuracy of current measurements.

1.4 THE UNIVERSE ACCELERATES

A remarkable refinement of the paradigm in cosmology came with the adoption of evidence for a nonzero *cosmological constant*. The cosmological constant is a notion dating back to Einstein's original work on a cosmic picture consistent with general relativity, in which he found that maintaining a static Universe required introducing a new term Λ in his field equations. This term represents a phenomenon whose effect is cosmic repulsion. The idea was dropped soon thereafter as Hubble made the case for an expanding Universe, leading to Einstein's oft-quoted remark about this being the biggest blunder of his career.[2] Still, the idea has been resurrected from time to time as a way to reconcile apparent problems in cosmological data. This time, the evidence comes from observations of high-redshift supernovae. Supernova explosions are so luminous that we can now detect them, with targeted searches, to redshifts $z \approx 2$. A subset of these outbursts, type Ia supernovae, have sufficiently consistent peak luminosity to make them useful distance tracers. These type Ia supernovae are modeled as resulting from the accretion of mass onto a white dwarf in a close binary system until it is pushed past the Chandrasekhar mass limit, beyond which gravity overcomes the white dwarf's internal pressure. This collapse will generally be asymmetric enough to result in an explosion, destroying the star completely and releasing a luminosity comparable with an entire galaxy for a few days. SN Ia are attractive as cosmological probes because they all have nearly the same peak luminosity, and those variations which are present from one to another can be largely correlated with the independently measured decay timescale for the light. Furthermore, essentially all the heavy elements in such a system were produced internally during the white dwarf's previous nucleosynthetic career, making the initial chemical composition virtually irrelevant. Two major surveys for high-redshift supernovae concluded that the redshift–luminosity distance relation for these "standard bombs" requires a nonzero cosmological constant (Figure 1.4). For most astronomers in the field, the weight of the evidence and the status of the investigators have been convincing enough to make the cosmological constant an essential piece of the default cosmological model. Earlier work using gravitationally lensed quasars had bounded the cosmological constant from above; a larger value of Λ gives us more gravitational lensing for quasars, as it puts a relatively longer path length for light from high redshifts.

The evidence for acceleration does not rest solely on the supernova results. It is also indicated by the angular structure of the cosmic microwave background and by structure in the galaxy distribution, particularly the so-called baryon oscillations (Chapter 7). As pointed out by Neta Bahcall *et al.* (1999), it is crucial to have not only different pieces of evidence for cosmological pictures, but different *kinds* of evidence. Much use has been made of their "Cosmological Triangle", including

[2] The evidence that Einstein actually said this traces no farther than a remark by George Gamow, who could seldom resist letting a colorful phrase tune his descriptions. However, Einstein did later describe the cosmological constant, in print, as "theoretically unsatisfactory".

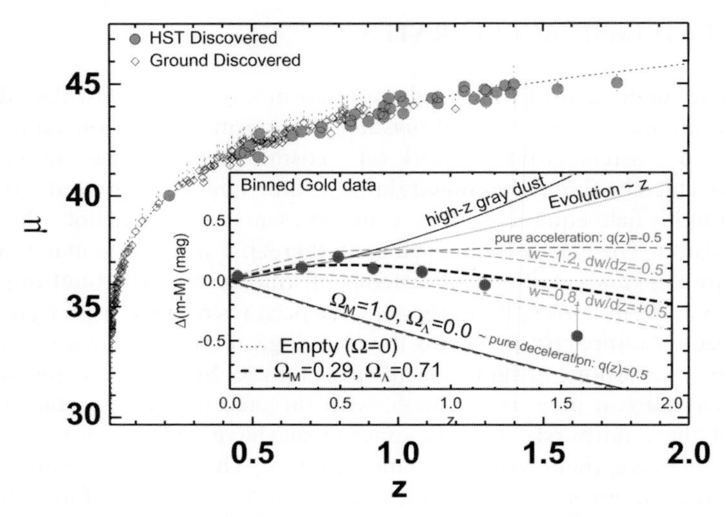

Figure 1.4. The evidence for cosmic acceleration, as found from type Ia supernovae, is summarized in this plot of distance modulus μ (apparent minus absolute magnitude) versus redshift for distant SN Ia. For clarity, the inset compares the values with those calculated for a hypothetical empty Universe. The uppermost curve represents the behavior if some kind of non-reddening dust, or evolution in SN luminosity, accounts for the intermediate-redshift behavior (large points with error bars), and the open gray curve indicates a Universe dominated by a cosmological constant. These diverge at high redshifts, where the redshift-binned mean values in the inset as well as the single available data point at largest z (far right, SN 1997ff at $z = 1.7$) is most consistent with the cosmological prediction (as shown, particularly if dust reddening is a factor). (Figure courtesy of Adam Riess.)

the microwave background, distant supernovae, and the mass distribution of galaxy clusters.

New expressions, beyond those we've used for $\Lambda = 0$, will be needed to relate redshift to the various distance measures, depending on the history of Λ. The cosmological constant introduces, in principle, a new degree of freedom in the history of cosmic expansion. For world models with $\Lambda = 0$, it has long been held that geometry dictates destiny. For a nonzero cosmological constant, and particularly since we have no direct evidence about the cosmic history of its value, the field of allowable cosmic histories is once again open. Under these conditions, as L.M. Krauss and M.S. Turner (1999) write, "there is no set of cosmological observations we can perform that will unambiguously allow us to determine what the ultimate fate of the Universe will be." While these uncertainties, based on what we now observe, do not propagate fast enough to give us serious uncertainties on the timing of galaxy formation, they do allow a remarkable range of cosmological origins and fates. For example, the behavior of the scale factor $R(t)$, even for a constant value of Λ, could allow the Universe to float at nearly constant R for a significant time before finally embarking on its present asymptotic expansion, or may give multiple epochs with quite different values for the Hubble constant.

At our current state of knowledge, it is more accurate to describe the SN Ia and microwave-background results as giving evidence for cosmic acceleration than as indicating a cosmological constant. We do not yet know whether the cause of the acceleration has an amplitude which is constant with cosmic time or changes in some way. Various theoretical prescriptions are described in the literature as the cosmological constant, dark energy, quintessence ... These notions differ in their predicted temporal behavior, usually denoted as the ratio w of an "equation of state" relating the effective pressure and density of the Universe: $w = p/\rho$. Einstein's cosmological constant amounts to $w = -1$. We cannot yet rule out significant changes in w with z, or important departures from this value. Available SN Ia samples constrain w roughly to the range -0.80 to -1.25. Proposals for addressing possible variations with redshift have revolved around large and statistically well-bounded supernova samples, or measurements of the baryon-oscillation signature in the galaxy distribution at large z (Chapter 7).

In light of the *Hubble* determinations of the cosmic distance scale, *WMAP* measurements of structure in the cosmic microwave background, and photometric data on high-redshift supernovae, a remarkable consensus has emerged on the best-fitting parameters of a standard Big Bang cosmology. In this cosmic view, the Hubble constant is $H_0 = 71 \pm 7\,\mathrm{km\,s^{-1}\,Mpc^{-1}}$, spacetime is flat ($\Omega = 1$), all forms of matter contribute a fraction $\Omega_m = 0.3$ of the critical density needed to make the Universe flat, and a cosmological constant (or perhaps a time-varying equivalent) contributes the remaining 0.7. It remains unclear just what processes give the cosmological constant, and hence what its time history is. Assorted notions have been floated as to what kind of field could have this effect, and how it might vary with cosmic expansion and thus epoch.

1.5 BIBLIOGRAPHY

Books

Christianson, Gale (1996) *Edwin Hubble: Mariner of the Nebulae* (University of Chicago Press). A recent biography of Hubble, dealing with his early life as well as scientific accomplishments.

Guth, Alan (1997) *The Inflationary Universe: The Quest for a New Theory of Cosmic Origins* (Addison-Wesley). A good introduction to cosmic inflation, involving one of the major protagonists.

Hubble, Edwin (1936) *Realm of the Nebulae* (Yale). Hubble's work on galaxies, described in this work in his own words (there is a more recent paperback edition).

Linde, A. (1990) *Particle Physics and Inflationary Cosmology* (CRC Press).

Misner, Charles W.; Thorne, Kip S.; and Wheeler, John Archibald (1973) *Gravitation* (Freeman). An exhaustive text on all applications of general relativity in cosmology, for the really curious. Includes development of the tensor calculus used throughout.

Peacock, J.A. (1999) *Cosmological Physics* (Cambridge). A very detailed treatment, at the graduate level and including a great deal of the relevant particle physics.

Peebles, P.J.E. (1971) *Physical Cosmology* (Princeton). This is a textbook introduction to the "standard" treatment of cosmology, and the whole issue of applying general relativity to cosmology.

Rees, Martin (2000) *New Perspectives in Astrophysical Cosmology* (Cambridge University Press). Worthwhile musings on the state of cosmology at the end of the twentieth century.

Silk, Joseph (2001) *The Big Bang* (Freeman). A fairly painless introduction to the standard cosmological picture, presuming very little background.

Weedman, Daniel W. (1986) *Quasar Astronomy* (Cambridge University Press). This volume includes some of the practical details in applying the equations of standard cosmology to astronomical measurements.

Weinberg, Steven (1972) *Gravitation and Cosmology: Principles and Applications of the General Theory of Relativity* (Wiley). A rigorous treatment of relativity and cosmology, at the graduate level.

Journals, etc.

Bahcall, N.A.; Ostriker, J.P.; Permutter, S.; and Steinhardt, P.J. (1999) "The Cosmic Triangle: Revealing the State of the Universe", *Science*, **284**, 1481–1488.

Disney, Michael (2000) "The Case against Cosmology", *General Relativity and Gravitation*, **32**, 1125–1134. A deliberately provocative and refreshingly contrarian view of how much we really know, and how much we can know, in cosmology.

Faber, S.M. (2003) "Conference Summary—Observational Cosmology", in W.L. Freedman (ed.), *Carnegie Observatories Astrophysics Series, Vol. 2: Measuring and Modeling the Universe* (Cambridge University Press) (also *http://arXiv.org/abs/astro-ph/0302495*). More than a simple conference review, this article makes some very provocative analogies as to how seriously we should take arguments from the anthropic principle, suggesting that such arguments may constitute a reason to take multiple-Universe ideas seriously.

Krauss, Lawrence M. and Turner, Michael S., (1999) "Geometry and Destiny", *General Relativity and Gravitation*, **31**, 1453–1459. Describes the wide range of possible cosmic histories allowed by current (and foreseeable) data for a nonzero cosmological constant, in an essay written for the Gravity Research Foundation. The authors point out that when a cosmological constant is allowed, geometry does not equal destiny for the Universe.

Levshakov, S.A.; Molaro, P.; Lopez, S.; D'Odorico, S.; Centurion, M.; Bonifacio, P.; Agafonova, I.I.; and Reimers, D. (2007) "A new measure of $\Delta\alpha/\alpha$ at redshift $z = 1.84$ from very high resolution spectra of Q1101-264", *Astronomy and Astrophysics*, in press (astro-ph/0703042). Recent limits on variations in the fine-structure constant α over cosmic time, derived from comparing absorption lines from very different species over a wide wavelength range.

Mattig, W. (1958) "Über den Zusammenhang zwischen Rotverschiebung und scheinbarer Helligkeit", *Astronomische Nachrichten*, **284**, 109–111. Derivation of the closed-form relations between redshift and various distance measures (such as luminosity and angular diameter) in the traditional cosmological models, as functions of the Hubble constant and deceleration parameter.

Percival, W.J.; Baugh, C.M.; Bland-Hawthorn, J.; Bridges, T.; Cannon, R.; Cole, S.; Colless, M.; Collins, C.; Couch, W.; Dalton, G. *et al.* (2001) "The 2dF Galaxy Redshift Survey: The power spectrum and the matter content of the Universe", *Monthly Notices of the Royal Astronomical Society*, **327**, 1297–1306. The dropoff in galaxy clustering at large scales, perhaps marking the long-sought approach to cosmic uniformity, has been found from the 2dF survey project at the Anglo-Australian Telescope. A complete description of this project, and its data releases, can be found at *http://www.mso.anu.edu.au/2dFGRS/*.

Sandage, A.R. (1988), "Observational tests of world models", *Annual Review of Astronomy and Astrophysics*, **26**, 561–630. An extensive treatment of the classical cosmological tests as applied to galaxies, and where they fail because of uncertainties in our knowledge of galaxy evolution. This review is particularly helpful in tracking distinctions between properties of the world model (a cosmological picture) and the world map, its representation in observed quantities.

Slipher, V.M. (1915) "Spectrographic observations of nebulae", *Popular Astronomy*, **23**, 21–24. The first published report of substantial redshifts for several galaxies.

Spergel, D.N. *et al.* (2007) "Wilkinson Microwave Anisotropy Probe (WMAP) Three Year Results: Implications for Cosmology", *Astrophysical Journal*, in press (astro-ph/0603449). This paper reviews the constraints on cosmological parameters based on the second release of WMAP data on the microwave background, falling firmly in the consensus cosmology. The authors note that the parameters from WMAP data overlap those estimated from various combinations of other data (high-z supernovae, galaxy clustering, and light-element abundances).

Internet

http://www.astro.ucla.edu/~wright/CosmoCalc.html For specific models and redshifts, Ned Wright's Javascript calculator provides the conversions from observed quantities to distance, luminosity, and size. This works for models with nonzero cosmological constants, going beyond the "classical" Friedman models for which Mattig derived the often-used analytic expressions.

2

Galaxies today

2.1 PATTERNS IN THE MIST

In the local Universe we see an enormous variety of galaxies. They differ in size, mass, stellar content, and structure. However, there are important kinds of order among galaxies. It is a largely unspoken scientific mantra that "patterns have reasons". It's hard to gaze upon Half Dome in Yosemite Valley or the oxbow lakes along a meandering river without envisioning a history. Likewise, among galaxies, there are clear relationships between disparate properties, which must result either from the way that they formed, or from their subsequent evolution. In our quest for the processes of galaxy formation, it is important to distinguish properties which reflect initial conditions in some way from those which result from long-term evolutionary processes independent of the details of galaxy formation.

The most widespread way to classify galaxies traces back to Edwin Hubble's work. He promulgated a system of describing a galaxy's morphology—more specifically, its appearance on photographs taken using the blue light that was then easiest to record. This classification turns out to correlate strongly with the stellar and gaseous content of galaxies, so that it is useful as an indicator of a galaxy's overall history. Most bright galaxies could be naturally grouped into the broad categories of elliptical, spiral, and irregular (Figure 2.1). The spirals were first separated into barred (SB) and "ordinary" (S) spirals, depending on whether the center is circularly symmetric or shows a more or less straight, broad distribution of stars cutting right across it. Each spiral category was further divided into subtypes a, b, c, depending jointly on how prominent the central bulge appears and the texture and tightness of the spiral arms. Elliptical galaxies seem at first glance to be much of a piece, and Hubble distinguished them only by their apparent shape—E0 for one appearing circular, E5 for one with an apparent axial ratio of 2:1, on to the flattest genuinely elliptical images at E7. Irregular galaxies were in most cases small and low in luminosity, rich in young stars, rather like spiral disks that couldn't quite organize a spiral pattern.

Figure 2.1. The contrasting properties of elliptical and spiral galaxies are illustrated by the color-composite image of the galaxy pair NGC 4647/9 in the Virgo Cluster. The spiral NGC 4647 shows a blue disk with bright clusters, the loci of current star formation, and a redder central bulge. The elliptical NGC 4649 (Messier 60) shows a much redder overall color and smooth texture, testifying to its lack of current star formation and old stellar population. Some members of its rich globular-cluster population appear as faint starlike objects superimposed on its outer regions. See also color section. (This image was made from two-band CCD observations by the author and R.E. White, III, with the 2.1-meter telescope of Kitt Peak National Observatory.)

Hubble himself noticed that the types fall naturally into a tuning-fork arrangement (Figure 2.2), and that near the branching of the fork was a "missing" type, which he denoted S0. Such galaxies would have the disk and bulge of spirals, perhaps with a bar, but no significant star formation giving them the colors and stellar content of ellipticals. Such galaxies were indeed found later, especially in rich clusters, and their relation to spirals and ellipticals has turned out to be a frustrating question.

This system has been extended by other investigators, notable by Sidney van den Bergh (1998) and Gerard de Vaucouleurs (1959). These revisions key on aspects not

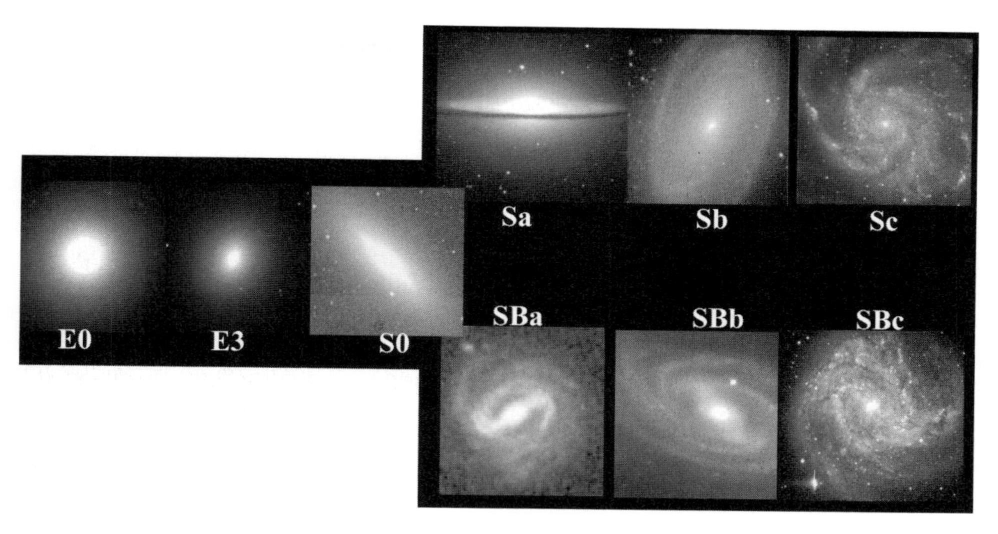

Figure 2.2. Hubble's tuning-fork diagram, filled in with galaxies. This collection of galaxies spans the classical Hubble types, illustrating the changes in bulge-to-disk prominence and spiral pattern along the sequences. Beyond this original set of types, and not shown here, are even less organized Sd/SBd spirals plus their corresponding irregular galaxies. These are optical images, close or identical to the photographic bands used to define Hubble's classifications. (Data by the author, from Kitt Peak and Lowell; from Greg Bothun; and taken from the Digitized Sky Survey, taken by the Oschin Schmidt Telescope on Palomar Mountain, with compression and distribution by STScI.)

given much attention in Hubble's original descriptions. The de Vaucouleurs system recognizes subtypes among elliptical and S0 galaxies based on the slope of the intensity profile and presence of dust. Among spirals, it distinguishes not only barred and nonbarred, but transitional spirals with weak bars, and adds codes for spiral patterns arising in a central spiral, ring, or blend. And for stages along the Hubble sequence, the de Vaucouleurs system keys specifically on the prominence of the central bulge, rather than using the texture of the spiral arms (this was perhaps a relic of the days when so many classifications on the Hubble system came from plates taken at Palomar when that was the only really large optical telescope in existence). This system introduced spirals beyond Sc, namely Sd plus "Magellanic" spirals Sm (shading into the Magellanic irregulars Im in ways that mostly depend on the imagination of the classifier). The additions by van den Bergh took into account the character of the spiral pattern. Even at a fixed stage along the spiral sequence, the arms may be either long, thin, and continuous, in a "grand-design" spiral, more patchy and discontinuous, or very difficult to trace at all. He found that these kinds of spiral pattern correlate, statistically if not uniquely, with the galaxy's luminosity (and generally linear size as well), so he introduced a *luminosity classification* based on the spiral structure. Spirals with the best-organized structure would be, for example, Sc I, less well-organized ones would be Sc II, and so on, to the least regular patterns at Sc V.

This last classification bin may be empty; only the very latest types Sd and irregular types seem to come in luminosity class V.

For many workers, these schemes have proven compatible enough that they feel free to mix and match features of the various definitions, to the irritation of the purists who try to keep each one as a fundamental reference set based on specific galaxies classified in a particular way against a particular set of standard objects. This works, most of the time, because the various classifications correlate for most galaxies. Classifying a spiral based on the arm texture (i.e., degree of resolution into bright stars and associations) and on the bulge-to-disk ratio or the pitch angle of the spiral pattern usually gives concordant results (though the galaxies which land in different classes from these various techniques are clearly interesting, not least in telling us what assumptions in classification don't always apply). The refinements of luminosity class and internal ring/spiral structure are largely independent of how the main type (or stage) is determined among spirals. Thus it is that Sandage (1961) adopted van den Bergh's luminosity classes for the types in the *Revised Shapley–Ames Catalog of Bright Galaxies*. As we entered the digital era, a particularly useful, and in hindsight obvious, refinement was de Vaucouleurs' introduction of a numerical equivalent to the stage along the Hubble sequence, which made correlation and regression analyses easier and more reproducible. This was originally done to make it easier to produce the *Reference Catalog of Bright Galaxies* directly from 1976-vintage computer output, and the practice has proven very convenient with the growing role of computers in data handling over the ensuing 25 years. On this scheme, Sc galaxies have a type $T = 5$, Sbc have $T = 4$, through to elliptical galaxies from -3 to -5 and with the latest-type irregular galaxies at $T = 10$. We will see this again in looking at how the physical properties of galaxies vary along the type sequence.

What is, perhaps, most remarkable about this general scheme for morphologically classifying galaxies is that a shorthand system for describing ordinary photographic images turns out to correlate very well with other, physically interesting, properties of galaxies. This is almost as miraculous as if Linnaeus had been able to use field observations of bird plumage to deduce their chromosomal patterns. As quantitative data on the colors, stellar content, and gas content of galaxies accumulated, these quantities turned out to change radically along the sequence of Hubble types. Elliptical galaxies are, with only a few interesting exceptions, uniformly red with a very narrow color spread and lacking in both the neutral and molecular gas phases that fuel star formation, and indeed they also lack the infrared and emission-line indicators of ongoing star formation (Chapter 4). Irregular galaxies are very rich in interstellar gas, and have accordingly high rates of current star formation; many seem to have maintained a constant rate of star formation for their whole histories. Spirals have various intermediate properties; going along the Hubble sequence Sa–Sb–Sc or SBa–SBb–SBc, we find increases in the star-formation rate and relative amounts of interstellar gas. While recent work has shown exceptions to these trends among S0 and Sa galaxies, the statistical properties are inescapable.

It is not guaranteed that the Hubble classification system is all-inclusive, especially as we can look into regimes very different than the bright, prominent, and nearby galaxies from which it was defined. Indeed, by number, most galaxies in the local Universe do not fall easily into its bins. The offenders are of several kinds—small and dim dwarf galaxies, galaxies of such low surface brightness that we could not find them at all until recently, and galaxies which have been altered by environmental effects. On top of this, the Hubble classification was set up to describe the appearance of galaxies as seen in blue light, as recorded by traditional photographic emulsionsA galaxy's appearance may change radically as we look in different parts of the electromagnetic spectrum and see the different components of its stars and interstellar material (Figure 2.3).

This change of a galaxy's appearance with wavelength may become important as we compare galaxies in the distant Universe, usually observed in visible light and thus by their emitted ultraviolet light, with nearby galaxies, often observed in visible light. This effect has been called the *morphological K-correction*, by analogy with the K-correction for galaxy brightness used to account for our measuring different parts of the galaxy's spectrum at various redshifts. This has driven several projects aimed at defining the ultraviolet properties of nearby galaxies, so that comparison with high-redshift galaxies can rely on data rather than expectation.

Our census of galaxies might possibly be complete only in our very close neighborhood. There, and in the nearby groups where we can search very deeply, we find that faint (usually small) galaxies far outnumber luminous ones, just as happens for stars. This is quantified by the *luminosity function*, the space density of galaxies expressed for various luminosities. Like many other important facts about galaxies, this was noted by Zwicky, who was so bold as to define dwarf, pygmy, and gnome galaxies. At any rate, dwarf galaxies are still with us, somewhat loosely defined as having absolute B magnitude fainter than -18. By this definition, although the Magellanic Clouds are not dwarfs, the term does apply to our close neighbors the Seven Dwarfs (as listed in the title of a famous paper by Aaronson, Hodge, and Olszewski (1983): Leo I, Leo II, Ursa Minor, Sculptor, Fornax, Carina, and Draco) and their siblings throughout the Universe.

Dwarf galaxies don't fit comfortably in the Hubble sequence. Most could be shoehorned into the elliptical or irregular classes, but the traditional sequence loses discriminatory power when stretched so far beyond its original domain. There are genuine dwarf ellipticals, and then there are dwarf spheroidals, which have a different relation of surface brightness to luminosity. Dwarf irregulars certainly exist, and there are also more regular dwarfs with the same kind of average intensity profile. There are also bursting dwarfs, or blue compact galaxies, whose evolutionary relation to the other kinds is still under investigation. From detailed, star-by-star investigations of nearby dwarfs, and statistical studies of the overall properties of more distant examples, a rich picture has emerged of systems which have been strongly shaped by interactions with their environments. Several of the dwarfs near our Galaxy have undergone multiple episodes of star formation, plausibly triggered by gravitational encounters with the Milky Way. Dwarfs undergoing bursts of star formation are often found to host global winds ("superwinds") driven by the energy input of the stars' radiation, winds, and supernova explosions. For dwarf systems, these winds are likely to sweep the galaxy free of gas, as traumatic an event in its history as

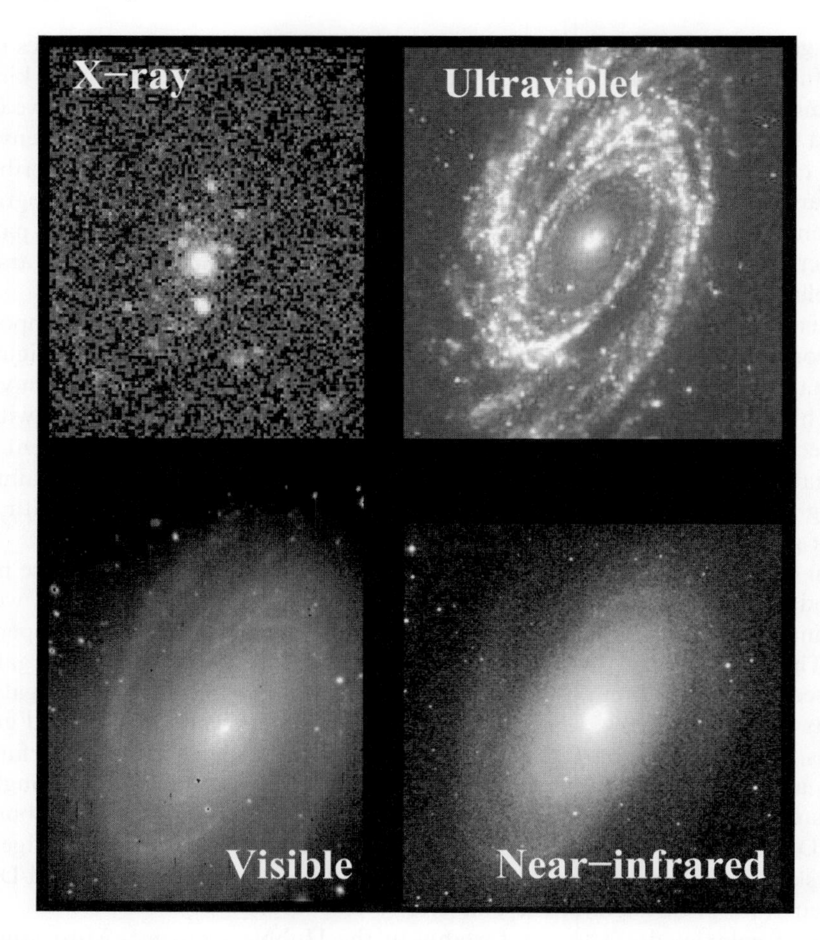

Figure 2.3. The changing appearance of galaxies in different parts of the electromagnetic spectrum is illustrated by this series of images, showing the nearby spiral galaxy M81. In X-rays, only the active nucleus and accreting compact objects in binary systems (white dwarfs, neutron stars, and black holes) appear, with no hint of the rich stellar structure. In the ultraviolet, only the hottest (unobscured) stars are bright, so that the spiral pattern can be traced via star-forming regions, but the central bulge of old stars has almost vanished. This appearance resembles some spiral galaxies seen at large distances, where the redder bulge light has shifted out of the optical bands. The familiar visible-light image shows both the spiral pattern and the old bulge population, while in the near-infrared, the spiral pattern and star-forming regions are much more subdued. This change in apparent structure with wavelength has been dubbed the "morphological K-correction". X-ray data are from ROSAT, extracted from the HEASARC archive at NASA's Goddard Space Flight center. The UV image is a combination of observations obtained by the GALEX satellite at wavelengths 1,500 and 2,300 Å, used courtesy of NASA. The optical images are reproduced courtesy of Greg Bothun. (The infrared data are from the 2-Micron All Sky Survey (2MASS), a joint project of the University of Massachusetts and the Infrared Processing and Analysis Center/California Institute of Technology, funded by NASA and the NSF.)

the preceding starburst. In a few cases, there is even evidence of dwarf galaxies being carved out of larger galaxies today, when bits of tidal debris from a string gravitational interaction become gravitationally bound. This notion, of the incidental production of dwarfs during encounters between massive galaxies, is another of the many legacies of Fritz Zwicky.

The Swiss–American astrophysicist Zwicky (1898–1974), who worked for many years at Caltech, still stands as a towering and lonely figure in 20th-century astrophysics. He essentially discovered dark matter, recognized dwarf galaxies as a class and their vast numbers, drew attention to galaxies with high surface brightness, foresaw the use of gravitational lensing to extend the range of our telescopes, and played a key role in postulating the existence of neutron stars and in defining the phenomena of supernova explosions that proved to be their progenitors (as he proposed in a remarkable 1933 lecture). He oversaw the first reasonably uniform and deep effort at cataloguing galaxies and clusters across the northern sky, which gave crucial evidence that clustering is a universal feature of the space distribution of galaxies. Along the way, Zwicky managed to alienate virtually every other important astronomer except (temporarily) Walter Baade, with predictable results. The introduction to his *Catalogue of Selected Compact Galaxies and of Post-Eruptive Galaxies* (alias Zwicky's Red Book) makes some characteristically pithy observations from his view of astronomy and astronomers. Within a mere three pages, he writes of "sycophants and plain thieves", refers to the "comparatively rare incidents in the USA in which the gentlemanly spirit upheld by so many of our great predecessors" was exhibited, the "useless trash in the bulging astronomical journals", and the fate of "lone wolves who are not fawners and apple polishers". Still, it is striking to note how often in the history of ideas about galaxies we find that Zwicky got there first.

The vast numbers of dwarf galaxies, and their typically low surface brightness, were some of the factors inducing Arp (1966) and Disney (1976) separately to consider whether our picture of galaxies is shaped largely not by what exists, but by selection effects—our doors of perception. Both workers noted that catalogued galaxies span a rather limited range of surface brightness (suitably averaged from catalog values), and that we would miss both small galaxies of high surface brightness (photographically appearing much like stars) and galaxies of low surface brightness, which would be lost in the diffuse glow of the natural night sky. Indeed, they showed that some of Zwicky's compact galaxies skirted the "starlike" line, while the Sculptor dwarf galaxy has a surface brightness that should have rendered it invisible—and indeed, it showed up first in photographs that showed its individual stars, rather than the average diffuse glow that would be seen at greater distances. These ideas led, eventually and indirectly, to the identification of galaxies of low surface brightness (or LSB galaxies). They have been identified by such techniques as photographic amplification, teasing the very faintest smudges out of photographic plates, examination of modern sky surveys taken using improved photographic emulsions, or smoothing long-exposure CCD images. Most turn out to be, indeed, small and intrinsically dim dwarf galaxies. However, there also exists a completely unsuspected class of giant LSB galaxies—systems which are as bright as typical spirals, but so physically large that their surface brightness lies several orders of magnitude below

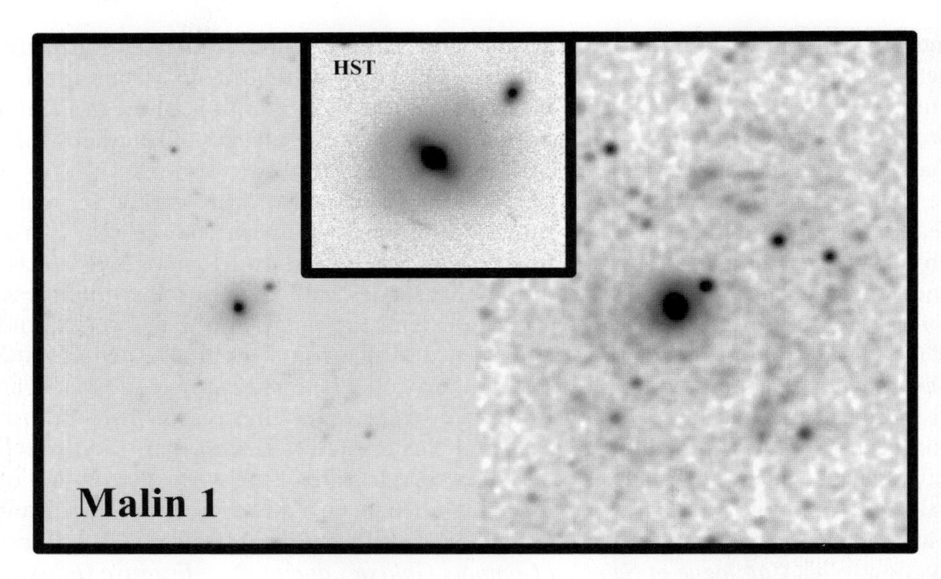

Figure 2.4. The prototype giant low-surface-brightness (LSB) galaxy Malin 1. A typical exposure depth (*left*) shows only the nucleus and bulge of the galaxy. Smoothing the image and stretching the intensity by a factor of ten (*right*) reveals the extensive and patchy spiral structure which extends over a much wider area, visible over a diameter of 200 kiloparsecs and well into the massive disk of neutral hydrogen. (The inset is from a Hubble image in red light, showing the distinct bar and spiral structure marking a "normal" disk at the center of the huge LSB structure. The wide-field blue-light image was obtained by Greg Bothun at the 1.3-meter telescope of MDM Observatory. HST image from the archive, from a program with Chris Impey as PI.)

the natural skyglow from the ground or even low Earth orbit. The first such example known, Malin 1, turned up when David Malin's photographic image-amplification techniques showed a very large LSB object in the direction of the Virgo Cluster (Figure 2.4).

Redshift information showed that this system is well in the background, at about 15 times the Virgo core distance. This galaxy showed interesting and paradoxical properties. Its central bulge, the only part that shows up on images of typical exposure, contains a modest AGN. The disk starlight is blue, but the overall star formation rate (SFR) has to have been very low for its entire history. This low SFR is not due to any lack of gas, as Malin 1 has a total H I mass among the highest ever measured for any galaxy. Further surveys—notably work by Schombert and co-workers starting from the second Palomar survey—have uncovered additional galaxies of this kind. These discoveries expand the bounds of systems which we can now find (Chapter 4, Figure 4.10).

Recent analysis of HST images has shown that Malin 1 does have a "normal" disk, but one of such small angular size as to have blended with its bulge in the discovery images (thereby making the extrapolated central surface brightness of the

Figure 2.5. The spectacular extended-ultraviolet (XUV) disk structure of the interacting spiral galaxy NGC 1512. Only the central disk is optically prominent. The filamentary nature of the outermost arms is common for these XUV disks, as is their large radial extent compared with the better-known inner disk. In this case, interaction with the bright companion NGC 1510 is likely responsible for much of the star formation seen in NGC 1512. (This image is a composite of near- and far-UV bands, roughly centered at 1,500 and 2,400 Å, from the GALEX mission—NASA and the GALEX team.)

disk artificially low). Its enormous outer disk may then be related to a family of structures brought to wide attention by the GALEX UV survey (although seen occasionally in earlier data)—the so-called XUV (extended ultraviolet) disks (Figure 2.5). A significant fraction of bright spirals show filamentary spiral arms extending well beyond the pattern mapped in typical optical imagery, whose contrast is by far strongest in the UV bands. Hα may be weak or undetectable in these patterns, which indicates either an initial-mass function deficient in the most massive stars (B0 and hotter) or an unusual history for star-forming gas in these regions.

The new surveys do seem to indicate that these LSB galaxies, and LSB structures surrounding "ordinary" galaxies, cannot dominate the overall galaxy luminosity density, so that we may be correct in concentrating on the more obvious galaxies when trying to explain their history. Still, any theory of galaxy formation should be able to tell us why the whole range of surface brightness and history exists. Ideas in which these systems have never interacted with other systems to trigger extensive star formation, or started in such low-density regions that they barely collected into galaxy-like masses of ordinary matter, have been discussed.

2.2 GALAXY COMPONENTS AND QUANTITATIVE CLASSIFICATION

There has been much recent interest in more quantitative forms of galaxy classification. While the familiar Hubble sequence correlates with interesting physical quantities (perhaps better than we would have any right to expect), such properties as bulge:disk intensity ratio, disk scale length, and color of the stellar population are both more directly relevant to models of galaxian history, and more likely to be recovered at high redshift than the visually estimated Hubble type. This trend was pushed further by HST images of high-redshift galaxies, with the realization that many of them do not fit comfortably within the traditional classification bins.

There is at this point no single universal set of numerical, objective galaxy classifications. Those using image structure rely, to one degree or another, on the fact that galaxies contain components with different kinds of intensity profiles, and which are associated with different stellar populations and formation histories. The disks of spirals and S0s, and in many cases irregular galaxies, are reasonably well described in surface brightness by an exponential disk:

$$I(r) = I(0)e^{(-r/r_0)}$$

modified by the orientation of the disk with respect to our viewing direction. In this case r_0 is the scale length of the disk, a characteristic measure of its size.

In contrast, the central bulges of spirals, and the overall light profiles of elliptical galaxies, are more centrally concentrated. Various empirical forms have been used to parametrize them:

$$\text{de Vaucouleurs or } R^{1/4}: I(r) = I_e \exp(-7.67((r/r_e)^{1/4} - 1))$$

$$\text{Hubble or Hubble–Reynolds law}: I(r) = \frac{I_0}{(1 + r/r_0)^2}$$

$$\text{King model}: I(r) = \frac{I_0}{1 + (r/r_0)^2}$$

each modified by the apparent axial ratio of the system. Furthermore, high-resolution measurements of many late-type spirals show that their bulge components have a brightness profile which is more nearly exponential than any of these forms, leading some to distinguish them as "pseudobulges" to stress the possibility that they may have formed differently from ellipticals and the more elliptical-like bulges of early-type spirals. In particular, some models suggest that these pseudobulges formed gradually as secular processes scattered stars from an initially thin disk into a thicker but still exponential structure.

It is not clear whether any of these formulae are much more appropriate than the others. Among them, only the King model has a particular theoretical foundation, and has a simple analytical form for the radial distribution of starlight in three dimensions:

$$\rho(r) = \rho_0(1 + (r/r_0)^2)^{-1},$$

but it does not fit the cores of many galaxies very well. Sérsic (1968) has introduced a more general form

$$I(r) = I(0)\exp[-b_n[(r/r_e)^{1/n} - 1]]$$

for galaxy components, in which the exponential and $R^{1/4}$ laws are special cases (with $n = 1$ and $n = 4$, respectively). The index n from this fit is related to the concentration indices derived from global profile fitting, a factor of special interest when examining images of high-redshift galaxies where the information content of a galaxy image may charitably be described as "modest". When more image pixels are available, many systems are best fit with multiple, nested Sérsic profiles, often corresponding to the intuitive disk/bulge distinction.

These parametrizations can be used to simplify the decomposition of a galaxy image onto the contributions coming from a disk and bulge. A global fit may be done to the whole image, varying the parameters of the bulge and disk functions to improve the result, or a comparison of major- and minor-axis profile can be employed, using the three-dimensional shape difference of disk and bulge populations as a starting point. The disk:bulge intensity ratio shows a systematic variation with Hubble type (Figure 2.6), so that it has similar behavior with gas content and color, although disk:bulge ratio measurement is fundamentally a different process than assigning a Hubble type since classification includes characteristics of the spiral pattern as well as bulge prominence.

As we examine galaxy images which are less and less well resolved, containing less information, the last information we can retrieve is some index of the image's concentration—for example, the second moment of intensity about the center, or one of a variety of similar indices comparing the total intensity and central surface brightness. Such indices can, with proper control sets, be recovered even in the face of substantial image blurring, as demonstrated by such tests as measuring galaxies from ground-based images and using HST data as "ground truth". These kinds of results give us the most remote hints of what the mix of galaxies is doing in the early Universe. One set of indices that has seen particularly wide use is the "CAS" triple, which measures concentration, asymmetry, and clumpiness. As particularly championed by Conselice, the concentration $C = 5\log(r_{80}/r_{20})$ where r_{80} and r_{20} are the radii enclosing 80% and 20% of the light in a radially averaged intensity profile.

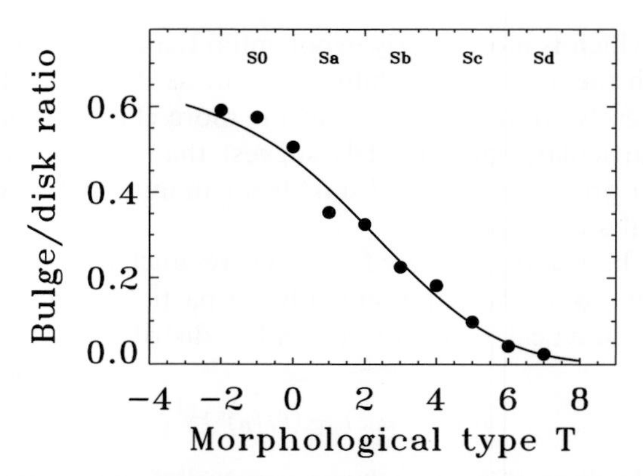

Figure 2.6. Hubble type versus quantitative parameters. The bulge-to-disk luminosity ratio in blue light, as derived by F. Simiens and G. de Vaucouleurs (1987), versus numerical morphological type T. Error bars come from the scatter seen among members of each type. The curve is their approximate interpolating function, illustrating that the behavior is smooth and monotonic. The Hubble-type equivalents for several of the values of T are indicated at the top. The Hubble sequence is, in large part, one of changing bulge-to-disk ratio. The value for a given morphological type will depend, as well, on observed wavelength.

A measures the asymmetry, using the simple mathematical notion that any function may be split into its even (symmetric) and odd (asymmetric) parts, so their ratio (in this case weighted by total luminosity) quantifies an image's asymmetry. The definition is

$$A^2 = \frac{\sum (I_0 - I_r)^2}{2 \sum I_0^2}$$

in which I_0 is the intensity in the image, I_r is the intensity at the same location rotated by 180° about the brightness peak, and summation is taken over the pixels above the brightness threshold of interest in the galaxy image. These parameters will change with the resolution of the image, in the sense that their dynamic range across the galaxy population decreases for poorly resolved images; simulations, often based on resampled images of nearby galaxies, are required for particular data sets to understand how to compare various results. To quantify the clumpiness of a galaxy's light, the S parameter makes use of a version of the images smoothed to some reduced resolution σ, so that

$$S = 10 \sum \frac{I - I_\sigma - B}{I}$$

where B is the background value, the summation over galaxy pixels excludes the nucleus, and the minimum numerator is set to zero. Empirically, these parameters have broad physical interpretations. More concentrated galaxies tend to be more massive (and are broadly of early type on the Hubble system). Asymmetry is one symptom of recent gravitational interaction, while high-frequency structure in optical images has long been known to correlate with ongoing star formation. This system

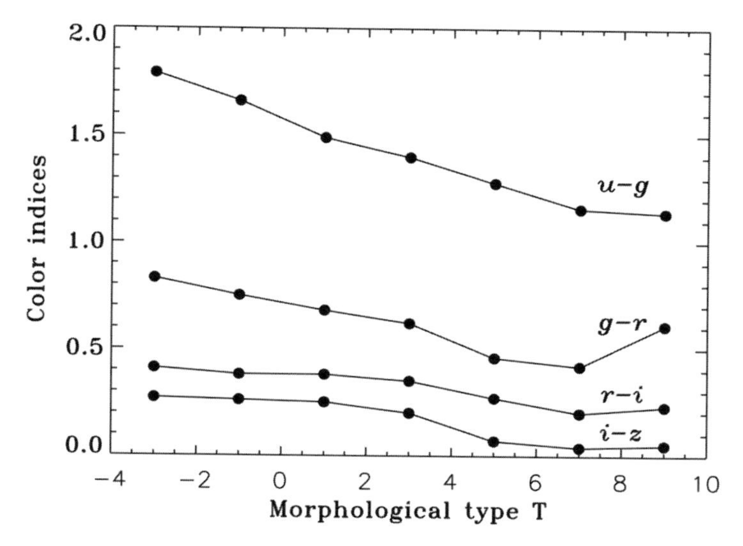

Figure 2.7. The color of galaxies along the Hubble sequence. The star-forming history of galaxies is traced by their broad-band colors. This is illustrated in these mean colors from a sample of 456 bright galaxies in the Sloan Digital Sky Survey, with the bands approximately at 3,500 Å (u), 4,800 Å (g), 6,250 Å (r), 7,700 Å (i), and 9,100 Å (z). Smaller differences among these magnitudes (color indices) denote a bluer spectral shape. The difference is most conspicuous at shorter wavelengths (as in $u - g$ color), since stellar evolution is most rapid for the more massive bluer stars. (Data from Shimasaku *et al.*, *Astronomical Journal*, **122**, 1238, 2001.)

has seen use in asking how local galaxy samples populate the *CAS* space, and in comparing this with samples at high redshift.

Such a pattern of grouping and quantifying galaxies through some small set of observable quantities, such as color, image concentration, and an index of image symmetry, takes us into the more abstract field of quantitative classification. For many years, a need has been expressed for fully quantitative and numeric ways to classify galaxies, not necessarily to supplant but more completely to supplement the classical morphological schemes such as the Hubble and de Vaucouleurs systems. Such parameters could, if we are fortunate, be linked directly to the dynamical components of galaxies, and be defined in ways that make them more robust to data quality and resolution than the usual visual classifications. As candidate systems have been proposed and evaluated, it is one measure of the lasting influence of the Hubble system that most of the quantities they derive have been shown plotted as functions of ... Hubble class. We may consider such interesting additional quantities as color of the stellar population, gas content per unit optical luminosity, normalized star-formation rate from Hα or far-infrared indicators, or mean rotational velocity, and in each case we see a clear correlation with galaxy type (as seen in sample quantities in Figure 2.7). The color dependence can be so strong that color is often used as the independent variable in studies of galaxy populations, as a proxy for the star-forming history of a system (Chapter 4).

2.3 DISKS, BULGES, AND DISSIPATION

A recurring feature of quantitative classification, and of our understanding of galaxy structure, is the distinction between bulge and disk components. This is widely understood to reflect the different processes by which these systems took shape. Exponential disks seem to result from dissipative processes, in which components can lose energy (such as to collisions between gas clouds, in which some orbital energy is given up to internal heat). The conservation of angular momentum in this situation leads naturally to a thin disk, which is dynamically cold—that is, it consists of material in circular orbits with a very small local dispersion in velocities. The same general considerations apply to forming rings around giant planets or accretion disks around compact objects. This fits with the basic history of the disk of our own Galaxy as inferred from the ages, motions, and makeup of its stars.

In contrast, elliptical galaxies and the central bulges of early-type spirals (collectively known as spheroids) seem to have resulted from a formation process in which dissipation was not important. Such nondissipative collapse would happen if the stars had already formed (leaving little cold gas for further generations of stars) before the final collapse to the configuration we see. The transition from an extended collection of stars to the compact configurations we actually see has to take place by a process known as violent relaxation. In classical dynamics the redistribution of energy among some group of interacting particles is known as relaxation, and the short time available for this process in galaxies (at least in terms of the systems' dynamical timescale) leads to this initially oxymoronic label.

Within galaxies, we see an interplay between stars and interstellar material so complex that it can fairly be called galactic ecology. It is the broad workings of this interplay that lets us associate stellar composition with time, and link the properties of gas in galaxies to their evolution. Stars form from interstellar gas, most immediately from molecular hydrogen with a salting of heavier elements. Stars then return the favor by ejecting gas back into their surroundings, through such diverse avenues as stellar winds, nova outbursts, planetary nebulae, and supernova explosions. In some of these, the returned material was once deep enough inside a star to have been a product of nuclear fusion, so that the interstellar medium becomes enriched in heavier elements with time. Which elements are enriched depends on the stellar source, with supernovae of different types giving differing relative amounts of oxygen, sulfur, and iron, for example. Even a single episode of star formation will give a changing chemical makeup in the interstellar gas as stars with different lifetimes return their portion to the surroundings. In turn, the enriched interstellar gas forms new stars, some of which are massive enough to further enrich the gas. The Sun is often described, loosely, as a "third-generation star" in this sense—not that there have been three distinct episodes of star formation, but that we would expect the solar abundance (of about 1% of its mass in elements heavier than helium) to be reached only after two rounds of such recycling.

Stars are excellent historical probes of the kind of material they formed from. Although nuclear processes in the stellar cores change the mix of elements there, the atmosphere—what we measure in spectroscopic analyses—does not reflect these changes until late in a star's red-giant phase, so that the exterior composition reflects the original makeup. For long-lived stars, this gives us a chemical window into the interstellar medium long past. As we shall see later, the compositions of various kinds of stars are a powerful way to excavate galactic history, and to connect what we see in our neighborhood with the early Universe as revealed by observations of galaxies at high redshifts.

2.4 BIBLIOGRAPHY

Books

Arp, H. (1966) *Atlas of Peculiar Galaxies* (California Institute of Technology). This set of remarkable Palomar photographs made it clear what bizarre forms could be seen among interacting, and otherwise peculiar, galaxies. It also set forth the role of surface-brightness selection in determining what kinds of galaxies we can detect to a particular sensitivity limit. The Atlas was published in a rather limited photographic edition to preserve the quality of the images; identical material appeared at reduced scale in *Astrophysical Journal Supplement*, **123**, 1–20 plus associated plates (1966).

Binney, J. and Tremaine S. (1987) *Galactic Dynamics* (Princeton University Press). An extensive and definitive treatment, at the graduate level, of galaxy dynamics, including both disk and spheroidal components, and many related astrophysical issues.

Buta, R.; Corwin, H.; and Odewahn, S. (2006), *The de Vacouleurs Atlas of Galaxies* (Cambridge University Press).

Müller, R. (1986), *Fritz Zwicky, Leben und Werk des grossen Astrophysikers, Raketenforschers und Morphologen*, (Verlag Baeschlin, Glarus, Switzerland). This German-language volume appears to be the only complete biography of Fritz Zwicky, including his work in rocketry and application of his "morphological approach to thought and action" to diverse fields of knowledge, as well as his astrophysical research.

Sandage, A. (1961) *Hubble Atlas of Galaxies* (Carnegie Institution) and Sandage, A. and Bedke, J. (1994) *Carnegie Atlas of Galaxies* (Carnegie Institution). These volumes contain the standard examples for classifying galaxies on the Hubble system. The Hubble Atlas defined the system for a generation of astronomers, and the two-volume Carnegie Atlas includes high-quality photographs of over 1,200 galaxies with Sandage's commentary on their structures and classification.

Sérsic, J.L. (1968) *Atlas de Galaxias Australes* (Observatorio Astronómico, Córdoba, Argentina). This atlas of galaxy photographs—produced at a time when there were no optical telescopes in the Southern Hemisphere as large as 2 meters and the many beautiful galaxies in this region were little known—also introduces the general Sérsic form for the light profile of galaxies.

van den Bergh, S. (1998) *Galaxy Morphology and Classification* (Cambridge University Press). This monograph discusses issues of galaxy classification, both qualitative and quantitative.

Zwicky, F. and Zwicky, M.A. (1971) *Catalogue of Selected Compact Galaxies and of Post-Eruptive Galaxies* (Zwicky: Gümlingen).

Journals

Aaronson, M.; Hodge, P.W.; and Olszewski, E.W. (1983) "Carbon stars and the seven dwarfs", *Astrophysical Journal*, **267**, 271–279.

Bothun, G.D.; Impey, C.D.; Malin, D.F.; and Mould, J.R. (1987) "Discovery of a huge low-surface-brightness galaxy—A protodisk galaxy at low redshift?", *Astronomical Journal*, **94**, 23–29. Describes the discovery and overall properties of Malin 1, the first known giant low-surface-brightness galaxy.

Conselice, C.J. (2003) "The relationship between stellar light distributions of galaxies and their formation histories", *Astrophysical Journal Supplement*, **147**, 1–28.

de Vaucouleurs, G. (1959) "Classification and Morphology of External Galaxies", *Handbuch der Physik*, **53**, 275–310 The original exposition of de Vaucouleurs' extension to the Hubble classification system. An online version is maintained at *http://nedwww.ipac.caltech.edu/level5/Dev/frames.html* Some of de Vaucouleurs' students have recently presented this system in an extensive and systematic way, based on digital imagery with controlled display properties (see Buta *et al.*, 2006).

Disney, M. (1976) "Visibility of galaxies", *Nature*, **263**, 573–574. A pivotal paper in quantifying the role of observational selection, and especially surface-brightness selection, in our view of galaxies.

Schombert, J.M. and Bothun, G.D. (1988) "A catalog of low-surface-brightness objects—Declination zone + 20 degrees", *Astronomical Journal*, **95**, 1389–1399.

Simien, F. and de Vaucouleurs, G. (1987) "Systematics of bulge-to-disk ratios", *Astrophysical Journal*, **302**, 564–578. A compendium of quantitative properties compared with morphological types of galaxies.

3

The fossil record in nearby galaxies

3.1 INTRODUCTION

In our own galaxy, and with recent instrumental advances in a growing number of our neighbors, there is a vast amount of historical information to be gleaned from the "fossil record"—the combinations of age, chemical composition, and orbits of the individual stars. Such data led to the first realistic picture of galaxy formation, and have now been able to trace the aftermath of individual events in our galaxy's history.

Perhaps the single most seminal concept was introduced when Walter Baade recognized distinct stellar populations during the 1940s. This distinction was easier to see when looking at other galaxies in the Local Group than in our own, since all the stars in another galaxy are at nearly the same distance and we can compare them without making laborious individual distance estimates. In an initial reconnaissance of these galaxies, Baade had found that the brightest stars in some regions (such as the spiral arms of the Andromeda spiral M31) were easily recorded on the blue-sensitive photographic plates then available, but that the bright inner parts of M31 and of its elliptical companion galaxies M32 and NGC 205 resisted any such resolution. Even using the 100-inch (2.5-m) Hooker telescope on Mt. Wilson, which was to remain the world's largest until 1948, his photographs showed only a continuous glow from vast numbers of stars too faint to distinguish individually. Star clusters in the Milky Way furnished a model which Baade found could be applied to these distant stellar populations. All the stars in a cluster have essentially the same age, making them excellent laboratories for studying the evolution of stars. The disk of our galaxy contains numerous *open clusters*, often including a few thousand stars, with clusters having a wide age range from less than a million (sometimes still forming new stars) to a few billion years. In contrast, the *globular clusters* are found grouped around the center of the galaxy, with 250 or so forming a nearly spherical distribution extending outward beyond the easily measured individual stars of the disk and halo. These clusters are very massive and rich, with some containing more than a million stars concentrated in a region only a few tens of light-years across. Stars in globulars are uniformly old, with cluster ages estimated in the range 12–13 billion years. The visible stars making up the clusters differ due to these different ages. More massive stars run through their life cycles more rapidly than lower-mass ones, using up all the core hydrogen (which is fused to helium to power the stars during their *main-sequence* phases), so that which mass ranges are still on the main sequence, which have advanced beyond that to become red giants, and which are gone altogether into stellar remnants (white dwarfs, neutron stars, or black holes) changes systematically with the age of a star cluster. Based on experience with clusters in the Milky Way, Baade conjectured that the inner regions of these galaxies might be composed of stars like those in globular clusters, instead of the younger open clusters (like the Pleiades). The difference would show itself most clearly in the colors of the brightest stars. In the younger open clusters, the brightest stars are quite blue, being hot stars at the massive end of the main sequence. In contrast, such stars have long died away in the older globular clusters, whose brightest stars are red giants, both fainter and redder than the brightest stars in younger assemblages.

Detecting such red giants in other galaxies posed formidable technical problems. They had to be separated from the diffuse glow of the greater numbers of stars which would remain too faint to distinguish, and they would be easiest to find in red light instead of the blue light hitherto employed. Red-sensitive photographic emulsions were barely up to the task, and would require exposures of many hours' duration. The required image sharpness was delicate enough to require refocusing the telescope during the exposures, without interrupting them or allowing the images to shift within the camera. Baade had taught himself over the years how to track the focus changes as the telescope structure cooled during the night, by inspecting the aberrated appearance of a star at the edge of the field of view that was being used to keep the telescope tracking precisely. He also benefited from wartime security measures. As a German citizen, he was confined to Los Angeles County—a region which includes Mount Wilson. Many of the other staff astronomers at Mount Wilson were away on war-related work in optics, navigation, or the nascent field of electronics, so telescope time was more plentiful than usual. Finally, the perceived risk of air attack led to the lights of Los Angeles being blacked out, offering conditions for deep probing from Mount Wilson which have been matched since only when very thick banks of coastal fog appear.

The combination of Baade's drive and skill as an observer and the exceptional conditions did indeed suffice to show the red giants—barely. When his work was published in the *Astrophysical Journal* in 1944, original prints of the photographs were bound into each issue, for fear that the images were too delicate to survive reproduction in printing. From his immediate observational result—that spiral arms have their brightest stars blue and the center of Andromeda plus its elliptical companions have their brightest stars red—he surmised, correctly, that he was dealing with two distinct populations of stars, best represented in our immediate neighborhood by the two kinds of star clusters. In a move setting nomenclatural tradition for astronomers, he denoted them Population I and Population II. It helps in keeping them straight that the Sun belongs to Population I.

The existence of the two stellar populations, and the fact that they could be distinguished by location and distribution within galaxies as well as by the kinds of stars they contain, were the keys to showing that galaxies have histories. Indeed, the 1963 volume summarizing a series of Baade's lectures was perhaps the first serious attempt at addressing together *The Evolution of Stars and Galaxies*. If Friedman made the Universe expand, and Hubble first navigated the galaxies, it was Baade who gave them a history.

Once their existence had been identified in other galaxies, the details of the stellar populations could be better probed in our own neighborhood, the Milky Way and its close companions the Magellanic Clouds. Doing so required the painstaking collection of precise magnitudes, colors, and apparent motions for huge numbers of stars, an effort so time-consuming that much of the art was in knowing which samples of stars would most quickly probe the right effects. In today's world of digital sky surveys, electronic data distribution, and highly efficient detectors on huge telescopes, it can be difficult to recall how expensive each of these measurements was in human effort and telescope time. Given these limitations, some surrogates had to be used for the quantities of greatest interest. Chemical composition, for example, could be measured directly from spectral lines in only a handful of the brightest stars, so a substitute was used involving colors measured by comparison of a star's brightness through standard filters. Because absorption lines from individual atoms are more numerous toward the ultraviolet, the amount of mismatch between the shapes of a star's spectrum in two wavelength intervals could substitute for detailed analysis of stars' spectra. In this case the brightness of each star was measured using a photomultiplier tube, through filters corresponding to the three standard bands denoted U (ultraviolet), B (blue), and V (visual). The difference in magnitude between two different spectral bands is a quantititive measure of color, which can be compared across different sets of filters to yield information on temperature, chemistry, or interstellar reddening. The temperature and abundance of heavy elements are mixed in the $U - B$ and $B - V$ color indices, but the composition has a much stronger effect on $U - B$. Therefore, astronomers could compare the two color measures and use the mismatch between the two, compared with what we would expect for stars composed of the same mix of elements as the Sun, to estimate the heavy-element fraction. From long tradition, astronomers often refer to all elements more massive than helium together as "metals". In discussing objects as composed of hydrogen, helium, and metals, we may stand in the tradition of the ancient Greeks' four elements: earth, air, fire, and water.

Similarly, in tracing the orbital excursions of stars perpendicular to the plane of the Milky Way, a proxy had to be used because the stars actually located far from the plane are too faint for such observations. Instead, stars were selected based on how far their velocity would take them from the galactic plane, selecting the ones now moving so rapidly through our vicinity that their orbits will eventually carry them to such heights. This selection used the local motions of the stars, derived from their radial velocities, proper motions, and estimated distances. These data allowed a prediction of how far the star's orbit carries it from the galactic plane, and thus whether each star is bound within the disk or belongs to the (nearly spherical) distribution of halo stars.

These shortcuts allowed a comprehensive comparison of stellar populations in our vicinity, based on both chemistry and motions, presented in a classic 1962 paper by Olin Eggen, Donald Lynden-Bell, and Allan Sandage (often known simply as ELS after its authors). Their work set the benchmark in the study of galaxy formation, and it is no stretch to say that every major work on the formation and early history of galaxies since has originated either in support of or reaction to ELS. Their data made it clear that chemistry and motions are linked for the two stellar populations. All metal-rich stars like the Sun orbit in the Galaxy confined to a thin disk, where the spiral arms and dense interstellar material are, and where star formation still proceeds. These stars have a wide range in age, including the very youngest stars we can find. The metal-poor Population II stars, in contrast, occupy a wide range of orbits, not particularly concentrated to this plane but ranging widely "above" and "below" it, and not confined to being nearly circular either. We now know, in fact, that some of these stars orbit opposite the direction of rotation of the disk, in retrograde paths. Since heavy elements are formed in stars and subsequently released into the interstellar medium to seed future generations of star formation, it is clear that a star's chemistry should be linked to its age. Furthermore, just as a star's atmospheric composition generally samples its initial makeup, the current orbits of stars within the Galaxy largely reflect the orbits that they were formed in. Interstellar gas clouds can be moved by forces other than gravity, such as supernova explosions or collisions with other clouds, but a star is so dense by comparison that its orbit is nearly frozen once it forms. Only the cumulative effects of gravitational encounters with other stars or massive interstellar clouds can slowly alter a star's orbit. Therefore, if Population II stars are on orbits which are collectively in a spherical distribution, the Galaxy must have had that shape when they were formed. Similarly, since Population I stars form a thin disk, that was the shape of the star-forming parts of the Galaxy when they were produced.

Table 3.1. Stellar populations in the Milky Way

Property	Population I	Population II
Velocity relative to Sun	$<20\,\mathrm{km\,s^{-1}}$	$<400\,\mathrm{km\,s^{-1}}$
Excursion from disk plane	<200 parsecs	Up to 10 kiloparsecs
Orbits	Coplanar, circular	Eccentric, any direction
Metallicity	Near-solar	Down to 10^{-5} solar
Age	0–10 Gyr	10–13 Gyr
Associated ISM	Cool and ionized gas, dust	None
Clusters	Open	Globular

This striking difference led to a picture of the Milky Way's formation in which an enormous cloud of gas collapsed from a nearly round initial state to a thin, orbiting disk, forming stars all the while. The first surviving stars, dating to a time so early that the collapsing assembly was still spherical, are in Population II, including the globular clusters. The conservation laws for energy and momentum quite generally guarantee that collisions between gas clouds will lead to a thin disk lying in the average orbital direction, much as happens for dust and ice particles in the rings of giant planets, gas and dust around newborn stars, and material orbiting accreting objects such as neutron stars or black holes. Population I stars have been forming continually in this thin disk down to the present time. This picture, applied literally to galaxies in general, suggests that there was a brief epoch of intense starbirth when Population II stars were created—an epoch which must have been short because the age spread among Population II stars is relatively short, less than a billion years out of an age of 12 billion years or so. Galaxy formation in a single brief event has become known as monolithic collapse—the transformation of an initial population of galaxy-sized clouds of gas into stars, with elliptical galaxies and the Population II portions of spirals being formed in a very quick initial event, during the initial collapse toward a flat shape. Applied to different kinds of galaxies, monolithic collapse suggests that elliptical galaxies formed stars so efficiently that essentially no gas was left over to make a thin disk, while irregular galaxies were so inefficient that virtually no stars were produced until such a disk had appeared.

As so often happens in research, further observations showed that reality was not as simple as this overall picture. Further studies over the following decades showed that the Galaxy hasn't been a "closed box" over all this time, and that its assembly has been more piecemeal than such a single, mammoth event.

The first of these complications comes under the name of the "G-dwarf problem". This arose when enough main-sequence stars could be observed in our neighborhood to assess what fraction had various levels of heavy elements (metallicity). If a set amount of material started out as pure hydrogen and helium, and its subsequent evolution consisted of some fraction being converted into stars with subsequent expulsion of enriched matter, the distribution of stars by metallicity can be calculated. The observed distribution, though, violates this expectation. The simplest conceptual model for the chemical evolution of gas in a galaxy is a one-zone closed-box model—one in which no gas enters or leaves the system and the gas is assumed to mix uniformly through the volume. We distinguish two classes of stars—those with lifetimes much shorter than the timescales of interest, and those that last longer. In this approximation, gas is recycled into the interstellar medium at a rate depending only on the rate of star formation and the relative numbers of stars formed with various masses, a distribution known as the *initial mass function* (which enters the problem through the ratio of stars formed with short and long lifetimes). This is a sensible first approximation since so much of the nucleosynthesis of heavy elements takes place in the most massive (and short-lived) stars. Under these conditions, there is a simple analytic relation between the metal abundance in the gas and the remaining gas fraction (independent of the time at which this gas fraction is reached). If R is the fraction of gas returned to the interstellar medium, and y is the yield, or fraction of stellar mass which is transformed into heavy elements, this simple model predicts a heavy-element abundance Z according to

$$Z = y \ln m - 1$$

and also that the fraction of stars formed so far with abundances less than some value Z will be

$$S/S_1 = [1 - m(Z/Z_1)]/(1 - m_1)$$

where m_1 is the current mass fraction. This simple model is designed for primary elements—those produced within the stars beginning with hydrogen. The behavior of secondary elements—those which require a seed abundance of some heavier nucleus—will in general differ. Furthermore, for a more realistic model and one which can now find application at high redshifts, one needs to take into account the different recycling timescales for various heavy elements. Different elements will be recycled with the highest yield from different stellar masses (particularly distinguishing supernovae of types I and II), so the instantaneous part of the approximation needs to be relaxed in a more realistic treatment. The name "G-dwarf problem" traces to the fact that it was for Sun-like G-dwarf stars that it was first recognized, although it applies to all other kinds of long-lived stars as well. The requirement for stars to trace the metallicity history of the Galaxy is obviously that we use a class of stars long-lived enough for us to see survivors from the earliest epochs of metal enrichment still existing as "witnesses" to that history. Spectral features in the overall light from other galaxies shows that the G-dwarf problem is not unique to the Milky Way. There are too few low-metallicity stars in the solar neighborhood for the present-day metal abundances. The same results persist in recent studies that incorporate a detailed accounting of our neighborhood, differences in stellar temperature at a given mass, and thus for differing lifetimes at constant spectral type, and winnow the sample to the stars that are not much younger than the disk itself.

The obvious answer to the G-dwarf problem is to break the closed box—perhaps it's telling us that the Galaxy wasn't made in a day, but that gas continued to fall into it even after star formation began in earnest. In retrospect, this may have been the first move toward the contemporary use of the phrase "galaxy building" to stress what a drawn-out process may have led to today's galaxies. This solution implies that there has been, and probably still is, substantial gas which has yet to be incorporated into galaxies, which might be found in such forms as 21-cm emitting clouds in our own vicinity (perhaps to be identified with the high-velocity clouds which don't follow the disk dynamics in the Milky Way) or as hydrogen absorption against the light of background quasars, which lets us look for its traces far back in cosmic time. Other solutions have also been pursued, since any process which breaks the closed-box assumptions can help. For example, a very short-lived generation of massive stars enriching the gas before any surviving stars were formed would go in the right direction, as would certain kinds of changes in the stellar initial-mass function or metal yield with time.

Quantities such as metal abundance are, naturally, not constant across a galaxy. It is quite common among both ellipticals and spirals to see abundance gradients among stars and gas, in which the metal fraction rises toward the center. In a general way, this is to be expected if the gas available for star formation is gradually concentrated to the center as star formation proceeds. This can even occur if star formation always occupies the same volume, since material ejected at high speeds from supernova explosions will be caught more efficiently deep in the galaxy's gravitational potential than near its outskirts. Barred spiral galaxies usually show no gradient within the radial range where the bar is strong, in line with calculations showing that gas flow along the bar will mix gas from various radii quite effectively.

Galactic stars have been found with [Fe/H] as low as −5.4 (for the current record-holding star HE1327−2326). In this notation, [Fe/H] represents the ratio of iron to hydrogen compared with the Sun, in logarithmic units, so that this star has only $10^{-5.4} = 0.000\,004$ times the Sun's iron content.[1] This contrasts sharply with the lowest gas-phase metallicity yet found in galaxies, abundances that would scale to [Fe/H] = −1.7 for constant ratios among metals. This points up the difficulty in doing an obvious match—asking whether the enrichment and star-formation history of the spheroid and stellar halo of the Milky Way tell the same story we see by tracing the whole galaxy population back in redshift. In principle, we don't need to study the chemistry of high-redshift galaxies in detail to check a global concordance, owing to a fortunate coincidence which seems to have been first exploited by Lennox Cowie and collaborators (1988). Virtually all the heavy atoms which have eventually been ejected into the interstellar medium were synthesized in fairly massive stars, above about 2 solar masses. These are the same stars which dominate the ultraviolet light output of galaxies (unless dust extinction is important), so there is a relation between the mean metal abundance today and the average ultraviolet surface brightness of all galaxies (neglecting AGN, where the energy output is not necessarily connected to nucleosynthesis). As given by Cowie *et al.* (1988), for smoothed mass density of metals ρZ, the mean ultraviolet surface brightness is

$$S = 2.1 \times 10^{-25} [\rho Z/(10^{-43}\,\mathrm{g/cm^3}]\ \mathrm{erg/cm^2\ s\ Hz\ deg^2}$$

The local estimates of ρZ are a few times higher than this fiducial number, for $S \approx 5 \times 10^{-25}$. There are sufficient UV photons, mostly from galaxies bright enough to see in data such as the Hubble Deep Field, to account for most of the metal production needed locally, and this is reflected in the limiting case which can be imposed on histories of star formation derived in other ways. In fact, taking account of the likely amount of dust absorption in star-forming galaxies pushes the integrated history of star formation uncomfortably close to this limit, unless there is a substantial reservoir of processed material outside bright galaxies. Recent data make it clear that such a reservoir does exist in the intergalactic medium, so this constraint may not be as simple as it first seemed.

[1] Some care should be taken in classifying stars on this basis; these stars are not nearly as poor in, for example, carbon as in iron when compared with the Sun. It is also noteworthy that no stars are yet known with [Fe/H] in the range from −4 to −5.

Attempts to connect the enrichment history of stars in the Milky Way with the abundances in gas in galaxies seen at early epochs, by looking to substantial redshift, have not fared much better. There are many estimates of metal abundances in the gas which is detected as it absorbs light from quasars, in many cases probably gas in galaxies or their extended gaseous halos. While the fractions relative to hydrogen aren't always available, the relative abundances among the various heavy elements can be derived. However, a variety of selection biases make it difficult to be sure whether we have a truly representative sample. In particular, gas with high abundances is most likely to occur in the inner parts of galaxies, regions in which there will be more dust and we will be less likely to see enough ultraviolet light from a quasar reaching us to do the measurement. At this point, we do not see the clear trend that might have been expected, of metal abundances decreasing as we look farther in redshift, which is to say increasing with cosmic time, as successive generations of stars have enriched the gas for future luminaries.

3.2 THE GALACTIC HALO: E PLURIBUS UNUM?

More detailed study of the Milky Way's stellar halo yielded evidence that its assembly was not the single smooth event assumed by ELS. This became especially clear from a major study of the chemistry of stars in our galaxy's halo by Leonard Searle and Robert Zinn in 1978. They analyzed the spectra of 177 red giants in 19 distant globular clusters, highly recognizable tracers of the Population II stellar halo in the Milky Way. Beyond the Sun's distance from the galactic center, they found no trend in the mix of abundances with distance, although a considerable range in chemistry was seen at each radius. Confronting these results with several conceptions of galaxy formation, they suggested that these clusters may have originated in independent systems which, after undergoing varying amounts of star formation and chemical enrichment on their own, were incorporated into the growing Milky Way. This notion replaced a monolithic event for the Milky Way with a more gradual buildup from smaller galaxy-like units, whose mode of formation is yet to be determined.

Nearby stellar census results also suggest a more complicated history. Below about 0.04 of solar metallicity, enough stars scatter off the main relations from ELS to prompt Bruce Carney and coworkers (1996) to write, "These results contradict the model that the metal-poor stars are a single population that is *only* the relic of the earliest stages of the Galaxy's collapse." Some very metal-poor stars have nearly circular orbits, and in detail there is no one-to-one correlation between metallicity and angular momentum for the entire population. They see evidence of kinematically distinct populations which could be the relics of distinct accretion events. Again, these data show that, while the main Population I/II distinctions are quite robust, further subgroups can be altered with events in galactic history.

Quite recently, wide-field telescopes and automated data reduction have enabled the remarkable discovery that the shreds of some of these infalling victims may still be

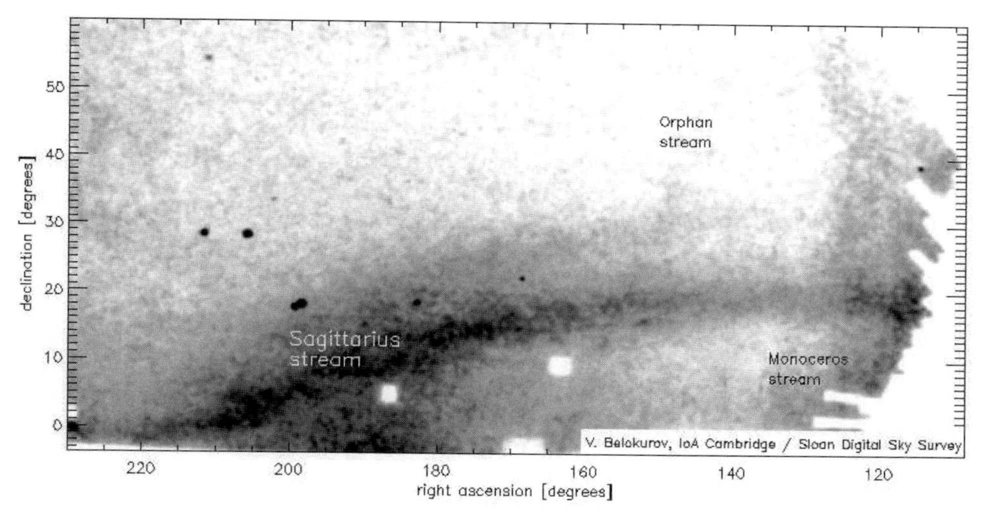

Figure 3.1. The "Field of Streams" map from the Sloan Digital Sky Survey (SDSS). This depicts the sky density of stars identified as part of the galactic halo, after rejecting the vast number of foreground disk stars based on their colors and apparent magnitudes. The prominent streams are identified as remnants of dwarf galaxies accreted by the Milky Way, whose remnant stars are being stretched into sky-spanning streams by the gravitational field of the Galaxy. (Vasily Belokurov, Daniel Zucker, and the SDSS Collaboration.)

identifiable in our galaxy's halo. The "Spaghetti Survey", led by Heather Morrison, has led to the identification of streams of stars which are distinct in both chemistry and kinematics, still identifiable after many billions of years as threadlike trails through the general background of stars. Their technique involves wholesale measurements of stars with a Schmidt telescope, through a set of filters designed to separate distant giant stars from nearby dwarfs of similar temperature, and so to allow immediate estimates of chemistry and distance for large groups of stars. This conclusion fits very nicely with the conclusions from Searle and Zinn's study, by showing that some of the final additions still leave discernible traces in galactic structure, instead of having been completely homogenized by orbital dynamics. Additional streams of stars in the galactic halo have been identified using data from the Sloan Digital Sky Survey, which allow a three-dimensional mapping (projected onto the sky in Figure 3.1). The number of accretion events we can identify in this way is important not only in understanding galaxy growth as such, but because straightforward models of dark-matter history predict many more dwarf-galaxy-mass condensations than we can associate with visible dwarf galaxies. The models might in fact be correct, if many dwarf galaxies have been assimilated by giant galaxies.

Perhaps the issue we should consider is how much monolithic and how much piecemeal growth occurred in reaching today's galaxies. As we will see, there are aspects of the galaxian bestiary which are best accounted for by each scenario.

3.3 GLOBULAR CLUSTERS: OUR COLLECTION

Globular star clusters (Figure 3.2) play an especially important role as fossils of galaxy history, not only in our own Galaxy but now in vast numbers of galaxies as distant as the Coma Cluster. This is, first, because they are so luminous, and easily identifiable and measurable; and, second, because the dense mass concentration of a globular cluster makes it very long-lasting against gravitational disruption and hence the best place to look for a record of star formation early in galactic history. It has not

Figure 3.2. The galactic globular cluster 47 Tucanae. This color-composite picture shows the color and luminosity contrast between red giants and the brightest main-sequence stars, characteristic of old Population II stars. See also color section. (This image was produced from two-band CCD data obtained by the author, R.E. White, III, and C. Conselice at the 1.5-meter telescope of Cerro Tololo Inter-American Observatory.)

proven easy to reconcile what we find in the demographics of globular clusters with the record of stellar populations in galaxies as a whole.

Globular clusters are excellent laboratories for understanding stellar evolution in old systems, being so rich that even brief evolutionary phases are well-represented. The obvious test against stellar-evolution calculations is estimating ages by fitting the observed and synthetic main-sequence locations, especially the turnoff marking core hydrogen exhaustion and the rapid pilgrimage toward red giants (Figures 3.3 and 3.4). While this exercise shows (in a robust way against a range of stellar models) that globular cluster stars are old, the oldest systems that we can identify wholesale in the Milky Way, it is still subject to errors of about 1 Gyr because of uncertainties that are difficult to address. Some subtle effects, that are second-order perturbations in other applications of stellar evolution, make a crucial difference in looking for the slow changes that characterize such old stellar populations: the exact initial fraction of helium, history of mass loss of red giants, and whether stellar rotation varies substantially among a single cluster's stars all turn out to matter at this level of precision.

The same kinds of fitting that are useful for age estimates can also yield photometric distances, though in globular clusters most of the same information comes from a census of the RR Lyrae stars by themselves. RR Lyrae stars are pulsating variables found among horizontal-branch stars, a phase of red-giant evolution in which all these variables have nearly the same luminosity and thus are very useful distance indicators. It was an initial application of such distance indicators that led Harlow Shapley to show that the center of the globular-cluster system (and thus the center of the Milky Way) is about 24,000 light-years away in the direction of Sagittarius. Globular clusters also let us assess the wholesale demographics of Population II stars, particularly by multifilter photometry and, more recently, using fiber optics to measure the spectra of many stars in a single observation. Such observations have turned up interesting, and sometimes puzzling, differences in the patterns of red-giant evolution among globular clusters and their populations of RR Lyrae variables.

One long-standing puzzle among the stars in globulars may have been resolved by recent observations, including Hubble Space Telescope imaging and additional ultraviolet data. For old stellar systems, there is a clear limit to how hot and bright stars on the main sequence should be, yet some globular clusters show a smattering of stars above this turnoff point. These "blue stragglers" must either be younger than the rest of the cluster or have been somehow rejuvenated. It now appears, from studies of their locations and frequency, that these are remnants of a vanishingly rare occurrence—the direct collision of two stars, yielding a more massive and faster-evolving product. Only in the dense cores of globular clusters, where the typical separation between neighboring stars is measured in astronomical units rather than light-years, do such events happen often enough to leave a signature we can find. Indeed, the fraction of blue stragglers in globular clusters is a strong function of local stellar density.

Red giants in nearby globulars are amenable to fairly detailed spectroscopy and hence abundance analysis. Such studies have been pursued for over 20 years,

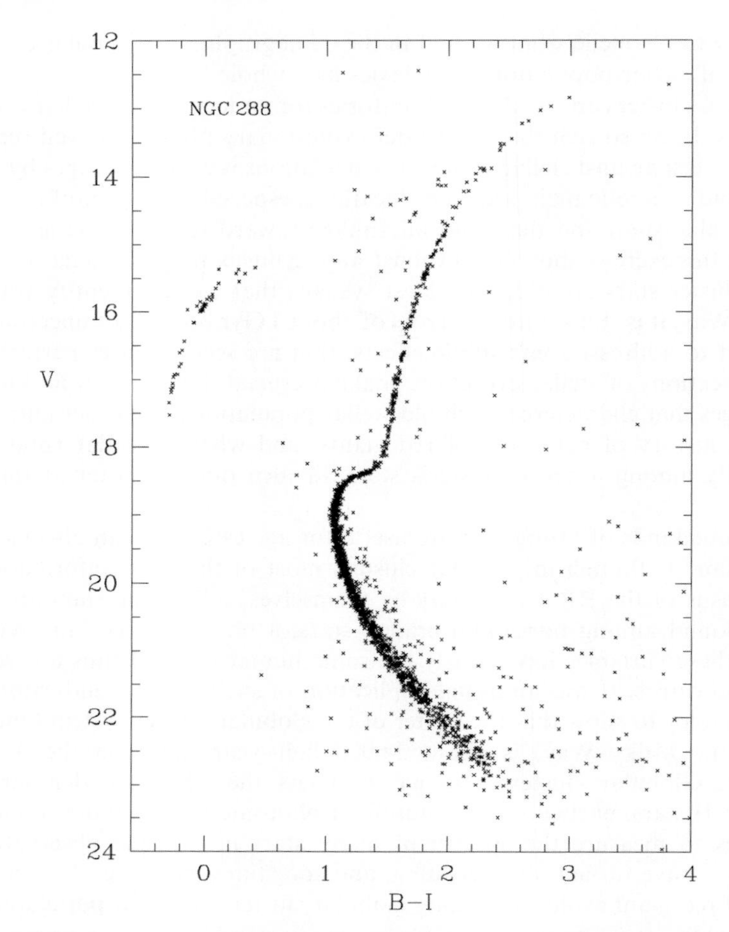

Figure 3.3. Hertzsprung–Russell (color–magnitude) diagram of the globular cluster NGC 288. This illustrates several features of stellar evolution which are important in the history of galaxies. Here, the B − I color index stands in for stellar temperature, and the magnitude scale is in the observed (rather than absolute) scale. The richly populated main sequence fades into observational errors at the faint end (*bottom*), and turns off from its diagonal course where stars are now evolving to become red giants (having exhausted their core hydrogen). The "horizontal branch" appears to the left, not being very horizontal in this color system for this cluster, marking those stars which are undergoing core helium fusion. The well-populated sequence climbing to the right is the red-giant branch, joined from the hotter (*left*) side by the asymptotic giant branch, showing stars on their final brightening before the core remnant becomes a white dwarf. The crowding of stars near the main-sequence turnoff, and their brightness, makes them an important contributor to the integrated light of galaxies with old populations. Also visible in this diagram is the sequence of binary stars; equal-mass binaries appear about 0.3 magnitude above the main sequence, particularly noticeable near magnitude $V = 20$. Scattered stars away from these sequences are mostly unrelated foreground stars, which thus appear at irrelevant magnitudes compared with cluster members. (Courtesy of Peter B. Stetson of the National Research Council of Canada and Peter A. Bergbusch of the University of Regina.)

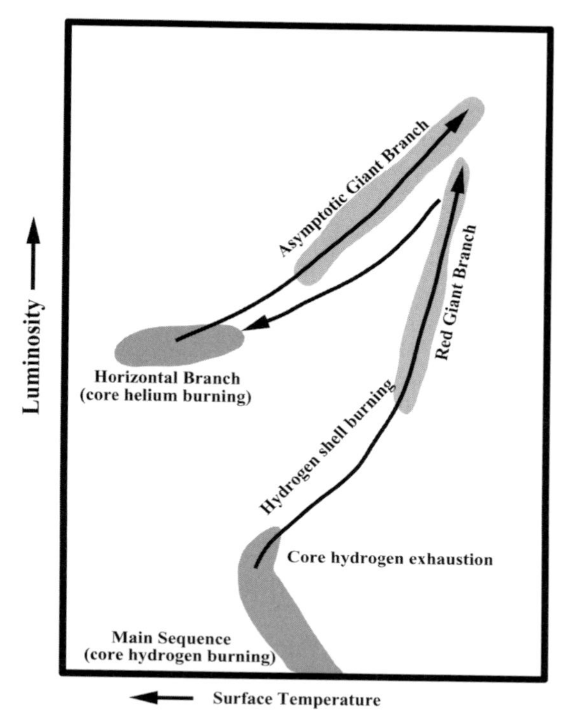

Figure 3.4. Schematic depiction of late stellar evolution. This is a sketch of the upper part of the Hertzsprung–Russell diagram for an old stellar population, as seen in globular clusters. The arrow traces the path of a single star, which is not exactly the same as the locus of existing stars at any one time. However, these phases are all so rapid compared with the main-sequence timescale that the relative numbers of stars in each region are a useful measure of the rate of evolution through those phases. The surface chemistry ceases to reflect the original makeup of the star as it ascends the red-giant branch, once mixing starts to bring processed material to the surface. More pronounced mixing occurs on the second ascent, when the star is on the asymptotic giant branch. During this phase, shell burning of H or He is occurring, with instabilities eventually leading to very high mass loss (and expulsion of a planetary nebula, in some mass ranges).

and have shown some surprising variations in the chemistry of cluster members. One needs to be able to observe well down the red-giant sequence for this, since the brightest stars in the optical and near-IR are on the second ascent of the red-giant branch in the HR diagram. In this evolutionary stage the products of fusion earlier in a star's life are finally mixed by convection all the way to the surface, changing its surface composition significantly (Figure 3.4).

Evolutionary calculations show that these stars can bring material synthesized in the star's early history to the surface in one of several possible "dredge-up" episodes, so a spectroscopic analysis would reflect a contribution from the star's own nucleosynthetic history as well as its initial abundances. Apparently, some clusters did not form from chemically homogeneous material, perhaps reflecting transient enhancement from a small number of variously placed supernovae during cluster formation.

Typical globular clusters would have been unable to bind supernova ejecta, and therefore should not have enriched subsequent generations of star formation, making them attractive snapshots of early Milky Way abundances. There is some evidence that the brightest and most massive cluster known in our galaxy, ω Centauri, is an exception, with an unusually wide star-to-star spread in chemical abundances. It has been speculated that such massive systems (in the Milky Way and other environments such as Andromeda, with its very luminous cluster Mayall 1) could have originated as the nuclei of dwarf companions which were later accreted, leaving only these clusters as identifiable remnants. This process starts to blur with the piecewise construction of the galactic halo inferred by Searle and Zinn (1978).

Globular clusters in the Milky Way are reasonably well catalogued, though some objects near the galactic center may remain unrecognized due to foreground extinction. Current catalogs list 150 such clusters.

3.4 GLOBULAR CLUSTERS ELSEWHERE

Studies of globular clusters in other galaxies have pointed out that these systems put strong constraints on the history of the galaxies themselves, a fact pointed out in numerous papers by van den Bergh. The most notable difference between globular-cluster systems in various galaxies is seen in their specific frequency S_N. This quantity was defined by Harris and van den Bergh (1981) as

$$S_N = N \times 10^{0.4(M_V+15)}$$

That is, this represents the number of globular clusters identified per amount of starlight, in units corresponding to the luminosity associated with absolute magnitude $M_V = -15$. In a simple picture in which galaxies form a constant fraction of their stars into globular clusters, S_N would be a constant except for any differential evolution owing to different formation times of the stars and clusters. In fact, wide variations are found between galaxies, especially between Hubble types. Elliptical and S0 galaxies have systematically higher S_N values than do spirals, although there is considerable variation among these types. Dwarf galaxies also exhibit a wide range, with dwarf ellipticals running high. Central cluster galaxies can have extremely high values, so high that the cluster systems may be more properly considered to belong to the cluster potential than an individual galaxy (Figure 3.5).

These results are challenging to accommodate in a scheme which builds up elliptical galaxies by merging disk systems, or produces S0 galaxies by stripping the interstellar medium from spiral galaxies. Any transformation of galaxies has to satisfy the globular-cluster statistics, which often seem at odds with other morphological evidence. In a hand-waving sense, this may be accounted for if mergers of spirals form stars systematically in young globular clusters. There would have to be a weighting toward this mode of star formation, since merely forming a fraction of

Table 3.2. Globular cluster populations in nearby galaxies

Name	Type	M_V	N_{cl}	S_N
Fornax	dSph	−13.7	6	19.9
NGC 147	dE5	−15.0	4	4.0
NGC 185	dE3	−15.2	8	6.6
NGC 205	S0/E5	−16.5	9	2.3
SMC	Im	−16.9	2	0.3
LMC	SBm	−18.6	15	0.5
M33	Sc	−19.2	30	0.6
M81	Sb	−21.2	210	0.7
NGC 1399	E1	−21.3	3,600	11.9 (cluster center)
Milky Way	Sbc	−21.3	160	0.5
M31	Sb	−21.7	270	0.6
NGC 5128	S0 + Sp?	−22.0	1,700	2.7
NGC 3923	E4/S0	−22.1	4,300	6.2
NGC 4649	S0	−22.2	5,800	7.6
M87	E0	−22.4	16,000	17.5 (cluster center)
NGC 4472	E1/S0	−22.6	7,400	6.7
NGC 3311	E0/cD	−22.8	13,700	10.4 (cluster center)

stars in these clusters in the same fraction as the galaxies had initially would not increase S_N. From van den Bergh's 1995 paper, these challenges include (1) the luminosity function of globular clusters having a very different form than that seen for young luminous clusters in merging systems such as NGC 4038/9, which would require preferential dissolution or destruction of the fainter clusters, (2) the relative numbers of metal-rich and metal-poor clusters in nearby giant ellipticals appear to have more metal-poor clusters than one would expect from merging fairly luminous spirals, and (3) the internal abundance ratios for globular clusters are not what we would expect for nucleosynthesis in a disk population. The α-elements[2] are relatively more abundant than iron-peak elements (as compared with solar values), which makes more sense for star formation directly from very gas-rich systems (or subsystems). This constraint could be satisfied by mergers from disk systems, but only if the disks were still largely gaseous rather than stellar at the time of the merger.

Several recent, key observations have given renewed emphasis to the role of globular clusters as historical markers.

Many starburst galaxies, particularly interacting and merging systems, contain very rich, luminous "super star clusters" (SSCs) that could be young globular clusters. A few had been identified from ground-based observations (such as the compact and luminous clusters in the nearby starbursting system NGC 1569), but

[2] α-elements are species (magnesium and heavier) with even atomic number. They are produced by helium nuclei (α-particles) bombarding lighter nuclei, particularly during the explosion of a type II supernova. Compared with elements near Fe, mostly produced in type Ia supernovae, α elements would build up more rapidly when star formation is active.

Figure 3.5. The rich systems of globular clusters seen around some ellipticals, particularly in cluster environments, are illustrated by this R-band Keck image of the luminous elliptical galaxy NGC 6042 in the Hercules cluster, Abell 2151. Deep as this exposure is, it reaches only the most luminous clusters, only hinting at the richness of the cluster system. (Image courtesy John Blakeslee.)

their full significance became clear only when HST imagery showed how compact they must be and how common these are in violent galactic environments.

The colors of these clusters can be used to estimate ages, based on starburst models. There is a tendency for clusters in a single interacting galaxy to clump in age, and the ages are in reasonable accord with the dynamical status of the system—the clusters appear to form during the period of strongest tidal disturbance. The cluster ages are important for a key question: Are these actually as massive as nearby globular clusters, so they will be similarly tightly bound and long-lasting? The light we can see is dominated by much more massive stars than the ones dominating the mass, and at early times the mass-to-light ratio evolves strongly, so that both the age and IMF are needed to derive meaningful masses and tell whether these are in fact newborn globular clusters. We can come closest to answering this for the nearby SSC in the Large Magellanic Cloud, R136 at the center of the Tarantula Nebula. HST imaging can reveal stars down to about 2 solar masses in this cluster before the cumulative confusion from the light of brighter stars swamps their detection, so we know that the IMF continues to rise to at least this level. There are also dynamical measurements in the form of velocity dispersions from spectra of SSCs in several nearby galaxies. Using 8- and 10-meter telescopes, several groups have now measured stellar velocity dispersions of 15–33 km s^{-1}, which can yield cluster masses using the virial theorem and sizes from HST imagery. They derive total masses of 3.3×10^5–1.5×10^7 solar masses, implying that (some) SSCs are systems fully comparable with globular clusters, and likely to remain gravitationally bound as they age. If the Milky Way has forgotten how to make them, other galaxies still remember. If we can find out just what conditions are needed to form such clusters today, this might tell us what conditions were like when the Milky Way formed its globular-cluster population long ago.

HST data have also dramatically extended the census of globular clusters in more distant elliptical and S0 galaxies, as regards not only number and luminosity function, but optical colors. The luminosity function of globular clusters is as nearly constant from galaxy to galaxy among these systems as the data will show; if the SSCs really age into globulars, dynamical effects must change their luminosity functions. The luminosity function of globular clusters has a shape close to Gaussian and a fairly narrow width (standard deviation about 1.1 magnitude), making it a reasonable distance indicator if we can detect clusters close to the peak luminosity (at about absolute magnitude $M_V = -7.4$, with some evidence that this depends on metallicity at the 0.25-magnitude level).

Besides the luminosity functions, the color distributions of globular clusters hold interesting clues to the early history of many galaxies. To first order, the color of globular clusters is an age indicator, with metallicity playing a second-order role within a given galaxy. For both elliptical and S0 systems, it is common to have a bimodel color distribution, which is widely taken as evidence for two important epochs of cluster formation. At one point it was tempting to use this as evidence for formation of the present galaxy in a merger, with the bluer clusters formed during the merger. More detailed examination, however, shows that both of these cluster subsystems are statistically linked strongly to the properties of the galaxy as measured now. What kind of early history would give multiple epochs of globular-cluster formation, but have the number of these clusters proportional to the galaxy's present-day stellar luminosity? Brad Whitmore and coworkers invoke a mechanism that appears increasingly important in the early history and formation of galaxies, one which seems like swimming upstream during an epoch of collapse—gaseous outflows sweeping the galaxy clean of gas and temporarily interrupting cluster formation. In this picture the first clusters represent the initial epoch of active star formation, one result of which was to power a global galactic wind which swept the galaxy nearly free of dense gas. When much of this gas eventually fell back, a second star-forming epoch set in. This notion suggests several additional patterns—galaxies of both high and low mass should be less likely to have multiple cluster populations, since massive systems would be less likely to host a global wind and low-mass galaxies would be less likely to recapture any gas once a wind was set up. Such a scheme also satisfies most of the constraints from statistics and metallicities of globular clusters in nearby galaxies. The low-metallicity globular clusters may even date to the formation of the small "protogalactic" objects which are expected in hierarchical formation models, in which case they would have fallen along with their parent systems into today's more massive galaxies.

3.5 DWARF GALAXIES

Dwarf galaxies are very numerous, so it is obviously important to understand their history if we are to claim we know about galaxies in general. In some respects, dwarfs might have had very different histories than giant galaxies, since they represent lower masses and in some cases smaller fractions of normal to dark matter. These might mean that they came together more slowly or later than luminous, massive galaxies. Since dwarfs often accompany giant galaxies, we need to know how they relate and interact. Such interactions have been seen in the halo of the Milky Way, in evidence for disruption and accretion of former dwarf satellite galaxies (Figure 3.1). Even dwarf galaxies far enough from the Milky Way to be safe (so far) from this kind of gravitational mistreatment have apparently been affected by our galaxy, to a degree that appears to increase with proximity. The stellar populations of local dwarfs can now be studied in some detail, showing that several of them formed their stars in several very distinct episodes. The Carina dwarf galaxy shows particularly strong evidence of such episodes. Exquisite color-magnitude diagrams obtained by Tammy Smecker-Hane and collaborators, and Denise Hurley-Keller and coworkers, using telescopes at Cerro Tololo, show that most of the stars in this system formed not early on as might be expected for a spheroidal system, but in two quite distinct episodes about 7 and 3 billion years ago. The first such episode is not a complete mystery, if tidal stress from a close passage to the Milky Way disturbed a population of molecular clouds, but the second burst is harder to understand. If the first episode was terminated by either running out of gas or ejecting it in a galactic wind, where did the gas for the second episode come from?

There is a connection between the star-forming history of the numerous dwarf galaxies at the faint Universe. Star-forming dwarfs can appear temporarily bright during such an episode, and could appear in significant numbers as faint blue galaxies depending on how frequent and how long-lasting episodes of active star formation might be.

3.6 BIBLIOGRAPHY

Books

Baade, W. (1963) *The Evolution of Stars and Galaxies* (Harvard University Press). Baade's 1958 lectures on stellar populations and their implications, as edited by Cecilia Payne-Gaposchkin. A more recent reprint edition is also available.

Osterbrock, D.E. (2002), *Walter Baade: A life in astrophysics* (Princeton University Press). A detailed biography of Baade, including his contributions to our understanding of stellar populations and galaxy distances, and his sometimes-working relationship with Zwicky.

Journals

Audouze, J. and Tinsley, B.M. (1976) "Chemical evolution of galaxies", *Annual Reviews of Astronomy and Astrophysics*, **14**, 43–79. Describes the basic closed-box recycling approach to the distribution of stellar metallicity, and illustrates ways in which real galaxies must be more complicated.

Brodie, J.P. and Strader, J. (2006) "Extragalactic globular clusters and galaxy formation", *Annual Review of Astronomy and Astrophysics*, **44**, 193. Recent results on globular-cluster systems and their implications for the early history of galaxies, including the possibility that low-metallicity clusters may predate the assembly of massive galaxies.

Carney, B.W.; Laird, J.B.; Latham, D.W.; and Aguilar, L.A. (1996), "A Survey of Proper Motion Stars. XIII. The Halo Population", *Astronomical Journal*, **112**, 668–692. The status of dynamical studies of metal-poor stars in our vicinity, and the star-forming history they imply. The data show that many halo stars, and much of the halo structure, do not date back to a single primordial collapse event.

Cowie, L.L.; Lilly, S.J.; Gardner, J.; and McLean, I.S. (1988) "A cosmologically significant population of galaxies dominated by very young star formation", *Astrophysical Journal Letters*, **332**, L29–L32. Sets out the connection between the combined ultraviolet light from galaxies and the mean metal abundance in today's Universe.

Eggen, O.J.; Lynden-Bell, D.; and Sandage, A. (1962) "Evidence from the motions of old stars that the Galaxy collapsed", *Astrophysical Journal*, **136**, 748–766. The classic paper connecting the chemistry and kinematics of stars to the history of the Milky Way. Virtually everything done on galaxy evolution for the following three decades either built on or reacted to this paper, commonly abbreviated as ELS.

Harris, W.E. (1991) "Globular cluster systems in galaxies beyond the Local Group", *Annual Reviews of Astronomy and Astrophysics*, **29**, 543–579. Reviews the frequencies of globular clusters in various types of galaxies.

Harris, W.E. (1996) "A Catalog of Parameters for Globular Clusters in the Milky Way", *Astronomical Journal*, **112**, 1487–1488. Describes a master catalog of globular clusters in the Milky Way. The updated electronic data set may be found at *http://www.physics.mcmaster.ca/Globular.html*

Harris, W.E. and van den Bergh, S. (1981) "Globular clusters in galaxies beyond the local group. I—New cluster systems in selected northern ellipticals", *Astronomical Journal*, **86**, 1627–1642. Defines the specific frequency of globular clusters, which has proven important in tracing the evolution and merging history of galaxies.

Hurley-Keller, D.; Mateo, M.; and Nemec, J. (1998) "The star-formation history of the Carina dwarf galaxy", *Astronomical Journal*, **115**, 1840–1855.

Searle, L. and Zinn, R. (1978) "Compositions of halo clusters and the formation of the galactic halo", *Astrophysical Journal*, **225**, 357–379. Used the distribution of chemical abundances among globular clusters to argue that the Galaxy formed in a more complicated way than a single smooth collapse.

Shapley, Harlow (1939) "A determination of the distance to the Galactic Center", *Proceedings of the National Academy of Sciences*, **25**, 113–118.

Smecker-Hane, T.A.; Stetson, P.B.; Hesser, J.E.; and Lehnert, M.D. (1994) "The stellar populations of the Carina dwarf spheroidal galaxy 1: A new color–magnitude diagram of the giant and horizontal branches", *Astronomical Journal*, **108**, 507–513.

van den Bergh, S. (1995) "Some Constraints on Galaxy Evolution Imposed by the Specific Frequency of Globular Clusters", *Astronomical Journal*, **110**, 2700–2704. A collection of ways that globular-cluster populations pose interesting challenges to schemes for galaxy history. The tabulated properties of globular-cluster systems in galaxies were taken largely from this paper.

Wheeler, J.C.; Sneden, C.; and Truran, J.W., Jr. (1989) "Abundance ratios as a function of metallicity", *Annual Reviews of Astronomy and Astrophysics*, **27**, 279–349. Discusses the observational data on chemical history of the Galaxy, and compares it with models incorporating realistic distributions of stellar lifetimes and recycling timescales.

Internet

http://smaug.astr.cwru.edu/heather/spag.html. The "Spaghetti Survey" at Case Western Reserve University, for remnants of assimilated dwarf galaxies in the halo of the Milky Way.

4

Measuring galaxies

As in any kind of astronomy beyond our own solar system, examining galaxies confines us to extremely remote sensing. This means gleaning whatever we can from the radiation reaching us across space, using whatever clues our knowledge of physics gives. The measurable quantities are limited: from a given piece of sky, we can specify the amount of radiation per second at each wavelength and its polarization state, and for an object resolved by the telescope, we can measure this for many points within it to build up an image. If the galaxy's distance can be estimated, and the effects of intervening material can be accounted for, this gives us a remarkable range of information on size, luminosity, mass, kinds, and ages of component stars.

4.1 GALAXY DISTANCES

Distance is obviously a key property, but one that becomes increasingly difficult to evaluate as we leave our own neighborhood. Within a few thousand light-years, we can use *parallax angles*, shifts in stars' positions as seen from opposite sides of the Earth's orbit, for trigonometric ranging to nearby stars, to provide the most fool-proof rung on the cosmic ladder of distance indicators. Beyond this, we must take more subtle cues, most of them depending on our knowledge of how light propagates through space. Until we span cosmological distances, the traditional inverse-square law is an accurate description of how light disperses through three-dimensional space. Light from a source of luminosity L will spread out so that the intensity I measured at some large distance D from the source is given by

$$I = \frac{L}{4\pi D^2}.$$

This may be further modified by such factors as absorption of light by intervening dust grains, which we can often measure by the changes it produces in the source's color. To estimate the distance of a star or galaxy, we need to measure some property which is invariant to its distance from us. Examples include the pulsation period of variable stars, orbital velocity of stars within a galaxy, separation of images by a gravitational lens, or fraction of microwave background radiation absorbed by hot gas in a cluster of galaxies.

Stepping outward from the nearby stars whose distances we can measure with parallax across the Earth's orbit, we can broaden the demographics of stars with known distances by using star clusters. Stellar evolution gives distinct patterns in the temperatures and luminosities of stars of a certain age, which we can use by taking the stars of kinds we understand well from nearby examples and use the fact that other members of the same cluster are at the same distance. We need this to get large calibrated sets of the brightest stars, which we can then use to larger distances—bridging the gulf to other galaxies. This is particularly important in getting us to the next rung on the distance ladder—Cepheids.

Traditionally, the most reliable measures of the distances to nearby galaxies have relied on the use of *Cepheid variable stars*. Cepheids are a class of supergiant star whose combination of temperature and luminosity does not allow a stable structure, oscillating regularly in size and temperature about their averages. This class is named after the first discovered example, δ Cephei, but the brightest example is well known for another reason unrelated to its mild variability—Polaris. The utility of Cepheids in distance measurements stems from a strong relation between the pulsation period, easily measured from the associated light variations, and the stars' luminosity. This was initially discovered by Henrietta Leavitt, analyzing photographic data on the Large Magellanic Cloud. This nearby galaxy is an excellent laboratory for the study of stellar populations, being close enough to resolve many kinds of stars individually and offering a large sample of stars all at practically the same distance, so that patterns can be discerned without the laborious star-by-star distance measurements often needed within the Milky Way.

One of the goals of the Hubble Space Telescope was using Cepheid variables to measure distances to a useful number of galaxies at the 10% level, tightening up this technique's contribution to the overall extragalactic distance scale. This was done largely through a so-called "Key Project", involving an international team of dozens of astronomers. It proved possible to locate and measure Cepheids in galaxies beyond the prominent Virgo Cluster (Figure 4.1), so that the remaining issues in going from these distances to the expansion rate of the Universe—the Hubble constant—now hinge on how fast we are moving relative to the Virgo Cluster and how accurately the local Cepheid calibration is known.

In recent years, several less traditional, but effective, distance indicators have come into use, allowing even ground-based telescopes to measure distances to galaxies as distant as the Virgo Cluster. These rely on the statistical properties of the stellar populations in various ways. For galaxies with large numbers of globular clusters, we use the fact that their distribution in luminosity—the *luminosity function*—is remarkably constant from galaxy to galaxy (in groups and clusters, where we can compare them directly). These are the brightest subsystems in elliptical and S0 galaxies, making them attractive in these galaxies that contain no young stars and hence no

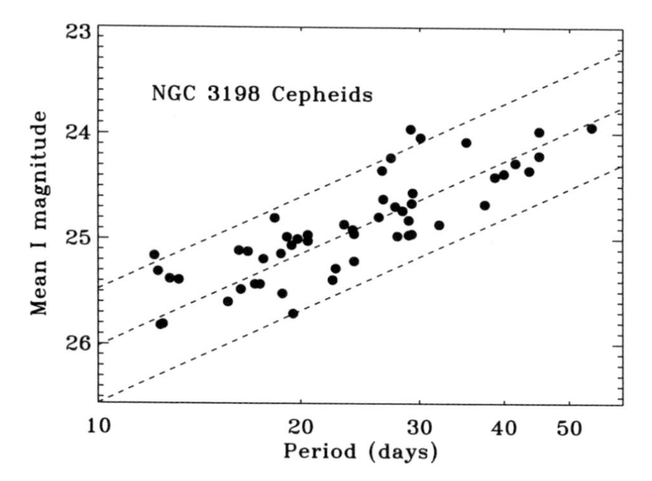

Figure 4.1. The period–luminosity relation for Cepheid variables, as seen in the Hubble Key Project data for the spiral galaxy NGC 3198. Near-infrared I magnitudes are shown, time-averaged over the pulsation periods, to reduce scatter produced by dust extinction. The central line indicates the mean relation, with the upper and lower ones showing the width of the band expected from stellar evolution (through the finite width of the instability region in the Hertzsprung–Russell diagram). Distance is measured from the brightness ratio (that is, vertical offset) with respect to some system of well-known distance, often the Large Magellanic Cloud. (These data were published by D. Kelson *et al.*, *Astrophysical Journal*, **514**, 614, 1999.)

supergiant stars to be Cepheid variables. Another possibility for these early-type galaxies is the use of *surface-brightness fluctuations*. These arise because each piece of a galaxy's image contains the light from a finite number of stars, and for a fixed surface brightness that number of stars depends on the galaxy's distance (Figure 4.2). In the ideal case of identical stars, the fluctuations between adjacent regions would reflect Poisson statistics, having a standard deviation relative to the mean of $1/\sqrt{n}$ when an average of n stars contribute in each part of the image ("part" being defined as a region which can be separately resolved). If we know the luminosity function of the giant stars contributing much of the light in elliptical galaxies (or in principle the bulges of spiral galaxies), a measurement of the statistical fluctuations between neighboring points in the image can give a distance estimate. Finally, as for globular clusters, the luminosity functions of planetary nebulae in galaxies are almost constant from one galaxy to the next. Images through filters that transmit only a very narrow band including the intense emission line of O^{++} at 5007 Å are very effective at detecting planetary nebulae and, together with images at other wavelengths, can distinguish them from other emission-line sources such as star-forming regions and supernova remnants. Ground-based imaging had ranged as far as Virgo for these detections, including a population of planetary nebulae in the space between the individual galaxies.

All three of these techniques furnish an interesting contrast to the use of Cepheids, because they can be applied to kinds of galaxies that do not host Cepheids.

Figure 4.2. Measuring distances using surface-brightness fluctuations. This shows a small region of the Local Group elliptical galaxy M32 (*top*). The brightest individual red-giant stars are visible. Successive panels show the effect of observing it at distances greater by factors of 2, so that the amplitude of observed fluctuations for a given surface-brightness level changes. This occurs as the mean number of stars per resolution element changes. (Images made with data retrieved from the NASA Hubble Space Telescope archive; R.M. Rich was the original principal investigator.)

They can, of course, be extended by the use of space observations to higher resolution and sensitivity, and such work is in progress.

Among stellar standard candles (more accurately "standard bombs") the brightest are supernova explosions (Figure 4.3). These come in a bestiary of types, some of which are much more standard and better understood than others. The distinctions among various kinds of supernovae—denoted types Ia, Ib, Ic, II, and perhaps others—can be made from their spectra and pattern of fading after the outburst. Type II supernovae, which have the confusing property of being associated

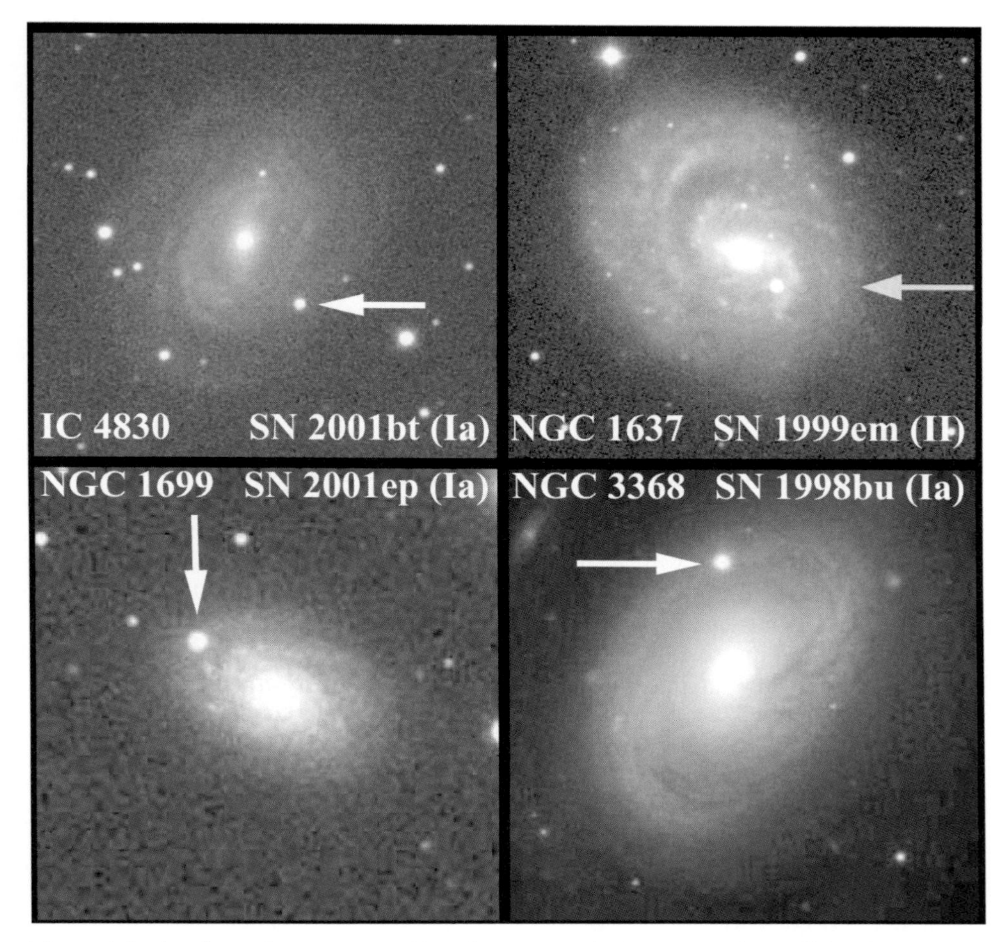

Figure 4.3. A gallery of supernovae. These images, made close to maximum light, demonstrate how bright supernovae can become, which is one of the factors making them important distance indicators. Three of these are type Ia explosions, which have sufficiently consistent peak luminosity to make them useful in probing cosmology. (Images courtesy of Nicholas Suntzeff, Cerro Tololo Inter-American Observatory.)

with Population I stars, arise from the death of massive stars. When the stellar core can sustain no more energy-producing fusion reactions, the star's gravity collapses the core inward, yielding a neutron star (and perhaps, sometimes, a black hole immediately). This transformation drives an immense shock wave and flood of neutrinos which blows the star apart. Such an event was seen as Supernova 1987A, the self-destruction of the star Sanduleak -69° 202 in the Large Magellanic Cloud. However, type II supernovae are not very useful as distance indicators, since their progenitor stars can have a significant range of masses and evolutionary states, exemplified by the fact that SN 1987A itself was substantially less luminous than was thought typical for type II events. In contrast, type Ia supernovae have proven to be very respectably standardized. These events can be recognized by their spectra, free of hydrogen lines and dominated by strong features of such heavy elements as sulfur, calcium, silicon, magnesium, iron, and nickel. These supernovae show a consistent peak luminosity, and the consistency can be made even better by removing a correlation between peak luminosity and rate of fading, making these excellent standard sources for measuring great distances across the Universe. The evidence for an accelerating Universe—a positive cosmological constant—came from samples of high-redshift type Ia supernovae.

When suitably calibrated, there are several global relations which can be used to estimate galaxy distances. Any of the Faber–Jackson, Tully–Fisher, and Fundamental Plane relationships described below can be used in this way.

Finally, at the greatest distances and for the largest numbers of galaxies, we are forced to estimate their distances solely from the Hubble law and a value of the Hubble constant (supplemented at large redshifts by a cosmological model linking redshift and distance over such large spans that the linear redshift–distance relation will not hold. This technique has the built-in systematic error that uncertainties in H_0 propagate in a systematic way across the whole sample.

The list above forms the rungs of a traditional interlinked ladder of distance indicators, in which each one helps calibrate subsequent techniques useful to greater distances. Such an approach, however carefully executed, means that the final and largest distances have error sources which have built up over all the calibrating techniques. As a check on all this, there are several more exotic distance indicators, useful in special situations, which rely more or less directly on physical principles to give galaxy distances directly from observations. Several of these provide very interesting alternate paths to extragalactic distances.

For an expanding (or alternately expanding and contracting) object, a combination of spectroscopic and brightness measurements can give its distance. In the *Baade–Wesselink* approach, applicable to such objects as Cepheids and the expanding envelopes of supernovae, one uses the Doppler shifts of spectral features integrated over time to ask how much, in linear units, the envelope has expanded over that time. Meanwhile, the difference in brightness and temperature, from photometric measurements, plus either blackbody physics or more detailed simulations, tells by what factor the object has expanded or contracted. The derived distance is the one at which that linear change corresponds to the right ratio and gives a consistent total brightness. This has been applied with reasonable success to both Cepheids and supernovae, giving confidence that the usual calibrations and assumptions of uniformity are sensible.

Again making use of supernovae, we can also use their *light echos*. The brief and intense flash of radiation from a supernova explosion illuminates the surroundings. When we include the peculiarities of seeing the illumination subject to the speed of light, what we will see is a ring of light scattered from material at a particular (increasing) distance in front of the supernova, with clouds of matter at various distances appearing as rings of different radius. As was shown in the context of light from nova outbursts by P. Couderc in 1939, these rings will expand from our vantage point at speeds well above the speed of light, purely because of the viewing geometry and the boundary condition that we see only those locations where the excess time taken for the light's detour before we see it corresponds exactly to the time since we observed the explosion. The combination of angular size and expansion speed constrains the allowed distance. This has worked most famously for SN 1987A. This approach may be extended, in principle, to older remnants in more distant galaxies by imaging in polarized light so that the scattered radiation has improved contrast.

In a similar vein, SN 1987A provided an independent measure of the distance to the Large Magellanic Cloud through unintentional observations of the way its explosion illuminated material the star had earlier blown off as a ring (Figure 4.4). Ultraviolet spectra taken within the first few weeks after we observed the explosion showed a strong temporary peak in emission from diffuse gas, which faded rapidly thereafter. Imaging from the *Hubble Space Telescope*, launched over three years later, showed that the star had been surrounded by a ringlike nebulosity, elliptical in projection. The strong line emission was produced as the initial energetic radiation ionized this material. Thus, the peak time of emission tells us, via light travel, how far the ring is from the supernova, and the angular size now tells how far the supernova is from us.

Gravitational lenses offer a way to measure very large distances independent of assumptions about the Hubble constant. This makes them attractive since we can see gravitationally lensed objects to very large redshifts (currently to at least $z = 4.7$; additional candidates have multiband fluxes consistent with redshifts as large as $z = 10$ but are still pending spectroscopic confirmation). The relevant point in this context is that the light travel times for various images of a lensed quasar (Figure 4.5) will differ, by times from weeks to years for various cases. If the mass of the lensing galaxy is well known—for example, from a velocity dispersion measurement—the delay between the paths scales with the distance to the source (and hence inversely with the Hubble constant, if the result is expressed in that way). There have been numerous programs to seek the time delay by correlating the variability patterns of lensed images, most successfully for the "original" doubly-imaged QSO 0957+561.

Since there are many structures in galaxies that approximate disks in circular rotation, there has been discussion of combining Doppler shifts and proper motions to yield geometrically determined distances. Very small sources which can be observed with radio interferometry, such as masers in star-forming regions or the atmospheres of evolved stars, can show proper motions due to their orbital motion. If these are in a circular structure, we can use Doppler shifts of similar sources at the apparent ends of the disk to measure the characteristic orbital velocity, and derive a distance by asking how far the object must be for this to match the proper motion seen for the tracers best placed to see their transverse motion. This has actually been carried out for masers in a disk close to the nucleus of the nearby active galaxy NGC 4258, with results that gave not only a distance, but a mass estimate indicating a central black hole.

Interaction between the cosmic microwave background and the hot particles between the galaxies in clusters can also yield distance estimates. In the *Sunyaev–Zeldovich effect*, the microwave background photons traversing a cluster will be

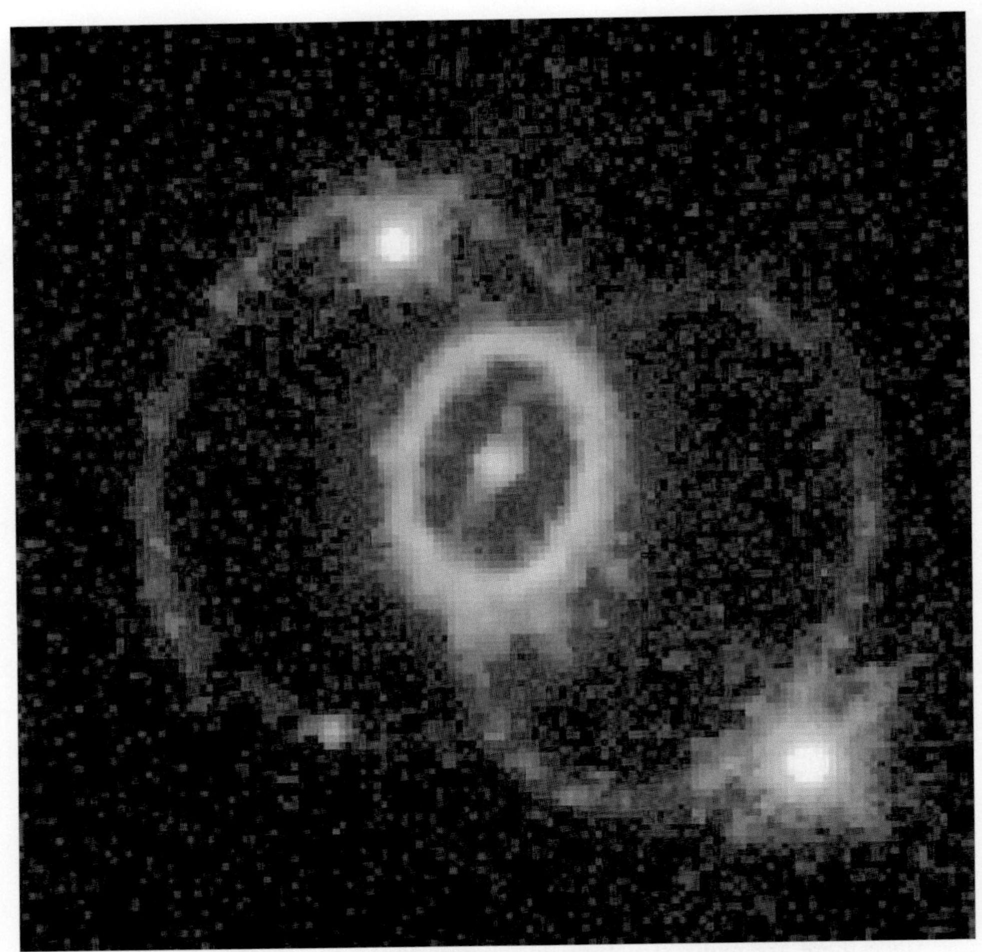

Figure 4.4. The mass-loss ring around SN 1987A in the Large Magellanic Cloud, as imaged with the Hubble Space Telescope in 1994. The flash of emission as the supernova's ionizing radiation reached this ring had been observed spectroscopically shortly after the explosion was seen. Comparison of the time required for the radiation to reach the ring and the angular diameter of the ring gives a geometric distance estimate for the Large Magellanic Cloud, independent of other steps in the distance ladder. Measurement of the ring turns out to be somewhat time-critical after the explosion, since the expanding explosion ejecta are now observed reaching the ring and heating it, which will probably destroy the emission-line gas. (Image courtesy of NASA, ESA, and C. Burrows of STScI.)

boosted to slighly higher energies by scattering off the energetic electrons in the intracluster gas. This is observed as a "hole" in the microwave background, at frequencies where more energy is lost than gained. Since the intracluster gas is itself observable in X-rays, its temperature and density profile are independently known,

Figure 4.5. The double quasar Q0957+561 and its lensing galaxies. The two bright images are from the quasar, at redshift $z = 1.41$. The lensing galaxy is partially visible close to the southern (lower) quasar image, with additional associated galaxies (at $z = 0.36$) throughout the field of view. QSO 0957+561 was the first identified example of strong gravitational lensing, and has offered a rich field for study with the radio structure of the quasar providing unique constraints on the lensing mass distribution. (This image was produced from multicolor WFPC2 images in the NASA/ESA Hubble Space Telescope archive, from data obtained with G. Rhee as the principal investigator.)

and such hot material is likely to be smoothly distributed simply because internal structure would smear out rapidly at these temperatures. The distance can be calculated by requiring the amount of gas measured in emission and absorption to be the same. This exercise has been done for a number of galaxy clusters, despite the microwave absorption being so minute (at the level of one part in 10,000) that for years this was almost the standard of comparison for an observation that was not worth trying. Its detailed application has been limited by the discovery that many clusters are dynamic environments with previously unsuspected levels of structure, but it is conceptually important in confirming the overall distance scale using completely different physical processes that any other technique.

4.2 GLOBAL PROPERTIES—SIZES, LUMINOSITIES, AND MASSES

Most astronomers have had, at times, to contend with physicists who like to remark that astronomy is the "science" where the error bars go on the exponents. Against this background, it is slightly galling that we cannot claim to have very precise measurements of the masses or brightness of any galaxies, and that estimates of their distances are only slightly better. Galaxies do present formidable obstacles to what would normally be good laboratory measurement practice, so astronomers can perhaps be forgiven this lapse. The radiation from galaxies comes from extended areas on the sky, without sharp edges, and is measured against various kinds of foreground and background radiation. Lacking sharp edges, they will not allow us to draw a circle around some region and say that it includes all the light.

Despite this fuzziness, we need some ways to describe the sizes of galaxies. There are several useful approaches, tied to various extents either strictly to observable quantities or theoretical concepts. It is straightforward to measure the brightness within some particular aperture size, so that there are measures of galaxy magnitude which integrate the flux within either a particular isophote or within an elliptical approximation to this isophote. Holmberg introduced a system integrating the light within an isophote corresponding to photographic (essentially blue) magnitude 26 per square arcsecond. This level proved too faint for routine photographic photometry, so de Vaucouleurs adopted a system based on B_{25}—the magnitude integrated above a surface brightness of $B = 25$ per square arcsecond. These systems have the advantage of being completely empirical, relying on no assumptions beyond our ability to calibrate the data properly. This is, in fact, not really trivial for bright galaxies, since we always see galaxies superimposed on foreground and background "sky" light which must be subtracted, and there is no clear place where we can guarantee that a galaxy's own light has dropped to absolutely zero. The best we can do is estimate the sky light from locations far enough away that the galaxy contribution is likely to fall below some small fraction of the likely sky error, while working to avoid systematic problems with the sky intensity from such causes as vignetting in the optical system, scattered light from bright stars, and actual changes in the sky brightness with position across large enough pieces of the sky.

These isophotal magnitudes have an obvious bias with galaxy surface brightness, missing an increasing fraction of the light for galaxies of lower surface brightness (a fraction as large as "all" for galaxies whose surface brightness nowhere exceeds the threshold level). In an effort to produce flux data which are more properly comparable for a wide range of galaxy properties, and less vulnerable to systematic bias from foreground absorption by Galactic dust or cosmological dimming, serious effort has gone into defining effective radii and magnitudes derived from flux within the effective radius. As defined by de Vaucouleurs, and used in the three successively larger Reference Catalogs, the effective radius is the radius within which half of the light from a galaxy originates (as projected in the sky; this differs from the projection of the three-dimensional radius enclosing a volume within which half of the light arises). This quantity will be directly comparable for all kinds of galaxies. The obvious problem, though, is that some theoretical input is needed to guide an extrapolation from the outermost observed piece of the galaxy to "total" flux. It is safe to assume that the surface brightness continues to decline monotonically outward, and the extrapolation is generally guided by the brightness profile where we can measure it. For example, in elliptical galaxies which follow an $r^{1/4}$ profile closely, it is reasonable to assume that this continues outward. For spirals, some sort of combined profile, including an exponential distribution of light from the disk, is taken to derive an extrapolation rule depending on Hubble type, as was done in the Reference Catalogs by de Vaucouleurs and collaborators (1991). The upshot is that the effective radius can be well determined, since the allowed range of extrapolation of the brightness profile won't change it much. The luminosity within the effective radius is only half as uncertain as the total luminosity (though having the same relative uncertainty), and, again, this approach allows fair comparison across galaxies with very different surface brightnesses. For particular purposes, other measures of size have proven useful as well. One notable example is the *Petrosian radius*, defined as the radius within which the mean surface brightness is a specified multiple of the surface brightness at that radius. This has proven important in measuring *Tolman dimming*, a direct prediction of cosmological models in which space is expanding, as distinct from those in which some other process produces the redshifts of distant objects.

Using any of these measures, galaxies have a vast range of luminosity. The faintest dwarf galaxies in our neighborhood (such as the Draco dwarf) have blue-light luminosities only about 250,000 times that of the Sun, while the largest cD galaxies in the centers of some rich clusters (such as the extremely luminous central galaxy in the cluster Abell 1413) range as high as 2×10^{12} solar luminosities, a range of 10 million to one. Similarly, we can detect the starlight in the smallest dwarf systems to perhaps 1,000 parsecs from the center (as poorly defined as that point sometimes is) and in the largest galaxies as far as 2 megaparsecs from the nucleus. Galaxies thus span an enormous range in size and stellar content. The mass range of galaxies greatly exceeds the thousandfold range in mass of individual stars, though their brightnesses do not differ quite as much as the brightest and faintest individual stars because of stars' strong relation between mass and amount of light produced per unit mass. Even among spiral galaxies, which occupy a moderate span of size and luminosity among galaxies, similar-looking galaxies can in fact differ in size by a factor of 10.

The distribution of galaxy luminosities is not only very deep, it is also strongly weighted toward dimmer systems, as expressed in the luminosity function Φ. This is the number density of galaxies (usually per cubic megaparsec) as a function of luminosity or absolute magnitude, strictly analogous to stellar luminosity functions. Summed over all galaxy types, the form of this function is well-determined except perhaps for the faintest dwarfs, and the shape is closely constant over a wide range of environments from the "field" to rich clusters. Its shape is well described by the Schechter function

$$\Phi\, dL = \Phi_0 (L/L_*)^{\alpha} e^{-(L/L_*)} (dL/L_*)$$

in which L_* is a fiducial luminosity and α is the asymptotic slope at the faint end in the (log number–absolute magnitude) plane. L_* is often taken as the typical luminosity

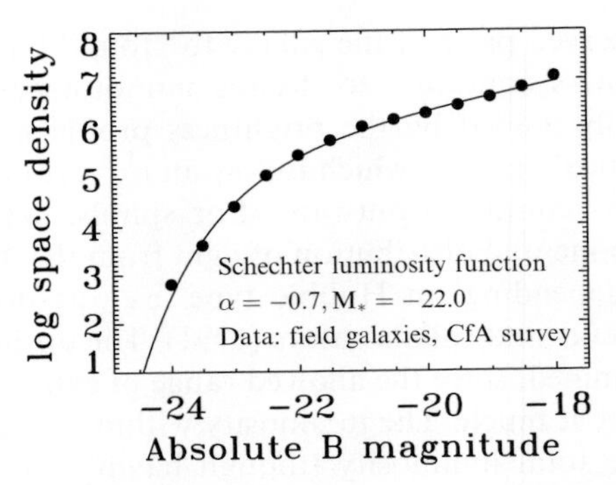

Figure 4.6. The luminosity function of galaxies, shown as the log of the space density (as expressed in Gpc^{-3}) versus galaxy absolute magnitude. Dots represent the space density of galaxies in various luminosity bins, as retrieved from analysis of the Center for Astrophysics redshift survey for "field" galaxies outside rich clusters. The curve shows a fit of the Schechter function to these data. The free parameters are in the absolute normalization, slope of the faint end, and characteristic luminosity for the turnover at the bright end. The shape parameters are remarkably robust for large samples of galaxies, and provide useful measures of the characteristic mass and luminosity scales for galaxy formation.

for bright galaxies (Figure 4.6). Any viable theory of galaxy formation should be able to reproduce the luminosity function, as well as the global relations described below.

4.3 TRACERS OF STAR FORMATION

Star formation is an important facet of galaxy evolution caught "in the act". Much of the history of a galaxy is written through changes in its stellar population, inextricably linked through the processes of starbirth and stellar evolution. We can ask whether the current level of star formation is comparable with its past level, or has increased or decreased markedly. There are several disparate tracers from various wavelength regimes which give insight to the rate of star formation averaged over different timescales. In fact, the optical and near-infrared regimes are distinguished by being the only parts of the electromagnetic spectrum in which we can easily see any but the youngest stars in galaxies.

4.3.1 Emission lines: optical and infrared

The first tracer of star formation to be developed and extensively used involves optical emission lines. The recombination lines from hydrogen have a particularly straightforward relationship to the number of ionizing photons from the hottest stars, which yields a star-formation rate (SFR) when coupled with stellar models and an initial-mass function.

In equilibrium, the number of photoionizations over the whole volume of gas ionized by a star will equal the number of recombinations as the liberated electrons become bound to protons again. This generally happens only after many weaker Coulomb encounters, which have the helpful effect of making the velocity distribution of the electrons thermal (Maxwellian) even though it would have been quite different immediately upon their ejection from neutral atoms. During recombination, most electrons will be initially captured into an excited (high-n) state, and decay to the ground level by a cascade of photon emissions. The number of recombinations can thus be measured starting with the intensity of some emission line. This is usually one of the optical hydrogen lines (Hα, Hβ), although instrumental advances have started to allow significant surveys using the infrared lines from transitions between higher pairs of n-values. These lines are intrinsically weaker in both energy and photon flux, but have the enormous advantage of being much less sensitive to dust extinction.

Using the notation in Osterbrock and Ferland's treatment, a balance between recombination and ionization requires that the number N_{LC} of ionizing photons per second satisfy

$$N_{LC} = \int_{\nu_0}^{\infty} \frac{L_\nu}{h\nu} d\nu = \int N_e^2 \alpha \, dV$$

where ν_0 is the frequency at the Lyman limit of hydrogen and α is the recombination coefficient, calculated including effects of resonent scattering which can keep essentially all atoms in $n = 2$ in extensive nebulae. For comparison with observations, we use the calculations of what fraction of decay cascades leads to a given emission line, which may either be derived from a full cascade matrix incorporating all processes between the various levels, or using emissivities so derived for various lines at the relevant electron temperature. An effective recombination α_{eff} rate can be defined, such that the emission rate of photons in the relevant emission line is simply $n_p n_e \alpha_{eff}$. (For example, at a typical electron temperature 10^4 K, each Hβ photon stands in for 8.5 recombinations, while an Hα photon results on average from nearly half of all recombinations.) The line luminosity will be this quantity, integrated over the nebular volume, as diminished by distance (via $4\pi D^2$ for small enough distances to be Euclidean):

$$L = \frac{\int n_p n_e \alpha_{eff} \, dV}{4\pi D^2}.$$

A set of young stars will give off a number N_{LC} of photons below the Lyman limit, and if the surrounding gas has sufficient column density, all these will be absorbed and eventually lead to recombinations (the *ionization-bounded* case). If we have a model of the stars' spectra in this region (which is unobservable because of this very absorption), we can calculate the number of ionizing phtons per second (also known as Q) as

$$N_{LC} = \int_{\nu_0} \frac{L_\nu \, d\nu}{h\nu}$$

in which the spectrum is converted to photons from the more usual energy units.

A widely-used relation was derived by Kennicutt (1998), using models for the ionizing continuum of hot stars. Extrapolating to the entire mass in stars with a

Salpeter (1955) initial mass function of power-law form, it is:

$$SFR = L(H\alpha)/1.12 \times 10^{41} \text{ erg/sec}$$

in solar masses per year, while counting only those stars with masses above 10 solar masses, which actually ionize surrounding hydrogen, it becomes:

$$SFR(>10M_{\odot}) = L(H\alpha)/7.0 \times 10^{41} \text{ erg/sec}$$

4.3.2 Ultraviolet continuum from hot stars

Since young stellar populations are distinguished by hot, massive, and short-lived stars, it is natural to seek their direct signatures in the emitted ultraviolet range. Especially shortward of about 2000 Å, only O and B stars are bright, so that a particular mix of stellar mass and age directly implies a star-formation rate averaged over less than 10^8 years. With an adequate spectrum, we can make progress in untangling the temperature mix of these stars, and estimate their abundances from the strengths of stellar-wind features. If the star-formation rate is constant over the lifetimes of the stars contributing in the far-UV (around 1500 Å), the UV spectral shape is close to a power law with $F_\lambda \propto \lambda^{-2}$. This translates to flux per unit frequency F_ν nearly constant with wavelength, so the star-formation rate is measured in a robust way using frequency-linked quantities:

$$SFR = 1.4 \times 10^{-28} L_\nu \text{ erg s}^{-1} \text{ Hz}^{-1}$$

again in solar masses per year.

The weakness of ultraviolet measurements lies in the necessary corrections for extinction and scattering by dust grains. Since star formation is aided by grains, and current star-forming regions are generally also dusty, the very galaxies which would otherwise be brightest in the UV are also the most susceptible to having their starlight absorbed by grains. Most of the stars' radiation is emitted in the wavelength range where absorption is most efficient. The abundance of grains depends on metallicity; empirically, there is a broad correlation between the slope of the observed UV spectrum and the metallicity as derived from emission lines. This fits with the dust/gas ratios found in nearby systems of various levels of chemical enrichment (particularly using the Magellanic Clouds for comparison), and has been used beginning with work by Daniela Calzetti and collaborators to estimate the effective extinction from the shape of the UV spectrum in star-forming galaxies. This approach has figured prominently in efforts to retrieve the overall history of star formation from galaxy statistics in, for example, the Hubble Deep Fields.

4.3.3 Far-infrared dust heating

The basic principle in retrieving star-formation rates from the far-IR is that whatever energy in starlight is absorbed by grains is reradiated at wavelengths corresponding to the equilibrium temperature of the grains. This statement needs some modification—small grains can be transiently heated by single UV photons so strongly as to give misleadingly high temperatures, not all the starlight is from young stars, grains are too small to be good blackbodies in the far-infrared—but it suffices to illustrate the basic energy balance. Let us consider a grain of albedo a and radius r. For incident starlight of intensity I in units of energy/(time × area), it will absorb $\pi r^2 I$ per unit time. In equilibrium, this will match the blackbody radiation from the surface $4\pi r^2 \sigma T^4$ where $\sigma = (2\pi^5 k^4)/(15c^2 h^3)$ is the Stefan–Boltzmann radiation constant.

The most important feature of this process is that, in equilibrium, luminosity is preserved (although, again, reality may introduce ugly complications along any given line of sight, due to structure in the stellar or dust distributions). The temperatures of the grains can be affected particularly by their location; active starbursts often have unusually hot grain populations, since the energy density can be high and the dust is often concentrated tightly around star-forming regions.

The relation between far-IR luminosity and star-formation rate is certainly affected by the detailed distribution of star and gas, as well as the initial-mass function of the stars. A relation such as

$$SFR = 4.5 \times 10^{-44} L_{FIR}\ (\mathrm{erg\ s^{-1}})$$

is representative, noting that there are assorted definitions of L_{FIR} in use (as detailed in Kennicutt's 1998 review).

Taking extinction and dust heating into account, a better overall star-formation tracer has been proposed which combines the emerging UV intensity with the amount of starlight absorbed and re-emitted in the far-infrared. At this point, sensitivity and resolution limits prevent us from applying this approach to any but the most luminous galaxies at high redshifts. Comparison of UV- and infrared-selected samples suggests that more luminous galaxies are dustier, as intuition suggests for objects with long and intense histories of star formation, and therefore the ones most effectively hidden from short-wavelength surveys. This is especially relevant to color-selected samples of galaxies, in which very dusty systems may be completely missed.

4.4 DYNAMICS AND DARK MATTER

Beyond the visible components which can be traced in increasing detail—stars, interstellar gas, and dust—abundant evidence shows that galaxies include vast amounts of material in some form that we cannot detect directly. This is the "dark-matter" problem.

The most straightforward sign of the dominant role of this material comes from the dynamics of spiral galaxies. Their thin disks of stars and gas in near-circular orbits allow a very clean demonstration of how to measure a galaxy's mass. In the approximation that the mass distribution is spherical (which only changes the outcome by 20% or so even for extreme galaxy shapes), the orbital velocity follows the generalized form of Kepler's third law

$$M = \frac{v^2 r}{G} = \frac{4\pi^2 r^3}{G p^2}$$

Figure 4.7. Direct image and aligned rotation curve for the Sb spiral galaxy NGC 5746. The Doppler shifts show a rapid rise in orbital velocity on either side of the nucleus, a dip upon crossing inner spiral structure, and roughly constant-velocity extensions on either side—that is, a "flat rotation curve". (The data were obtained by the author using the 1.1-meter Hall telescope at Lowell Observatory for the image, and the 2.1-m telescope of Kitt Peak National Observatory for the spectroscopic data.)

relating the mass M and orbital velocity v at a distance r from the center (or the equivalent incorporating orbital period p). The relevant mass is the total *inside the orbit*. Isaac Newton first showed that a spherically symmetric shell has zero net gravitational effect in its interior, and that theorem applies in this case. The most compelling initial evidence for unseen mass in galaxies came from rotation curves (Figure 4.7). A rotation curve is a set of measurements of the orbital velocity for various distances from the center, such as is obtained from an optical spectrum with the spectrograph slit along the major axis of a highly inclined galaxy, so that the Doppler shift differences we measure reflect most of the actual motion.

There were many attempts to derive total galaxy masses from photographic spectra, including a long series by Geoffrey and Margaret Burbidge using the 2.1-m Struve telescope at McDonald Observatory. In hindsight, the Universe was having one of its frequent laughs at astronomers' expense. The expectation was that mass would more or less follow starlight, with the exact ratio in each part of the galaxy telling us about the mass distribution of stars. That translates to a rotation curve which starts off slowly rising, as orbits enclose more and more mass, and eventually, when dealing with orbits outside most of the mass, gently rolling into the Keplerian falloff that we see in the Solar System, with $v \propto \sqrt{r}$. The sensitivity of photographic plates could trace emission lines from star-forming regions generally to a radius where the rotation curve was no longer rising, but clearly levelling off, so there were assorted attempts to fit total masses and mass profiles by assuming that this was where Keplerian motion finally set in. The joke was finally uncovered with the work of Vera Rubin and collaborators Kent Ford and Norbert Thonnard, who used a newly commissioned image-tube spectrograph on the 4-meter Mayall telescope at Kitt Peak National Observatory to record spectra of what had previously been undetectably faint outer regions of galaxies. Over and over, they found that the rise in rotation velocity did finally cease, and the rotation curves in that sense turned over, but in many galaxies the velocity does not decrease beyond that point. This is now conventionally described as a flat rotation curve. Much the same story was also being pieced together at that time from radio observations of interstellar clouds within the Milky Way. In this case the interpretation is more difficult because we generally do not know the distance of each cloud very well (unlike our bird's-eye view of other galaxies when measuring a rotation curve). One of two equally wild conclusions had to be drawn—either galaxies are imbedded in vast halos of invisible matter, whose mass completely dominates that of all the components we know how to detect by their radiation, or there is something important we don't know about how gravity works on galactic scales. By now, the concordance of such varying techniques as gravitational lensing and hydrodynamics of hot gas in galaxy clusters leads to the picture in which dark matter, whatever it is, is the dominent constituent of the Universe.

It is common to express the dark-matter issue in terms of the mass-to-light ratio, in solar units. This has the advantage of normalizing away scale differences between systems, and allowing easy comparison of galaxies and clusters of very different size and mass. Young stellar populations should have values well below unity, and old stellar populations have values 3–5 in the optical passbands. For galaxies as a whole, using the mass derived from the outermost measurable points, the M/L ratio may be a few tens, and for clusters it can be several hundred. Comparison of these numbers shows how dominant the unseen material is—more than 90% of the mass within the observable radius is dark matter in typical galaxies. Some dwarf galaxies have the smoothed matter density dominated by dark matter everywhere (shown particularly by the neutral hydrogen observations by Claude Carignan and coworkers).

Additional evidence on the role of unseen material comes from several other kinds of observations. Elliptical galaxies are more challenging than spirals for dark-matter measurements, since they generally lack the thin disk of emission-line gas which makes the kinematics of spiral disks so attractive. In a few cases observed in great detail, the velocity-dispersion profiles of stars in ellipticals require dark-matter halos at a level similar to those inferred in spirals. X-ray gas bound to some individual elliptical galaxies tells the same story. And, in those cases in which an elliptical has acquired a small, dynamically cold galaxy, producing a shifting set of stellar shells, their spacing also indicates a substantial dark component to the potential.

On scales of galaxy clusters, the presence of dark matter is shown by the velocity dispersion of galaxies and by the mass required to hold the hot X-ray gas in place. From velocity dispersions, the gross mismatch between estimated stellar masses in galaxies and the virial mass estimated from the galaxies' relative velocities and distribution within clusters was noted by Zwicky as early as 1937. This mismatch is universal among clusters of galaxies, and even larger than we see for individual galaxies; some of the dark matter is associated with the cluster potential as a whole rather than just individual galaxies. This picture is supported by observations of the distribution and temperature of intracluster gas, which must be in a potential much deeper than the galaxies alone could provide in order to stay bound. X-ray data can trace cluster dynamics much more finely than the galaxies' redshifts, in fact, since there is a limit to the number of galaxies, while the number of X-ray photons available to measure can be increased at will simply by lengthening the exposure time. Finally, gravitational lensing, in both its strong and weak versions, also traces the potential in clusters, and its results for the depth and shape of their potential wells are in fair agreement with the velocity dispersion and X-ray techniques. This is important because it relies on a completely different phenomenon, spacetime curvature, whereas the other two methods use the same physics, employing respectively whole galaxies and individual atomic nuclei as the test particles.

We do not know directly from observations of galaxies how much dark matter is present in the Universe as a whole. A uniformly distributed component can be detected only from cosmological arguments (e.g., from the density parameter q_0). Measurements of dynamics require observable tracers, such as ionized gas in spiral disks, stars in elliptical galaxies, companion galaxies, or X-ray gas. Global dark-matter conclusions must come from cosmological parameters, such as structure in the cosmic microwave background (Chapter 7), the rate of growth of large-scale clustering, or the history of cosmic expansion (Chapter 1).

4.5 GLOBAL CORRELATIONS

Big galaxies are big and small galaxies are small, a fact which dominates most correlations of global galaxy properties so strongly that the underlying physics can be hard to find. Still, and even given all these technical issues in measuring galaxies, there are still clear patterns that emerge and can give us powerful clues to how they got this way.

4.5.1 Dynamics and luminosity

The compelling evidence that galaxies are surrounded by dark matter (a more honest expression would be "visible galaxies are minority constituents of much larger and vastly more massive concentrations of something unseen") challenges the default assumption that mass traces light. It is not at all obvious that the luminosity of a

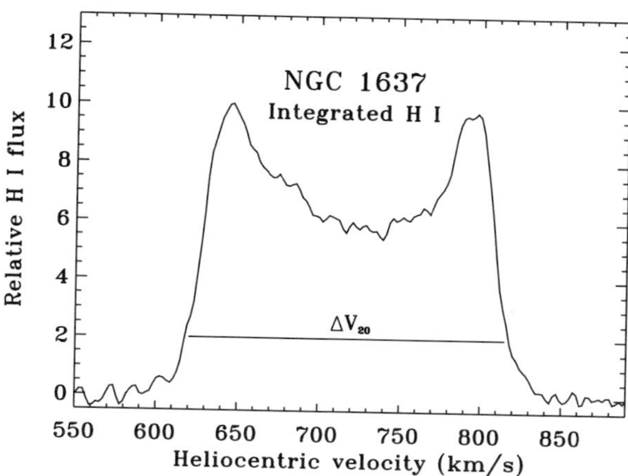

Figure 4.8. An integrated profile of the neutral hydrogen in a typical spiral galaxy, NGC 1637, observed in a beam large enough to encompass all its emission. A differentially rotating disk frequently gives this characteristic double-horned profile, as much of the gas on each side of the disk crowds into a fairly narrow range of radial velocity. The line indicates the 20% width, commonly used as an indicator in the Tully–Fisher relation, since this has been shown to scale with galaxy luminosity and its measurement is distance-independent. Data obtained with the 43-meter telescope of the National Radio Astronomy Observatory, and provided by Martha Haynes, as published in *Astronomical Journal*, **115**, 62 (1998).

galaxy (in whatever wavelength range) should tell us anything about its mass or mass distribution. However, just such relations actually hold, suggesting an intimate connection between the amounts of visible and dark matter. The original guises of mass-luminosity relations appeared separately for spiral and elliptical galaxies, and are so strong that they gave rise to a minor industry of their use for distance estimates.

The first-order indicator of a spiral galaxy's dynamics is the disk rotation velocity, which for a flat rotation curve can be quite well defined. This could be measured easily for hundreds of galaxies (by the late 1970s) using single-dish profiles of the 21-cm H I emission line. For a symmetric disk galaxy seen at any significant angle from face-on, the distribution of H I gas and the rotation pattern give rise to a characteristic double-peaked (or double-horned) line profile (Figure 4.8), from which the velocity separation of the peaks gives an indication of the rotation-curve amplitude when combined with a disk inclination to our line of sight as measured, say, from optical images. In groups and clusters, it turns out that these velocity widths are tightly correlated with galaxy luminosity, in the Tully–Fisher relation. This relation has been closely investigated because of its potential as a redshift-independent distance indicator. There are advantages to measuring the luminosities in the near-infrared bands, since internal dust extinction and recent star formation are less important there. The velocity amplitude can also be measured using optical spectroscopy, as was done in a survey of nearly 2,500 southern spirals from Australia in successive surveys by Mathewson, Ford, and Buchhorn.

An analogous relation holds for elliptical galaxies, but since many of them have essentially zero net rotation, a different indicator of the internal motions is needed. One can use the central velocity dispersion, derived now from the Doppler width of spectral lines as measured in a fixed aperture around the galaxy nucleus. In practice, this is easiest to derive by cross-correlation of galaxy spectra against stellar spectra or composites of the relevant types of stars. As shown by Faber and Jackson, there is a tight correlation between velocity dispersion and luminosity, roughly scaling as $L \propto \sigma_v^4$. As with the Tully–Fisher relation, this has seen extensive use as a distance indicator, since the velocity widths are distance-independent while the observed fluxes have a $1/D^2$ dependence.

Further study has shown that the Faber–Jackson relation is actually the projection on one plane of a more general three-dimensional construct, known as the fundamental plane. This was recognized by noting that the scatter of individual galaxies with respect to the mean Faber–Jackson relation was correlated with other quantities. Several parameters, evidently all correlated, can be used to measure the third dimension. A popular choice for the third axis is surface brightness within a fiducial radius (often the half-light radius), although metallicity as derived from absorption-line strengths has also been used. The fundamental plane for elliptical galaxies furnishes a strong constraint on any scheme that pretends to tell how they form and evolved—either only mass concentrations that follow this relation formed, or dynamics forced ellipticals into these patterns quickly. In particular, if many ellipticals are produced by merging, it must be shown that merger remnants obey the same Faber–Jackson relation as ordinary ellipticals.

The existence of the Tully–Fisher relation and the fundamental plane has taken on considerable significance with the recognition that galaxies are dominated by dark matter. These relations mean that the ratio of luminous to dark matter spans a limited range and varies in a highly systematic manner with luminosity and total mass. The normalization of these luminosity relations will generally depend on the cosmic epoch, since galaxies will show evolution even for the passive fading of systems lacking star formation. This has been observed by observations at substantial redshift, in which the luminosity for a given velocity scale is greater by amounts consistent with straightforward evolution of stellar populations.

There are also trends and scaling relations for internal structures of galaxies, which bear directly on their formation and history. In particular, many galaxies show pronounced gradients in metal abundance with distance from the core. Ellipticals show a uniform decrease in strength of metal absorption lines moving outward, which mirrors a similar trend for more luminous galaxies to be more metal-rich at corresponding radii. Color data suggest that globular clusters partake in these trends, as well as the general stellar populations. To first order, this fits with a variant of the classical picture for the formation of the Milky Way, with the star-forming material shrinking in all directions rather than forming a disk and being successively enriched by star formation as this proceeded. However, at a more subtle level there are wrinkles which aren't so easy to explain. For example, the simplest interpretation of the metallicity gradient as an implicit time sequence suggests that the ratios of various heavy elements should change with time and therefore radius, since different elements are produced by different kinds of stars (most notably the distinction between products synthesized by type II supernovae, which occur almost instantaneously on a galactic timescale, and type I supernovae, which take much longer from the initial formation of the stars). However, the relative abundances of key elements, such as Mg and F, are nearly constant within ellipticals. How could the products of disparate stellar sources become mixed without destroying the radial gradient? The notion of gas being lost preferentially in regions where the escape velocity is lowest has been introduced by a number of authors over the last two decades, and there is increasing evidence that global winds were ubiquitous in the early histories of many galaxies. This introduces an entirely new aspect to the chemical evolution of galaxies, one which connects directly to the intergalactic medium which we are finally becoming able to observe in some detail (Chapter 6).

4.5.2 Galaxy color sequences

Something similar may apply to the luminosity–metallicity relation, which is apparent not only in spectra of ellipticals but their broadband colors (since the accumulated absorption from metal lines is stronger for the emitted blue-light B and especially near-ultraviolet U bands). The color–luminosity relation for early-type galaxies is so consistent that high-redshift clusters with very small contrast in projected number density can be recognized because of the "red sequence" of ellipticals, becoming slightly redder for brighter galaxies. This correlation is very tight, with a dispersion of only about 0.02 in B – V color for low redshifts, a fact which tells us that the ages of the optically dominant stellar populations are closely synchronized for ellipticals (and most S0 galaxies). In a general way, deeper potential wells might have more generations of star formation before gas exhaustion and retain more or all of the ejected metals.

This red sequence of galaxies lacking star formation ("read and dead"), particularly apparent in clusters with ample populations of elliptical and S0 galaxies, has been known for some time. What was implicit in large data sets, and became clear with the very extensive and uniform measurements from the Sloan Digital Sky Survey and further amplified with the inclusion of GALEX ultraviolet fluxes, is that there is a surprisingly well-defined and analogous sequence of blue galaxies, in which star formation remains active. These two sequences are particularly clear in the color-absolute magnitude domain when integrating across whole galaxies (Figure 4.9), in which they remain well separated even though each population becomes bluer at lower luminosities, consistent with metallicity but likely also reflecting "downsizing" during galaxy evolution (Chapter 5). At each luminosity, the galaxy color distribution is strongly bimodal. The surprise is that the distinction is so sharp. Between the two sequences, there are distinct minima in a number of galaxies. This sharply limits the number of galaxies which are blue because of a burst of star formation superimposed on old, red stellar populations. The postburst color evolution of such a combination eventually becomes so slow that many such galaxies would violate the width and near-Gaussian symmetry of the red sequence. For most galaxies, such a transition can happen only once, so that we may inquire about the nature of a galaxy's shutting

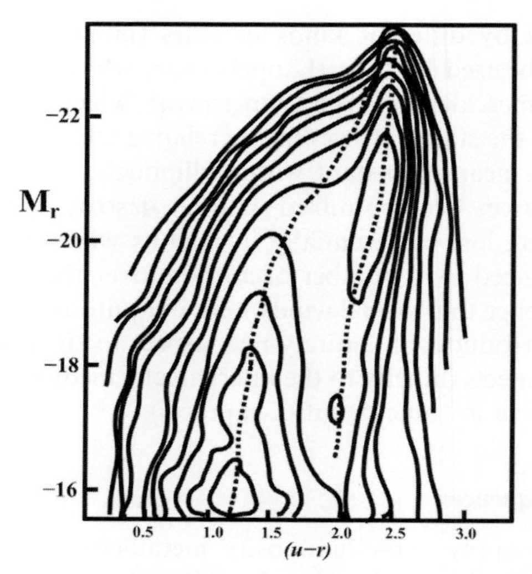

Figure 4.9. The red and blue sequences in galaxy color, from 207,000 galaxies catalogued by the Sloan Digital Sky Survey. Contours indicate the space density of galaxies, corrected for sample selection, and are spaced logarithmically. Thus, the highest contour maps a galaxy density 240 times higher than the lowest; the ridge lines are very well defined, and shown by the dashed curves. Missing contours indicate parameter values for which the correction for sample incompleteness leads to very large errors. Colors are from models fit to the entire galaxy spectral energy distribution, and absolute magnitude M_r is calculated within the Petrosian radius for $H_0 = 70$. (Figure taken using a smoothed version of the presentation in fig. 2 from Baldry *et al.*, *Astrophysical Journal*, **600**, 681, 2004; the ridgeline fits are also theirs.)

down its star formation. This might happen as a relatively sudden quenching, or a slow exhaustion of gas, but it must be very rare for galaxies to host multiple and powerful bursts of star formation. Certainly, there is global evidence of composite stellar populations, such as the fact that the most luminous blue galaxies have colors approaching the red sequence.

The properties of these sequences provide powerful constraints on galaxy evolution (Chapter 5), the role of merging, and the fueling of active galactic nuclei (Chapter 8). The color division also matches changes in the clustering properties of galaxies.

4.6 SELECTION BIAS IN GALAXY SAMPLES

Since all these scaling laws for galaxies are based on empirical samples, it is crucial to be sure that we aren't misled by correlations which come simply from the limits of our ability to find galaxies. As pointed out by Arp and Disney, there are obvious limits to what kinds of galaxies are easy to find, set by our environment. Galaxies that are too faint compared with the natural glow of the night sky from even the darkest sites will be extremely difficult to find. This situation changes somewhat when we consider

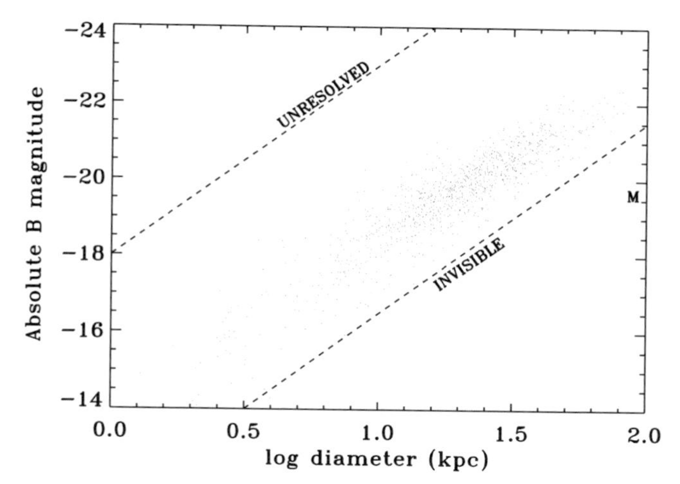

Figure 4.10. The relation between size and luminosity for detected galaxies, illustrated for 1036 galaxies in the Second Reference Catalog with well-measured values for both. The diagonal lines show constant surface brightness levels. The lower one is about 1% of the night-sky brightness at a typical ground-based site, so that galaxies below this line will be unseen in most surveys. The upper line indicates galaxies with an apparent size of 1 arcsecond, which would be confused with a star image on traditional photographs. The lack of points close to the "invisible" line indicates that low-surface-brightness galaxies have been missed in most catalogs to date. The letter M shows the approximate values for Malin 1, indicating that galaxies in these regions can now be found. Many high-redshift galaxies fall above the "unresolved" line, and can be seen as such only with space or adaptive-optics instruments.

observations from orbit, but not greatly—sunlight scattered from dust in the inner solar system, and general starlight scattered from dust throughout the Galaxy, make the V-band sky brightness from low orbit half of what a good ground-based site delivers. The situation improves considerably in the near-infrared and ultraviolet, but we have yet to see space instruments which really exploit this surface-brightness advantage. Diffuse sources are fairly easy to detect to 1% of the night-sky brightness, and become much more difficult below that. At the other extreme, a galaxy which is small enough could be mistaken for a star on typical images (as indeed happened with quasars because their cores are so bright compared with the surrounding galaxy). Considering galaxies that have been catalogued and well studied, we find that indeed all parts of the surface-brightness band that show up easily in our instruments are populated (Figure 4.10). It is more than a fair question to ask how many galaxies we haven't found which lie outside the combinations of apparent size and brightness that have been searched so far. Such systems certainly exist, and have been found when survey techniques have pushed outside this traditional envelope.

With appropriate techniques, it has now proven possible to look outside these surface-brightness boundaries. On the high-surface-brightness side, Zwicky began the recognition of compact galaxies, some of which appear much like stars on photographs from the Palomar Sky Survey. Most high-redshift galaxies appear starlike from the ground, with effective radii $< 0.3''$. There are not large numbers of compact galaxies nearby; for example, multicolor photometry of bright stars indicates that there are no 11th-magnitude compact galaxies masquerading as stars. By redshifts of $z = 0.1$, some starburst systems do have effective radii less than $1''$.

In the other direction, a whole class of low-surface-brightness (LSB) galaxies has been uncovered as more sensitive detection techniques became available (Chapter 2). At this point, they do not seem to be a dominant contributor to the starlight census of the Universe, but their origin and evolutionary status remain unclear, not least because they are extremely difficult to measure in most of the ways we would like to apply. Gregory Bothun has written that he started working on these objects after being advised by a colleague to "do something hard".

As we find ever more extreme kinds of galaxies—dwarf star-forming systems, luminous dust-shrouded objects—we appreciate that the wavelengths of our surveys introduce at least a statistical bias in what galaxies we include. Not surprisingly, we are most likely to detect objects in the spectral region of their strongest emission or strongest contrast against the background. Cross-correlating catalogs formed in very different parts of the spectrum—ultraviolet versus optical or infrared, for example—has provided important insight on how strong these biases are, and hence on the contribution of various kinds of galaxies to the overall cosmic budgets of star formation, luminosity, and stellar mass.

4.7 BIBLIOGRAPHY

Books

Block, David (1983) *Photographic Atlas of Primarily Late Type Spirals Printed as If Each Galaxy Were at the Same Distance* (University of Fort Hare, South Africa). This collection of photographs stresses the wide range in linear size for otherwise similar-looking galaxies, and thus the limitations of using morphological criteria alone as distance indicators.

de Vaucouleurs, G.; de Vaucouleurs, A.; Corwin, J.R.; Buta, R.J.; Paturel, G.; and Fouque, P. (1991) *Third Reference Catalog of Bright Galaxies* (Springer). Measured and derived quantities for increasing numbers of galaxies have been given in various catalogs, of which this is the most extensive available in print. With the explosive growth in available data and the ease of electronic distribution, data and literature references on individual galaxies are best obtained from such sources as the NASA Extragalactic Database (NED, *http://nedwww.ipac.caltech.edu*) or the Hyperleda service at the Observatoire de Lyon (*http://leda.univ-lyon1.fr/intro.html*).

Osterbrock, D.E. and Ferland, G.J. (2006) *Astrophysics of Gaseous Nebulae and Active Galactic Nuclei* (University Science Books).

Rowan-Robinson, M. (1987) *The Cosmological Distance Ladder* (Cambridge University Press). An overall (if slightly dated, pre-Hubble) review of the extragalactic distance scale.

Webb, S. (1999) *Measuring the Universe: The Cosmological Distance Ladder* (Praxis). A more recent survey of the cosmic distance scale, incorporating some of the Hubble Space Telescope Key Project results.

Journals

Arp, H. (1966) "Atlas of peculiar galaxies", *Astrophysical Journal*, Supplement 14, 1–20.

Baldry, I.K.; Glazebrook, K.; Brinkmann, J.; Ivezić, Ž.; Lupton, R.H., Nichol, R.C.; and Szalay, A.S. (2004) "Quantifying the Bimodal Color-Magnitude Distribution of Galaxies", *Astrophysical Journal*, **600**, 681–694. One of the earliest explorations of the distinct blue and red sequences in the galaxy population; Figure 4.9 draws on their presentation.

Calzetti, D.; Kinney, A.L.; and Storchi-Bergmann, T. (1994) "Dust extinction of the stellar continuum in starburst galaxies: The ultraviolet and optical extinction law", *Astrophysical Journal*, **429**, 582–601.

Côté, S.; Carignan, C.; and Freeman, K.C. (2000) "The Various Kinematics of Dwarf Irregular Galaxies in Nearby Groups and Their Dark Matter Distributions", *Astronomical Journal*, **120**, 3027–3059. A collection of dwarf galaxies with kinematics measured using neutral hydrogen. In many of these cases the dynamics must be dominated by dark matter even at the galaxy centers.

Couderc, P. (1939) "Les auréoles lumineuses des Novæ", *Annales d'Astrophysique*, **2**, 271–302. Worked out the properties of light echoes as seen from scattering material illuminated by nova or supernova outbursts.

Disney, M.J. (1976) "Visibility of galaxies", *Nature*, **263**, 573–575.

Faber, S.M. and Gallagher, J.S., III (1979) "Masses and mass-to-light ratios of galaxies", *Annual Review of Astronomy and Astrophysics*, **17**, 135–187. Summarizes the first results of extended galaxy rotation curves and the dawning realization that dark matter is dynamically dominant in galaxies.

Faber, S.M. and Jackson, R.E. (1976) "Velocity dispersions and mass-to-light ratios for elliptical galaxies", *Astrophysical Journal*, **204**, 668–683.

Jacoby, G.H.; Branch, D.; Clardullo, R.; Davies, R.L.; Harris, W.E.; Pierce, M.J.; Pritchet, C.J.; Tonry, J.L.; and Welch, D.L. (1992) "A critical review of selected techniques for measuring extragalactic distances", *Publications of the Astronomical Society of the Pacific*, **104**, 599–662. A comparison of several techniques for measuring galaxy distances, by practitioners of each.

Kennicutt, R.C., Jr. (1998) "Star formation in galaxies along the Hubble sequence", *Annual Review of Astronomy and Astrophysics*, **36**, 189. A comprehensive review of various ways to estimate rates of star formation in galaxies, and a consistently cross-calibrated set of expressions as adopted in this chapter.

Leavitt, Henrietta (1908) "1977 Variables in the Magellanic Clouds", *Annals of Harvard College Observatory*, **60**, 87–108

Leibundgut, B. (2000) "Type Ia Supernovae", *Astronomy and Astrophysics Reviews*, **10**, 179–209. A detailed review of progress in understanding type Ia supernovae, pointing to several kinds of observations that are badly needed to improve our models and better understand their utility for cosmology.

Mathewson, D.S.; Ford, V.L.; and Buchhorn, M. (1992) "A southern sky survey of the peculiar velocities of 1355 spiral galaxies", *Astrophysical Journal Supplement*, **81**, 413–659.

Petrosian, V. (1976) "Surface brightness and evolution of galaxies", *Astrophysical Journal Letters*, **209**, L1–L5. Defines the quantities leading to the Petrosian radius, whose use in photometric measurements cancels out uncertainities due to cosmological models and allows isolation of galaxy evolution in luminosity.

Salpeter, E.E. (1955) "The luminosity function and stellar evolution", *Astrophysical Journal*, **131**, 161–167.

Schechter, P.L. (1976) "An analytic expression for the luminosity function for galaxies", *Astrophysical Journal*, **203**, 297–306. Introduces the widely-applied Schechter expression for the luminosity function of galaxies.

Sunyaev, R.A. and Zel'dovich, Ya.B. (1972) "The observation of relic radiation as a test of the nature of X-ray radiation from the clusters of galaxies", *Comments on Astrophysics and Space Physics*, **4**, 173–178. Initial calculation of the flux decrease in the cosmic microwave background to be expected towards galaxy clusters, due to scattering by the hot intra-cluster gas.

Trimble, V. (1997) "Extragalactic Distance Scales: H from Hubble (Edwin) to Hubble (Hubble Telescope)", *Space Science Reviews*, **79**, 793–834. A recent review and comparisons of several specific techniques for measuring galaxy distances. A series of papers from a somewhat staged anniversary "debate" on the distance scale appeared in the *Publications of the Astronomical Society of the Pacific*, **108**, 1065–1097 (December 1996).

Tully, R.B. and Fisher, J.R. (1977) "A new method of determining distances to galaxies", *Astronomy and Astrophysics*, **54**, 661–673.

Zwicky, F. (1937) "On the masses of nebulae and of clusters of nebulae", *Astrophysical Journal*, **86**, 217–246.

5

Galaxy evolution

Important timescales from cosmology and stellar evolution overlap, which shows that galaxies must have evolved over cosmic history, demonstrating that their history lies before us to uncover in their propagating light. This point is well made in Figure 5.1, which shows how the lifetimes of various kinds of stars compare with the lookback time at a range of galaxy redshifts.

In retrospect, we might wonder why this recognition took so long, since stellar lifetimes certainly implied that major changes must have taken place over timescales of order 10^{10} years. This is close to the timescales for cosmic history suggested by the Hubble constant (through its inverse the Hubble time). For many astronomers, the key papers were Beatrice Tinsley's (1968) calculations of the effect of even passive evolution on galaxy colors and luminosity, showing that the uncertainties in our knowledge of galaxy evolution were so great that the classical tests for cosmological parameters, largely based on luminosity, could not be so applied. Beatrice Tinsley (1941–1981) was raised in New Zealand, and pursued her work first at the University of Texas and finally with a professorship at Yale capping her all-too-brief research career. She continued to work right up to the end of her life, cut short by melanoma, expressing this desire in a poem:

Let me be like Bach, creating fugues
Till suddenly the pen will move no more.

Robert Kennicutt, Jr., general editor of *Astrophysical Journal*, selected a paper based on her dissertation work for a centennial volume of the journal, and described its impact on astronomy by writing, "thereafter galaxy evolution, not cosmology, became the *raison d'être* for deep surveys of the universe."

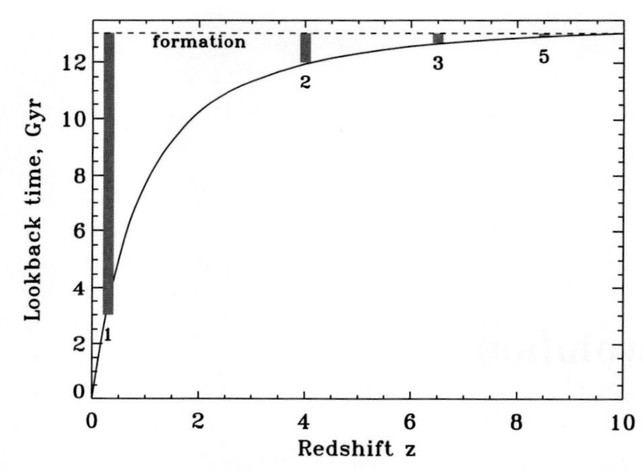

Figure 5.1. Timescales of stellar evolution and cosmology. The curve shows the relation between redshift and lookback time for a typical set of cosmological parameters. Galaxy formation is represented by the dashed line at the top. Shaded bars show how long stars of various masses formed at that time would survive. Thus, at redshift $z = 0.3$ we would see all the solar-mass stars from the initial burst. More massive stars are much shorter-lived, so we would need to observe at $z = 8$ to see stars of 5 solar masses from an initial short burst of star formation. (After a figure in the dissertation paper by Donald Hamilton in *Astrophysical Journal*, **297**, 371, 1985.)

5.1 THE CLOCKS OF GALAXY EVOLUTION

Galaxies can evolve in several ways, with these aspects not necessarily keeping pace in galaxies of all kinds. Most obvious (especially for observers conditioned by optical observations) is the history of star formation. This is reflected in the galaxy's luminosity at all wavelengths, simply as the sum of contributions of all its stars as affected by absorption and scattering from interstellar dust, and as a result we can in principle untangle the history of star formation from a spectrum of high enough quality spanning a wide enough wavelength range. In practice, we can answer certain broad questions about the history of most galaxies. This clock gives us one clue as to when we're approaching the time of galaxy formation observationally—truly young galaxies cannot contain old stars. There may be variation in how briskly they have begun forming stars, or how efficiently, but the starting time is fixed. This relates not only to galaxies at high redshift, but to searches for delayed galaxy formation still happening now.

As star formation and stellar lifecycles proceed, a galaxy's material must also evolve in *chemical composition*. Heavy elements, beyond beryllium in the periodic table, are produced only inside stars. At the outset, then, the composition of the gas available to make the first stars was that left over from primordial nucleosynthesis, as set out in Chapter 7. This means almost entirely hydrogen and helium. By the current epoch, slightly more than 1% of the mass in young stars in our part of the Milky Way consists of heavier atoms, "cooked" in preceding generations of stars. The evolution of the chemical makeup in a galaxy will depend on several factors. The history of star formation controls the input of heavy elements, as their sources, and the rate at which gas is both locked up in stars and expelled into the surrounding material. The distribution of these stars by mass, the initial mass function, sets which kinds of stars are made, and this in turn tells what elements will be produced when, since stars of different masses fuse different products in their core and release them into the interstellar medium at different times (or not at all). The galaxy's internal dynamics make a difference in whether these processes have local or global consequences, since the gas expelled from supernovae, planetary nebulae, and stellar winds may be mixed throughout a galaxy, through an annulus, or through only a small volume around the source. The disks of spiral galaxies often show gradients in abundance of heavy elements (metals, in astronomical parlance), reflecting a more active history of star formation and more effective retention of hot, enriched gas in the inner parts. However, such gradients are absent from barred galaxies, at least in the radial range covered by the bar. This makes sense in light of the dynamics of particles in a barred gravitational potential, which undergo substantial radial motion under the influence of the bar, so any gradient would be quickly smeared across the bar's full extent. Chemical evolution also gives us something to look for as a sign of galaxies just after formation—they would be very metal-poor, with minimal processing of hydrogen and helium to heavier elements since there has been little time for stellar evolution to take place. Just what pattern we would see, though, depends on details of what gas we observe and how well the earliest stars mix their products. Looking at a galaxy's emission spectrum alone can be misleading, since stars closely associated with the ones that enrich their environments light up the gas we can see. Depending on how rapidly gas is mixed through the galaxy, the initial enrichment (up to about the level seen in the Small Magellanic Cloud) can be so rapid that we would seldom expect to catch it in action.

Finally, and largely independent of the other two clocks, a galaxy will evolve *dynamically*, as its components exchange energy and angular momentum with each other and the galaxy's environment. The distribution of stars among orbits of various kinds may change, and such structures as disks, bars, and rings can form or evaporate. Even the structures of today's sturdy-looking elliptical galaxies are not, strictly speaking, stable, although the time required for further important changes in an isolated galaxy may vastly exceed the age of the Universe. Such evolution would have been much faster in the early Universe, as galaxies came together (in one or many major pieces) from more diffuse material or smaller units. By reaching high redshifts, we can hope to catch galaxies in these phases, when their initial dynamical conditions had not yet been erased, as long as there was some tracer we can observe to show us their forms. This might be the light of the first generation of stars, or some emission line from warm gas.

The dynamical process of a system approaching an equilibrium state, and particularly a state with equal sharing of energy among the constituent particles (stars, in this case) is known as relaxation. It must proceed by some kind of interaction between the particles, which, for galaxies, is expected to be two-body encounters between stars, which will occasionally approach one another so closely that substantial energy from their galactic orbits is transferred. However, the characteristic timescale for such relaxation is far too long (of order 10^{16} years for typical regions in our galaxy) to account for the regular structures of galaxies; unlike the much smaller globular star clusters, galaxies are not old enough for star–star encounters to have had an effect on their structure. Still, galaxies show enough structural symmetry to indicate that they have erased their initial conditions of formation, so that widespread redistribution of energy among the stars (and perhaps the gas clouds that predated them) must have taken place, and observations of galaxies at high redshift show that this *perestroika* took place in short order. These prompt interactions gave rise to what is known as *violent relaxation*, one of the more colorful phrases from dynamical astronomy. Interactions not only between individual objects, but between individual objects and features in the overall gravitational potential, can be important when that potential is changing rapidly, as during the collapse of a protogalaxy or a galaxy merger. Numerical and analytic results indicate that, indeed, such interactions can give galaxy-like profiles in a rather short time, only a few orbital timescales for most stars.

Once again, this gives us a sign of galaxies seen shortly after their formation—they would still be either quite diffuse or very clumpy compared with the symmetric and relaxed structure of galaxies today. Indeed, virtually every astronomer who has worked with deep Hubble Space Telescope images of faint galaxies has remarked that faint, high-redshift galaxies look less regular than their nearby counterparts. This observation by itself doesn't yet mean that we have seen dynamical evolution, since the cosmological redshift means that distant galaxies are usually observed by their own ultraviolet radiation, and spiral galaxies in particular are much patchier and less regular when seen in the ultraviolet than in visible light (Chapter 2).

5.2 PASSIVE AND ACTIVE EVOLUTION

One of the easiest ways to look for the evolution of galaxies is through their composite populations of stars, as reflected in the overall spectrum of each galaxy. Ignoring for the moment such complications as internal absorption by dust grains and the changing masses and temperatures of co-evolving close binary stars, the spectrum of a galaxy will be simply the sum of the spectra of its constituents. The galaxy spectrum $G(\lambda)$ can be expressed as

$$G(\lambda) = \sum_i n_i S_i(\lambda)$$

where the index i reflects various kinds of stars (or, indeed, other sources of radiation such as emission-line gas), n_i is the number of representatives of component i, and $S_i(\lambda)$ is the intensity of the ith component at wavelength λ. Using calculations of the evolution of stars with given mass and chemical composition, one can calculate the expected spectrum of stars in a given part of the Hertzsprung–Russell diagram, or the equivalent expressed in terms of the age and mass of the star. Thus, one can start with a history of star formation and predict the spectrum of the whole galaxy, an approach known as *population synthesis*. In practice, one might use either purely theoretical stellar spectra, or take empirical data on stars if they can be observed in a network spanning the ranges of temperature, luminosity, and composition that are needed. It is important that these entire ranges be covered; it can be shown, from the mathematics of the problem, that if some ingredient of the mix is lacking in the model, not only will the result be wrong, but it will contain no information as to what ingredient is missing. We also need to specify the relative numbers of stars formed as a function of mass, the *initial mass function* or IMF; while the IMF is reasonably constant in environments like nearby star-forming regions, there are signs that it may vary in regions of the most active star formation, and it was almost certainly weighted toward very massive stars ("top-heavy") in the early Universe.

This approach is especially effective for old stellar systems—elliptical galaxies and the central bulges of spirals. These systems have simple histories of star formation, with little or no recent activity, making them ideal laboratories to test our ability to synthesize stellar populations. Indeed, the models predict spectra in close agreement with observations for bursts of star formation seen 11–14 billion years after the fact, and very little subsequent starbirth.

Going beyond this, to infer detailed and complex histories of star formation, is more difficult. Not only are higher-quality data required, but degeneracies between various model parameters become more important. The role of dust, preferentially reddening and extinguishing the light from younger stars and from regions in the nuclei and spiral arms, becomes significant in spiral galaxies and in starburst galaxies. Still, there are trends with galaxy type that are robust to various ways of accounting for these effects. It is no surprise that spiral galaxies, in which star formation is still active, are found to have a more complicated star-formation history than ellipticals or their own central regions. The mix of original stars and more recent additions changes systematically along the Hubble-type sequence, with star formation most protracted for later Hubble types (Sc, for example). Here again we see a continuity along the type sequence, with the dynamical features producing the galaxy structure correlated with the entire history of star formation.

Models can be produced for any desired star-formation history, but it has proven fruitful to consider relatively simple combinations such as a single short burst of star formation, exponentially declining star-formation rate, constant star-formation rate, or an old population with a "frosting" of ongoing star formation. These are helpful not only in unravelling the history of present-day galaxies, but in suggesting what we should look for as we trace the evolution of galaxies by looking to high redshifts. A sequence of spectra predicted for an aging stellar population shows that much of the initial action takes place in the ultraviolet (Figure 5.2), as the hot massive stars die off first and disappear from the mix. For long times after that, up to the present age of the Universe, a very strong and useful spectral feature is the so-called HK break or 4000-Å break (Figure 5.3), a blend of absorption features just shortward of the twin absorption lines of Ca II (known ever since the time of Fraunhofer as the H and K lines). The amplitude of this "step" in the spectrum is a function of age, and depends only mildly on the composition of the constituent stars. This feature is also very strong and easy to observe in galaxies to beyond $z = 1$. It is thus a promising indicator of galaxy evolution; we expect the HK break to become weaker as we

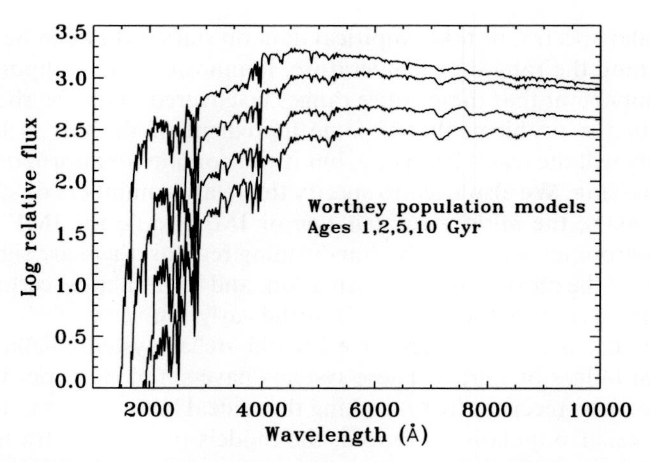

Figure 5.2. The evolution of a galaxy's spectrum. The calculated spectrum of an aging burst of star formation is shown at intervals 1–10 Gyr after the burst, showing the fading and reddening of the population over this time, as progressively less massive stars end their lives. At late times, the Balmer absorption lines weaken and metal lines, plus the break near 4000 Å, strengthen. Comparison with grids of such models can be used to derive the star-forming histories of galaxies, particularly in systems such as ellipticals with a simple history. As seen from the spacing here, galaxy luminosity is approximately logarithmic in its time development. (Data from models calculated by Guy Worthey and made available at *http://astro.wsu.edu/worthey/dial/dial_a_model.html*)

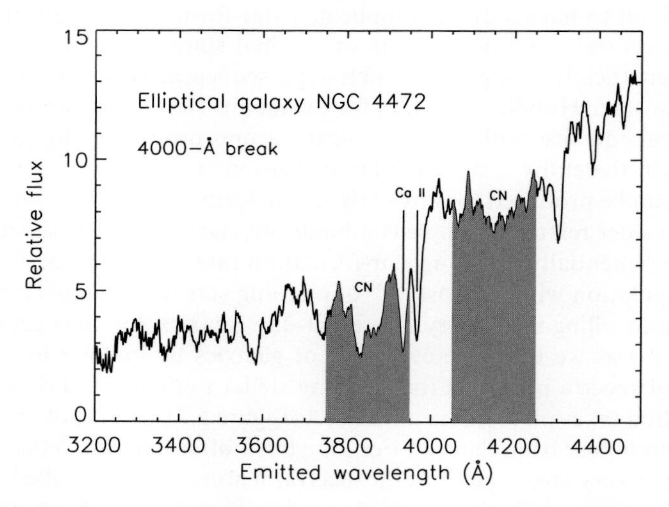

Figure 5.3. The 4000-Å break (sometimes called the HK break) shown in the spectrum of the luminous elliptical galaxy NGC 4472. A standard measure of its amplitude is the ratio of mean fluxes in the two shaded bands. The spectral break occurs near the two marked Ca II absorption lines, but its amplitude is strongly influenced by other species, such as the CN molecular absorption bands shown. (Spectral data by the author, using the 2.1-meter telescope of Kitt Peak National Observatory.)

observe galaxies at larger lookback times, when they were indeed younger. This test makes the most sense if we can preselect elliptical galaxies, to avoid mixing galaxies with very different histories. Several studies have now shown evolution in the amplitude of the break, either using selection on color (the reddest galaxies at each flux level will mostly be "red and dead" ellipticals) or from radio flux, since powerful radio galaxies are usually found in elliptical galaxies. The amplitude of the break and its slope with redshift in these samples can be fit if star formation did not completely end after the initial formation of the galaxies, but continued at a low rate until about $z = 1$.

Spectrum models also allow us to use another simple probe of galaxy evolution, a blunt instrument that is at once powerful and indiscriminate. For any scenario of how galaxies change in time as regards color and luminosity, both of which follow from the mix of star-forming histories that go into such a scenario, we can predict the number of galaxies we should see in a patch of sky for a particular color and brightness. Some of the first signs of galaxy evolution turned up in this way, as an excess of blue galaxies fainter than about magnitude 20. This in itself implied evolution in the galaxy population, though not just what kind of evolution it might be. Count analyses are very powerful in showing that *something* must be changing with redshift, but not exactly what it is that changes, since the count predictions involve integrations in both galaxy luminosity and redshift. Thus, the predicted counts at some color and magnitude level involve contributions from galaxies over wide ranges of distance and luminosity.

5.3 OUTWARD IN REDSHIFT AND BACK IN TIME

Being able to observe a large and systematically chosen set of high-redshift galaxies, so as to allow detailed comparison with the contemporary galaxy population and reveal how galaxies have evolved, has been an important goal for several decades. This is now being realized, so that we finally have a glimpse of the Universe at about one-fifth of its present age. Several strategies have led to this point.

5.3.1 Radio galaxies

Radio galaxies were the first systems (except quasars, whose relation to their host galaxies is so ill-defined that they don't tell us much in this context) to be observed in quantity at high redshifts ($z = 2$–4). There was a flurry of observations starting in the late 1980s to identify radio sources with high-redshift galaxies, as the fastest approach to measuring a sample of any kind of galaxy at high redshift. Particular subsets could be defined from the radio spectral index or optical flux, in efforts to bias the observed samples toward high redshifts. However, in applying their results to the more general questions of galaxy evolution, there has always been uncertainty as to how representative these objects are. If we are dealing with the most powerful radio sources in the observable Universe, objects which are exceedingly rare, how can we distinguish between features seen at high redshifts because of cosmic evolution and those seen at high redshifts because our targets are so rare that they are found only at distances corresponding to these long lookback times? For such rare objects, the effects of increasing rarity and evolution with redshift become degenerate. This made it all the more welcome when optical selection techniques were able to produce large numbers of high-redshift galaxies independent of radio properties.

5.3.2 Lyman α emission

In pushing galaxy samples to ever-greater distances, a cardinal difficulty has been distinguishing galaxies which are faint because they are very luminous and distant from the myriads which are faint because they represent closer, dwarf systems. Radio emission, with its pitfalls, has been one way to do this. Another, which has been tried for years but met with success only recently, has been the use of narrowband filters tuned to the emission lines expected for star-forming galaxies in specific redshift ranges. The strong Lyman α line of hydrogen, emitted deep in the ultraviolet at 912 Å, has been a favorite for this technique, since systems which are forming stars and still poor in dust should exhibit strong Lyman α, and no other strong emission lines in the same wavelength region. This effort succeeded only after many false starts, since the intensities where Lyman α galaxies showed up turned out to be fainter than once expected, and they may be strongly clustered so there is an element of luck in pointing to the right place at the right redshift. Clusters or rich groups containing these objects are known at $z = 2.4$–2.6 (Figure 5.4), and a few examples are known at $z > 4$. Although not quite as unusual as strong radio sources, these, too, may represent only a particular subset of galaxies at these redshifts. Theoretical considerations show that it is very easy for Lyman α photons to be trapped within the gas cloud for so many scattering events that they are likely to be absorbed by a dust grain or converted into some other kind of hydrogen emission line. The best environment for escape of Lyman α photons involves not only low metallicity to reduce the dust abundance, but a global outflow—a galactic wind—so that the emitting and absorbing atoms have a relative redshift and local absorption is dramatically reduced. Still, Lyman α emitters have provided a crucial set of young objects, some of them fainter (and thereby, perhaps, more common) than other techniques have delivered so far.

5.3.3 Lyman-break galaxies

A key breakthrough in recognizing "normal" galaxies at high redshifts came with the recognition that internal absorption from hydrogen will cause a complete break in the spectrum of any galaxy at the ionization edge of hydrogen, 912 Å, a feature known as the Lyman break. A high-redshift galaxy at redshift z can be recognized by blue colors across the red part of the spectrum, and a total absence of flux blueward of $912(1+z)$ Å (as shown for a nearby star-forming region in Figure 5.5). This specific technique was introduced by Steidel and Hamilton (1992), and has been applied with enormously fruitful results in numerous recent papers. It led to the immediate identification of many galaxies at $z > 3$ in the Hubble Deep Field, and by now thousands of such galaxies are known. Most of what we know about

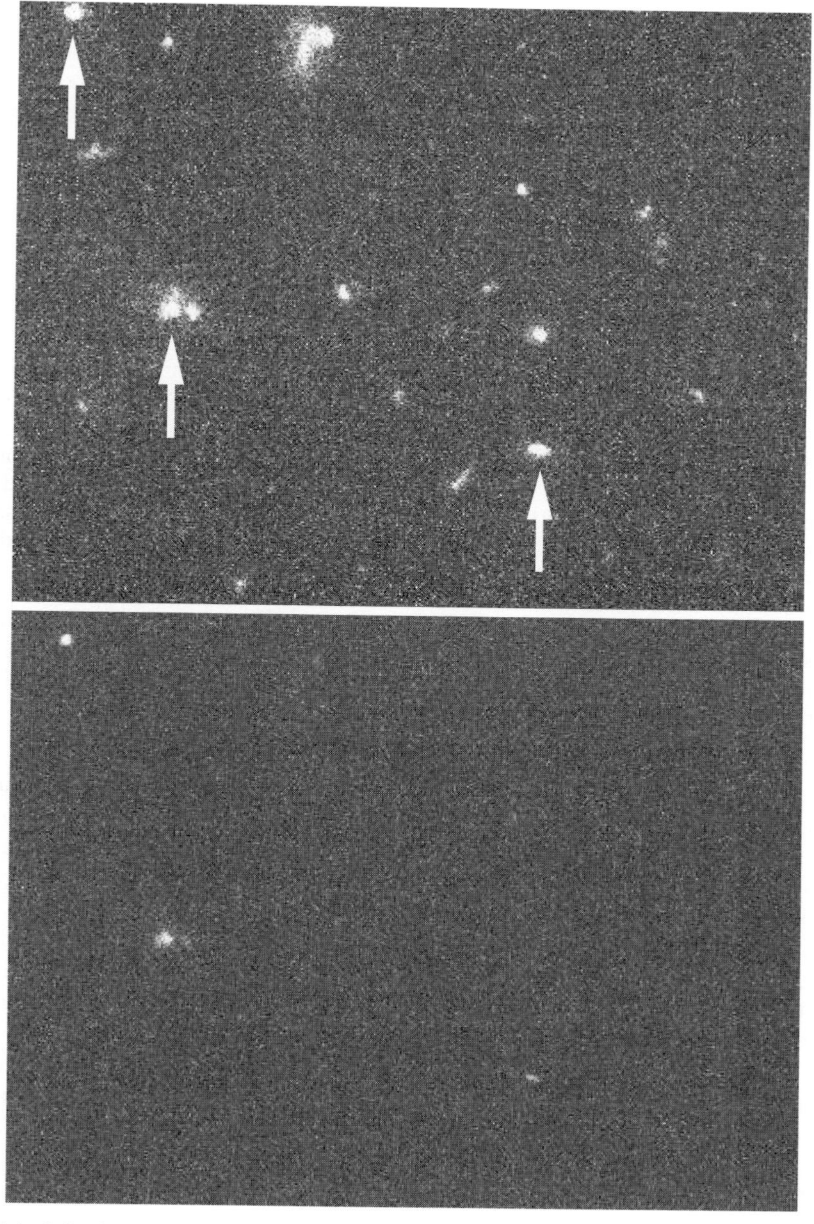

Figure 5.4. Selection of objects by Lyman α emission. This is part of the grouping near the radio galaxy 53W002 at $z = 2.4$, observed with the Hubble Space Telescope in a broad blue-light filter (*top*) and a narrow filter including Lyman α at this redshift (*bottom*). The arrowed objects show strong emission in the narrow band, and are indeed confirmed spectroscopically as active nuclei and star-forming objects at this redshift. (R. Windhorst, S. Pascarelle, W. Keel, and NASA.)

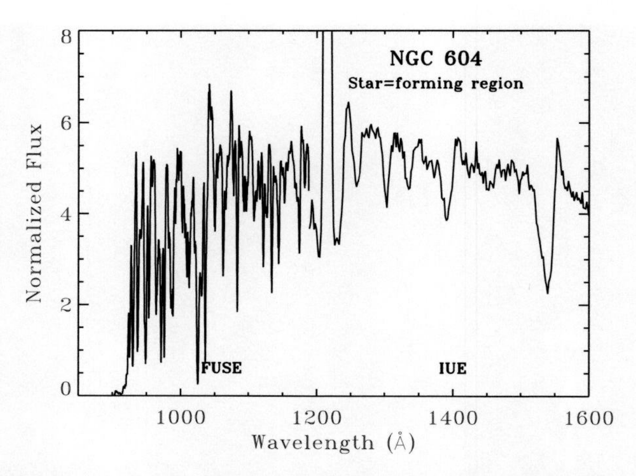

Figure 5.5. The Lyman break in the spectrum of a star-forming galaxy, shown for the brightest star-forming region in the nearby spiral galaxy M33. The continuum is very blue, and strong until close to the Lyman limit at 912 Å. It drops rapidly, due to blending of Lyman absorption lines, and eventually vanishes due to absorption by neutral hydrogen. Properly selected filters, using this behavior, can select star-forming galaxies in this way to high redshifts. In this spectrum, the Lyman α emission line at 1216 Å is dominated by incompletely removed emission from the Earth's escaping hydrogen, and not from the star-forming region itself. High-redshift galaxies may show this feature in either emission or absorption. (The longer-wavelength data are from the International Ultraviolet Explorer archive, and the short-wavelength data have been obtained recently with the Far-Ultraviolet Spectroscopic Explorer—FUSE.)

high-redshift galaxies comes from these samples, which is both good and bad. It is remarkable that we have come so far at this point, but we must recognize that there are biases inherent in finding Lyman-break galaxies. Only galaxies with active, recent star formation at the epoch we see them can appear, because if their ultraviolet luminosity is not high they will not be observed. Dusty galaxies may not be found, if the dust is positioned so as to absorb the crucial far-ultraviolet radiation. These issues will be important in asking how much of early star formation we can account for in these galaxies. Still, optical and near-infrared data on Lyman-break galaxies show that they are forming stars at a greater rate than any kind of galaxy in the nearby Universe, and give hints that they have lower heavy-element abundances than luminous galaxies have today. These are both important signs of galactic youth.

5.4 PHOTOMETRIC REDSHIFTS AND THE HISTORY OF STAR FORMATION

A generalization of this approach, tried in the 1980s by Loh and Spillar, but refined enough for widespread use only in the mid-1990s, is determination of so-called photometric redshifts. Galaxy spectra have only a limited range of properties, being to a good approximation a one-parameter family and to an excellent approximation a two-parameter family. Photometric measures of the galaxy's brightness through sets of broad-band filters, if they have high enough precision covering a wide wavelength baseline, can simultaneously determine the kind of galaxy spectrum and its redshift, as long as the galaxy spectrum is represented within the range of fitting functions. This works reasonably well for optical data for $z > 2$ and $2.8 < z < 7$, while near-infrared data are crucial for getting acceptable errors for $z = 2$–2.8 and and at very high redshifts when no optical flux remains due to Lyman absorption. Both the algorithms for retrieving photometric redshifts and the general confidence in the technique got a major boost from data collected in the Hubble Deep Field, in which a major international effort yielded a large set of redshifts of very faint galaxies to act as a training and testing set for the photometric technique. From photometric redshifts, we now have in hand samples of thousands of galaxies beyond $z = 2$ with known colors and luminosities.

With statistically significant samples of galaxies at redshifts up to $z = 4$, we can start to address some of the basic questions about galaxy evolution. One of the most important, with impacts ranging from the overall metallicity of galaxies to the timing of galaxy evolution and the ionization of the intergalactic medium, is the cosmic star-formation history. This may be expressed, for example, as the spatially averaged rate of star formation in solar masses per unit volume, as a function of redshift. The resulting plot is often known as a Madau diagram, or Madau–Lilly diagram, after the extensive discussion in a 1996 paper by Piero Madau and collaborators. That original derivation used photometric redshifts and UV luminosities from the Hubble Deep Field to imply a rapid increase in star-formation rate as we look back in redshift to about $z = 1$, a broad peak near $z = 2$, and a decline beyond that (while of course the actual behavior in time is the reverse of the redshift trends).

It was encouraging that the mean value of star formation over time from this work was close to the average expected from the metal content of present-day bright galaxies. Lennox Cowie and coworkers recognized as early as 1988 that a connection exists between the ultraviolet light emitted from galaxies and the rate of production of heavy elements, since the same stars that produce most of the heavy elements are important contributors to ultraviolet radiation. Thus, the typical metal content of the Universe today (as judged from bright galaxies such as our own) implies a total surface brightness, or sum from individual galaxies, in the ultraviolet. This total is slightly greater than the integral of the Madau *et al.* (1996) star-formation history, so at least we have a consistency check.

Other techniques, however, have indicated that the early history of star formation was more active and extensive than this. At issue is how much star formation was missed by the sample in the Hubble Deep Field, either because of internal dust absorption or because it took place in environments which were too dim to appear in the catalog. Several analyses, some starting from the same data, show star formation being either comparably active or actually more intense to $z > 4$ as compared with $z = 2$. Furthermore, recent results on the intergalactic medium show that the Cowie limit on cumulative star formation should be higher by a poorly known factor, since substantial masses in heavy elements were expelled by galaxies at early epochs and were not accounted for in that original value. This material is seen today as the metals in the hot gas in clusters of galaxies, and as the highly-ionized absorption seen from the intergalactic medium in O VI and similar absorption lines.

In view of its importance, and discrepancies in addressing this issue, a variety of techniques must be brought to bear in tracing the cosmic history of star formation. Infrared observations are particularly important, being insensitive to dust absorption and in fact relying on it by measuring the total amount of starlight which has been absorbed to heat interstellar dust. In our neighborhood, strong far-infrared emission is a hallmark of star-forming galaxies, and submillimeter observations have traced IR-selected galaxies to objects so faint that their redshifts cannot be accurately measured yet. Mid-infrared surveys of galaxies are an important goal of Hubble's successor, the James Webb Space Telescope (JWST), and the international millimeter-astronomy array ALMA will excel at detecting the dust from very high-redshift objects. In our vicinity the GALEX satellite has nearly completed its survey of the entire sky in the ultraviolet, for the first deep and unbiased census of unobscured star formation in the local universe (which is why the name stands for Galaxy Evolution Explorer). The current status of our knowledge of cosmic star-forming history is summarized in Figure 5.6, including results from UV- and IR-based surveys.

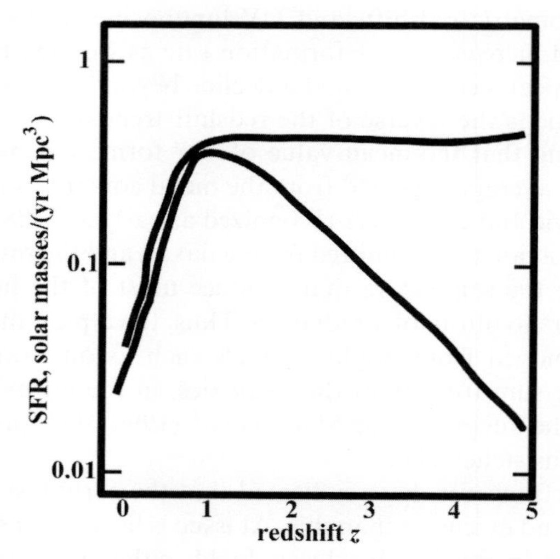

Figure 5.6. Cosmic star-forming history. This is a schematic version of the "Madau diagram", showing the comoving density of star formation over cosmic time. The lower curve is evaluated from the amount of star formation actually found from nearby surveys and the ultraviolet emission from high-redshift galaxies in the Hubble Deep Fields. The upper curve indicates the amount of star formation inferred from infrared observations and variously corrected analyses of the Deep Field data. In both cases, the inferred overall rate of star formation requires extrapolation upward by a factor near 40 to allow for the dimmer stars which are dominant by mass but unimportant in the luminosity of star-forming systems.

5.4.1 Host galaxies of gamma-ray bursts

In a more speculative vein, it may become possible to trace the history of star formation with such exotica as gamma-ray bursts. With substantial evidence that these represent beamed emissions from a species of supernova, they form an easy-to-detect subsample of massive star deaths which we can already see to large redshifts. As a newly applicable tracer of cosmic history, their story should perhaps be set out in detail.

One of the four Great Observatories launched by NASA, the Compton Gamma-Ray Observatory or CGRO marked significant advances in our ability to study the universe at high energies, in several ways. It carried detectors sensitive to hard X-rays and gamma rays over the thousandfold range from 30 keV to 30 GeV. Compton operated from 1991 to 2000, delivering on its promise to uncover new facets of the violent high-energy Universe. It detected numerous AGNs at gamma-ray energies, yielding new evidence of relativistic motion in their jets, and provided a census of repeating high-energy sources in the Milky Way. The mission also gave one of the key clues to the long-standing question of gamma-ray bursts, one which shows them to trace the history of star formation in an unexpected way.

The discovery of gamma-ray bursts was a byproduct of the Cold War. With the Limited Test-Ban Treaty, the nuclear powers of 1963 agreed to refrain from atmospheric (and space-borne) tests of nuclear explosions. Verification of compliance was a significant issue in ratification, to be carried out with what later became well known as "national technical means" (i.e., satellites). The USA launched the Vela series of satellites starting in 1963, on high 112-hour orbits which not only insured uninterrupted coverage of the eastern hemisphere, but would also detect clandestine nuclear tests on the lunar farside through emission from the expanding debris cloud. The second batch of Vela spacecraft, launched beginning in May 1969, carried improved detectors. Starting two months later, four of these Vela satellites detected unexpected bursts of gamma rays—not from Earthly nuclear arms, but from random directions in deep space.[1] The large orbits allowed crude directional determination, from the arrival times of the bursts as measured by different satellites. This allowed Ian Strong to analyze data collected by Ray Klebesadel at Los Alamos to show that they came from neither the Sun nor the Earth, thus representing a new and exciting astronomical phenomenon (Klebesadel *et al.*, 1973). Over a ten-year period, these satellites detected 73 cosmic gamma-ray bursts (with one retroactively found in 1967 data that had been insufficient on its own to warrant detailed analysis).

Models for the production of these bursts ran a wide gamut, since the only information available was the rate, total detected energy (fluence), and crude spectral shape of the bursts. Neutron stars were popular sites, since their deep potential wells could allow the liberation of vast amounts of gravitational energy. There were

[1] Vela 6911, far outlasting its design lifetime, may have finally done its designed job on September 22, 1979, when an optical flash was seen over the Indian Ocean. This has been widely rumored to be the test of a South African nuclear device, although open sources do not allow a firm conclusion.

calculations of the impact of asteroids with neutron stars, and of neutron-star mergers at the end of the orbital decay of a binary system. The general argument for burst production near neutron stars seemed compelling enough that by the launch of the Compton Observatory, most workers expected its results to trace the structure of neutron stars in the Galaxy.

Reality was to prove quite different. One of the most intriguing early results from CGRO was that the statistical properties of its large catalog of gamma-ray bursts confounded all expectations based on events involving galactic neutron stars. Its burst-detection system (BATSE, the Burst and Transient Source Experiment) incorporated eight detectors at the corners of the spacecraft structure, to record bursts from any direction not occulted by the Earth. The BATSE detectors could determine the position of each burst in the sky to only a few degrees' accuracy, but that was enough to show no concentration in the galactic plane or the galactic center. In fact, through the end of the nine-year mission, as a total of 2704 bursts were observed, the angular distribution remained as random as statistical tests could show. This in itself only meant that we were observing the sources within some volume that made them isotropic. In principle, sources among the nearest stars, the Oort Cloud, distant galaxies, or an encircling fleet of invading starships would satisfy this requirement. But even more telling was the distribution of bursts in intensity, in the same kind of $\log N$–$\log S$ diagram that had proven so powerful in establishing the evolution of quasars and radio galaxies. This time, we saw too few faint bursts compared with bright ones, a situation that implies that we are somehow centered in the distribution. This very non-Copernican situation could be satisfied by some kind of very large and hollow halo of sources centered on the Milky Way, so large that we would be nearly centered in it but not yet seeing any similar halo around Andromeda, or by a cosmologically evolving population, in which we would be central in time (equivalently in redshift). To extragalactic astronomers, the $\log N$–$\log S$ diagram fairly screamed "cosmologically distant".

Demonstrating this required an additional leap. It had been clear from early in the study of gamma-ray bursts that they do not have bright or otherwise obvious counterparts at other wavelengths, and faint counterparts cannot be found until the source's localization is better than the inverse of the number density of comparably faint sources (say, in the optical band). Individual detectors still have poor angular precision, and arrival-time triangulation from interplanetary spacecraft worked for too few bursts to be helpful (i.e., no clear counterparts emerged for classical bursts). This finally changed through the use of the X-ray tails of burst spectra. Strong bursts produce enough lower-energy radiation to be detected with imaging X-ray telescopes, easily bringing positional accuracies of an arcminute within reach. A field of view that small can be efficiently searched for variable (fading) objects, a strong hint that an optical object might be associated with transient gamma-ray emission. To carry this out successfully required the combined gamma-ray and X-ray detectors on the Italian/Dutch BeppoSAX satellite, launched on April 30, 1996. Its imaging X-ray telescope delivers $1'$ resolution in the 0.1–10 keV band. The burst observed on February 28, 1997 (known as GRB 970228) was the first to be located well enough to find an unambiguous optical counterpart, whose redshift proved to be significant

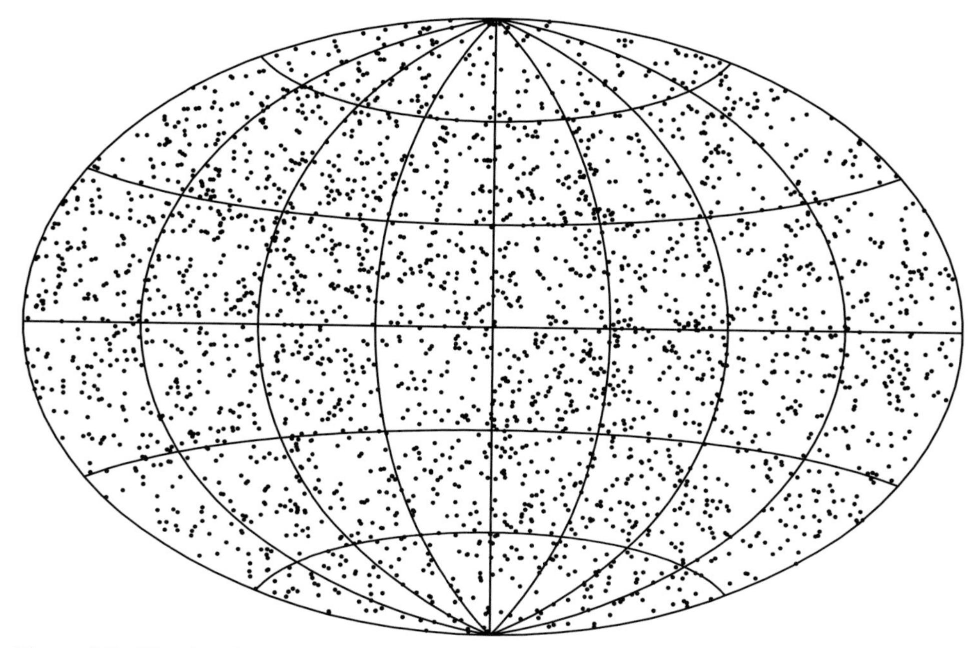

Figure 5.7. The sky distribution of gamma-ray bursts detected by the Compton Gamma-Ray Observatory, from the Fourth BATSE Catalog. This map is shown in galactic coordinates, with the galactic center in the middle. The equal-area projection makes it clear that the mean number per unit sky area is the same over the whole celestial sphere, with no concentrations toward the galactic center of the galactic plane, such as were expected if these bursts came from galactic stellar remnants. Individual bursts have a location error typically 2–3°, which may be compared with the grid lines spaced at 30° intervals. (These data are available from *http://gammaray.msfc.nasa.gov/batse/grb/catalog/current/*)

at $z = 0.695$. As of early 2007, there are 118 redshifts for counterparts of so-called long bursts (the short bursts might still be a different class altogether, since we have as yet only a handful of counterparts suggesting a different environment than the more familiar long bursts). The burst afterglows seen optically show no distinguishing emission lines, but redshifts can be estimated quite closely when the onset of the Lyman α forest appears against their continuum radiation, or a lower limit is set when there is a metal-line system from the gas around some galaxy along the lines of sight. Many of these redshifts properly pertain to the host galaxy, observed after the burst radiation has faded. After three decades of mystery, we finally know *where* gamma-ray bursts are. What can we ascertain about *what* exactly makes them?

The large redshifts for the host galaxies of gamma-ray bursts imply enormous radiated powers for these sources unless their radiation is relativistically beamed. In fact, production of such large amounts of gamma radiation almost certainly means that beaming is important in what we see, so the question is more precisely whether this beaming is confined to certain directions. If so, the energy requirements are relaxed accordingly. One model finding considerable favor with regard to physical plausibility involves a hypernova—the collapse of a high-mass star in which a central black hole forms, surrounded by a transient and very dense accreting disk of former stellar matter. This configuration would then launch relativistic jets along the poles, much as seen for accreting neutron stars and black holes from SS 433 to quasars, and a gamma-ray burst would be near the direction of each jet. Indeed, one gamma-ray burst seems to have been associated with the explosion of a peculiar supernova, GRB 980425 at $z = 0.0085$ which was associated with SN 1998bw in the galaxy ESO 184-G82. The connection with supernovae has been further strengthened by X-ray measurements of the afterglow of burst 011211, which detected spectral lines from magnesium, silicon, and other elements expected to be produced in a core-collapse supernova, blueshifted by about $0.1c$ with respect to the parent galaxy (itself at $z = 2.14$).

The gamma radiation from such a supernova would be associated with initial breakout of the jet through the photosphere, when we would see the shocked material, particularly internal shocks within the jet, Doppler-boosted by a factor $\gamma \approx 100$. The optical and IR afterglows in this model come as the jet material cools and is decelerated by entraining stellar and wind material. In a simple geometric picture, the fading pattern of these afterglows offers a way to distinguish between this jet picture and one in which some enormous explosion produces a relativistically expanding fireball. As relativistically outflowing material (in whatever geometry) decelerates, its beaming angle $\approx 1/\gamma$ increases, so that we observe radiation from a larger fraction of the whole outflow. For this reason, we will see a rate of fading which is the combination of each volume element sending us less radiation and seeing more volume elements. If the radiating region is a jet, the fading will accelerate once the beaming angle becomes comparable with the cone angle of the jet, since there is no additional volume to have radiation beamed from. Thus, in the simple jet scheme, we expect a break in the flux decline of optical and IR counterparts. There have been several (though not all) afterglows which do show such breaks in fading, with a much more rapid drop in flux starting a few days after the initial outburst was observed.

The high typical redshifts of the gamma-ray bursts, and the fact that these had to be relatively bright bursts to be localized and identified for redshift measurements, make these objects very attractive in studying galaxy evolution. This small sample already has a median redshift 1.0, a depth achieved for optical galaxy surveys only by going very deep (in fact, deeper than any existing spectroscopic samples outside the Hubble Deep Fields). Being able to measure even statistical properties of much fainter bursts could trace their occurrence far into the early Universe, and, if we can understand how they relate to the history of star formation, provide a completely new window on the earliest epochs of galaxy assembly.

The hypernova scheme implies that gamma-ray bursts trace the death rate of some kinds of stars. If these kinds of supernova are generally like type II, these must be massive and short-lived stars. On the other hand, if they require that a compact stellar remnant be pushed over its stability limit by accretion or merger with a companion, the progenitors could take cosmologically long times to reach the burst event. Some information on the kinds of stars involved comes from studies of the host

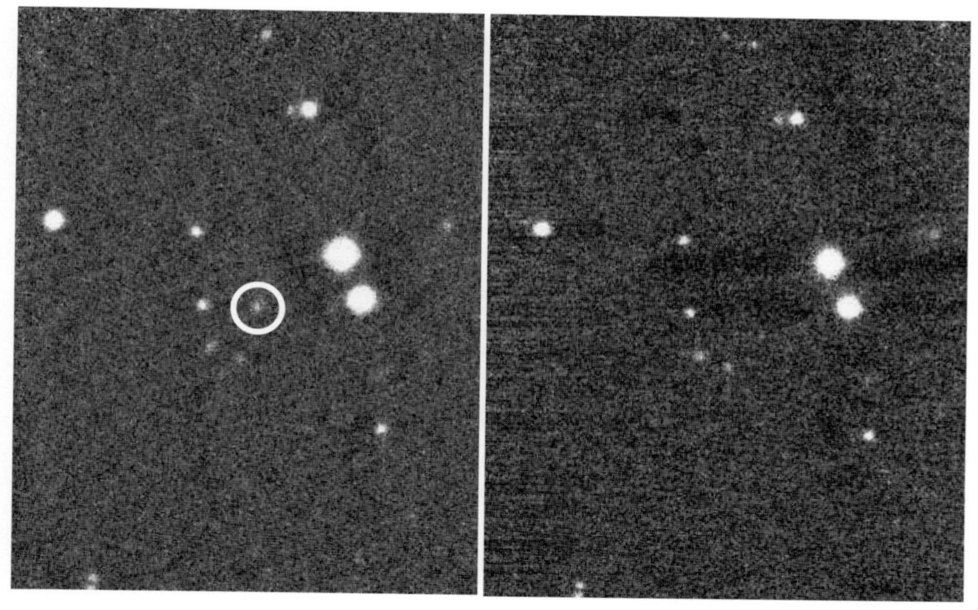

Figure 5.8. The fading afterglow of gamma-ray burst 991216, seen in these near-infrared (1.2–2.2 micron) images taken 2 and 4 days after burst arrival. This burst was found to have a redshift $z = 1.02$, and the very rapid fading of its optical/infrared afterglow suggested emission from a decelerating relativistic jet. (Data by the author using the 2.4-meter Hiltner telescope of the MDM Observatory, through the good offices of Ohio State University.)

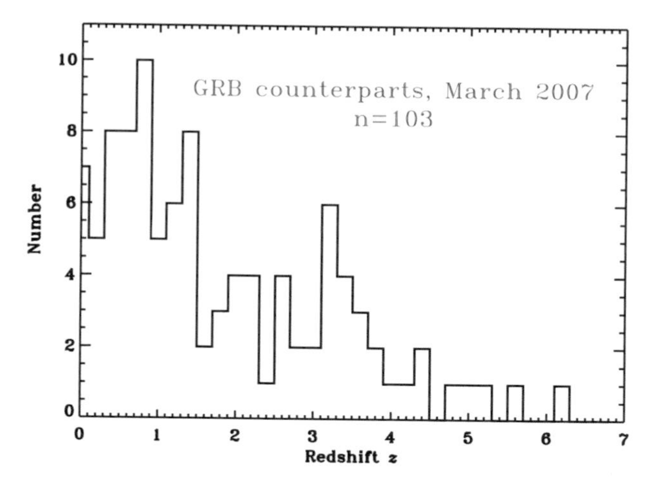

Figure 5.9. Histogram of secure gamma-ray burst redshifts, as collected by Jochen Greiner at *http://www.mpe.mpg.de/~jcg/grbgen.html*. Rapid identification and followup—made possible by the *Swift* mission—have been important in extending this to high values. No other known tracer has such a high characteristic redshift, or such a broad range, for flux-limited samples of the brightest examples.

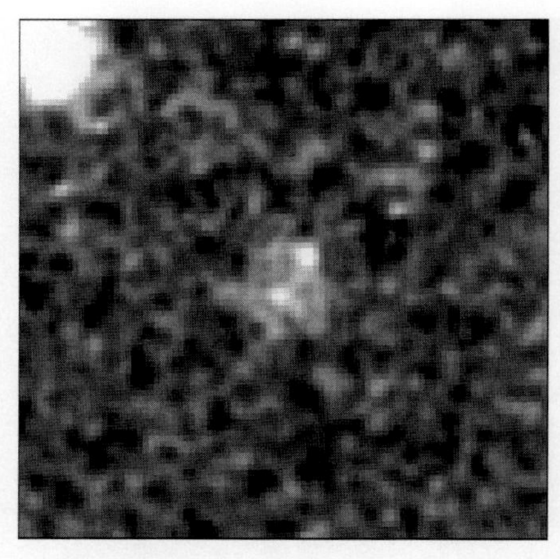

Figure 5.10. The host galaxy of a gamma-ray burst. This Hubble image of the fading afterglow of the first burst with an identified optical counterpart, GRB 970228 at $z = 0.695$, shows the host galaxy as well as the optical afterglow itself, on the edge of the galaxy image. The galaxy is detected over a region about one arcsecond in diameter, and could plausibly be a spiral with the burst projected about 3.5 kiloparsecs from the center. (This image is from data retrieved from the NASA Hubble Space Telescope Archive, with Andrew Fruchter as the original principal investigator.)

galaxies, especially from HST imaging. Most host galaxies, when one is detected, have a significantly clumpy structure and blue colors (Figure 5.10), linking the GRBs with star-forming regions.

Detailed study of the locations of afterglows in these host galaxies has shown that they roughly trace blue and ultraviolet starlight. This suggests that the progenitors were short-lived (as for core-collapse supernovae), not having had time to diffuse far through the galaxy since being produced in UV-bright star-forming regions. This supports the idea that GRBs can be used to trace the formation history of massive stars. Given the large redshifts found for relatively bright bursts, the fainter ones would be probing much higher redshifts, perhaps right up to the time of the first significant star formation. That would be a remarkable harvest indeed from the statistics of gamma-ray bursts.

Even more remarkable is the possibility that we are already observing gamma-ray bursts produced by the deaths of massive first-generation stars. These are predicted to explode in very powerful "hypernovae", and, if they produce narrow-beamed bursts at a relative rate similar to that required for type II supernovae in the later Universe, many of the bursts now being detected could come from these objects at $z > 10$. If this is so, we would expect to see their afterglows, as well as the hypernova emission itself, with deep observations further into the infrared than are typical for today. Lyman α and continuum absorption means that they are invisible at wavelengths shortward of $0.12(1 + z)$ microns. The flux levels we expect for the fading hypernovae are accessible for deep, targeted observations with existing equipment for $z < 17$, provided we have a position accurate enough to warrant spending large amounts of time on 8–10 m telescopes. At larger redshifts, spaceborne cryogenic instruments will be required, as atmospheric absorption, and especially emission, become rapidly worse with wavelength. Some bursts found at large redshifts (such as 050904 at $z = 6.3$) have been quite bright in gamma rays, indicating that we may already be detecting similar events to much higher redshift. There is a real possibility that the non-Euclidean dropoff in the counts of bursts versus intensity is a measurement of the first appearance of stars in the Universe.

A significant improvement in the localization rate, and hence statistics, of bursts came with the 2004 launch of *Swift*. This satellite's name derives from the goal of "catching gamma-ray bursts on the fly", generating identifications rapidly using on-board instruments for rapid response by ground-based facilities. This mission is deeply international, with co-investigators representing 40 institutions in 9 countries. Bursts are initially detected using a coded-mask detector (BAT) sensitive to hard X-rays (up to 150 keV). A coded mask, in which different parts of the detector see different subsets of the field of view, offers a fast and robust way to localize transient sources. This technique is very forgiving of partial detector failures, and allows fast algorithms for on-board estimation of the source location. This instrument yields immediate positions with a resolution of 22 arcminutes, accurate enough for the satellite to slew autonomously so as to place a burst within the field of view of an imaging X-ray telescope (XRT) and UV/optical telescope (UVOT). The XRT has $15''$ spatial resolution, and is sensitive over the softer energy range 0.2–10 keV. The UVOT by itself can yield superb locations for optical counterparts, with $0.5''$ pixels spanning a $17'$ field of view illuminated by a 30-cm primary mirror. A significant number of bursts will be seen serendipitously in the narrow fields of UVOT and XRT, independent of targeting from the gamma rays. Such an event had occurred prior to *Swift* only once in the history of GRB research, with the measurement of an optical flash by the ROTSE experiment in New Mexico while the gamma-ray burst was still in progress.

The evidence connecting gamma-ray bursts to supernovae is strong, although much remains to be done in understanding what kinds of supernovae from what kinds of progenitors. The data in hand indicate that these events may be the first to show us the earliest generations of stars, as our ability to observe them extends more deeply in other wavebands.

5.5 WATCHING GALAXIES EVOLVE

Recent years have seen remarkable progress in tracing how galaxies have evolved over cosmic time, driven both by technological advances and improved understanding of what we should be looking for. Using the approaches to finding very distant galaxies just described, we can start to piece together some aspects of galaxy evolution. The "Grail" in this endeavour would naturally be locating "protogalaxies", if we knew with any certainty just what they might look like and how to distinguish them from older galaxies.

Elliptical galaxies formed almost all their stars in a fairly brief, early episode. They may embody the Eggen/Lynden-Bell/Sandage collapse scheme best, although some of them seem to result from later mergers of disk galaxies. Their subsequent fading and reddening has been close to the idealized case of passive evolution, resulting only from the evolution of the constituent stars with no additional star formation. The scaling laws relating various properties of elliptical galaxies (from the fundamental plane) tell us that the structure of these galaxies was set closely by the scale or mass of the initial condensing material, and that subsequent developments have changed any of their properties only within this set of relations. Just how brief and how early the major star formation in elliptical galaxies might have happened is still uncertain. We do not yet see any population of objects numerous enough and bright enough to qualify as proto-ellipticals, which suggests that, if they were ever very luminous in the ultraviolet (instead of, for example, containing enough dust to absorb this part of the spectrum), their formation epoch must be at $z > 4.5$ to avoid utterly dominating the observed population of Lyman-break galaxies. The longer the time it took for initial star formation, the fainter a young elliptical would be to us, and the easier it becomes to "hide" this population. Likewise, if young ellipticals were as dusty as starburst galaxies are in our vicinity, they could elude searches for their ultraviolet or visible radiation, so that we would eventually find them in the infrared.

As discussed below, some ellipticals can be produced from mergers of spiral galaxies. Their rate of merging today can plausibly account for most ellipticals, but not explain easily why their stellar populations are so old. If most ellipticals come from mergers of other galaxies, there should be an age spread in at least their youngest stellar populations which reflects the overall history of galaxy mergers. Conversely, if most ellipticals are indeed extremely old and did not result from galaxy mergers, where is the remnant population for all the mergers which are clearly taking place? Part of the answer may lie in "dry mergers", where both progenitors were elliptical or similarly gas-poor S0 galaxies, so that the remnant would show the combined stellar and cluster content of both galaxies, but without an accompanying burst of star formation. Such mergers may be required to account for the most luminous elliptical galaxies, and for the clear distinction between the colors of star-forming galaxies and those with no ongoing starbirth. Faber and colleagues have particularly stressed that the evolutionary pathways allowed by the red and blue sequences in a color-luminosity diagram (Chapter 4) severely limit the ways the most luminous red galaxies can form. The fading and reddening—after a massive starburst—happen so slowly that mergers of gas-rich galaxies would be too blue for too long to make the most luminous galaxies (except, perhaps, in the rare case that internal dust happened to just cancel the effects of the blue stellar population in some particular set of passbands).

The red color sequence at high redshift, especially as seen in clusters, sets powerful limits on the typical formation time of elliptical galaxies as such. Data from the Spitzer Space Telescope, in particular, have shown that the red sequence is already distinct in clusters by redshift $z = 1.5$, and so narrow in color dispersion as to suggest that these galaxies had completed major star formation by a time corresponding to

$z = 2$–3. In part, this is a manifestation of the general "downsizing" pattern found in many tracers of star formation (Chapter 9).

Spiral galaxies have had a more involved history, one which is harder to trace because we don't see their structures in the same way at different redshifts. Their disks may have accumulated over a long timespan, from surrounding gas which itself might have already clumped into dwarf galaxies, and would have brightened dramatically as star formation took place at a slow (sometimes nearly constant) rate. Thus, the familiar spiral patterns may not have been common for billions of years; the most distant spirals clearly identified as such at this point are at $z \approx 1.4$. Our own galaxy furnishes ample evidence of slow growth of the disk, both chemically and structurally (Chapter 3). The bulges of some spirals show dynamical, as well as chemical, signs of having formed after a substantial disk was in place, so even this component of spirals may have had a protracted origin. Even as spirals are developing over cosmic time, their survival in that form depends on the local environment—as we shall see, the busy surroundings of a cluster of galaxies are unhealthy for spiral and irregular galaxies.

Studies of intergalactic material (Chapter 6) make it clear that young galaxies interacted strongly with their environments, ejecting vast amounts of enriched material into the surrounding gas. This means that we cannot treat galaxies as closed, isolated boxes, and that we must take the widest possible view to find all their products. And, beyond these trends, galaxy evolution can be forced, either in episodic or long-term ways, by external events.

5.6 ALTERED STATES AND FORCED EVOLUTION

Multiple lines of evidence show that interactions with other galaxies or the more general environment can drive galaxy evolution externally. We see this clearly in *starburst galaxies*. Starbursts are broadly defined as undergoing star formation so strongly that the available interstellar gas will be exhausted in a short time compared with the age of the galaxy, so that we must be seeing a transient event. Starbursts are statistically linked to galaxy interactions and mergers, and can be fostered internally as well via gas flow along bars. Clusters of galaxies show a pronounced change in their constituent galaxies with redshift, indicating that clusters can transform their galaxies. Both gravitational and gas-dynamical processes have been shown to contribute.

5.6.1 Galaxy interactions and mergers

Peculiar galaxies have long attracted special attention as extreme, unusual, and difficult to explain. Most of the objects in Vorontsov-Velyaminov's (1977) *Atlas and Catalog of Interacting Galaxies* and Arp's *Atlas of Peculiar Galaxies* display asymmetries, tails, plumes, and other kinds of structures which are now interpreted as the result of tidal encounters with other galaxies, sometimes leading to a merger of the galaxies involved. It was by no means obvious circa 1970 that gravitational effects alone could produce the kinds of long, thin, and straight plumes which are seen (some samples are illustrated in Figure 5.11). Early numerical work (following, belatedly, Erik Holmberg's analog calculations from the early 1940s) demonstrated that tidal effects in self-gravitating and rotating systems can have non-intuitive effects, and that well-known galaxy pairs with long tails could be reproduced in some detail. A turning point was the work by Toomre and Toomre in 1972, showing that even what we would now consider rudimentary simulations (test particles, galaxy masses concentrated at the nuclei) could reproduce many of the salient structures in the well-known systems M51, NGC 4038/9 (the Antennae), and NGC 4676 (the Mice).

Further work made it clear that star formation on a galactic scale, as well as events in galactic nuclei, could be controlled by tidal interactions. Larson and Tinsley (1968) examined the optical colors of galaxies from the Arp atlas, finding that the interacting Arp objects had colors which could be explained by recent bursts of star formation superimposed over "normal" more quiescent populations (represented in their study by inclusion in the Hubble Atlas). Similarly, many of the objects found by B.E. Markarian and colleagues in their survey of UV-excess galaxies, designed to detect Seyfert activity and unusual levels of star formation, are morphologically peculiar or obviously interacting. Likewise, using the far-infrared detections from the IRAS survey to trace these phenomena even when heavily shrouded in dust, we find that a very high fraction of galaxies detected in the far-IR shows such evidence of tidal distortion. As the far-IR luminosity rises, so does the fraction showing signs of interaction or merging, approaching 100% at a trillion (10^{12}) solar luminosities.

There has been no shortage of theoretical suggestions for just *how* interactions and mergers trigger bursts of star formation. Some of these have various amounts of support from various sets of data, but it is still unclear which (if any) actually account for what we see. Collisions between molecular clouds are a popular suspect, since clouds that formerly traced undisturbed orbits may now collide with each other if the galaxy is tidally disturbed at even a gentle level. The frequent occurrence of starbursts in a small, dense nuclear region has led to simulations showing how a bar can form temporarily after a tidal encounter, sap the angular momentum of disk gas very efficiently so as to drop the gas into the nucleus, and then change the mass distribution enough to evaporate the bar. Some calculations have suggested a role for pressure changes being transmitted from one phase to another in the interstellar medium, so that effects on hot gas in one part of the galaxy disk might be transmitted to molecular clouds in another part.

Independent of the details of just how interactions trigger star formation, on which there is still no clear consensus, the empirically established fact that such triggering does take place is important for galaxy evolution. In the stellar population and gas consumption that are major clocks for galaxy evolution, such a burst will mark a sharp step superimposed on the otherwise smooth and stately development expected for an isolated galaxy. Particularly in the early Universe, might this process have triggered most star formation? How many galaxies undergo interactions, and when is this an important factor in their development?

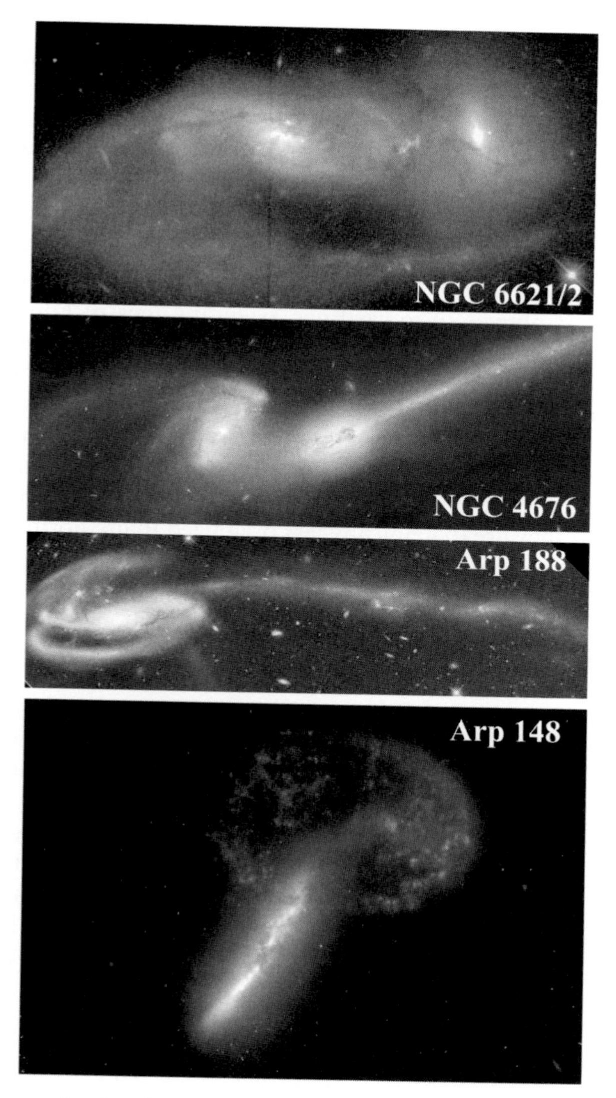

Figure 5.11. Galaxy pairs showing strong tidal distortions. NGC 6621/2 shows many young, luminous star clusters as well as a disturbed dust structure. NGC 4676AB (also known as the Mice) shows how tidal features can be quite different, as seen from various angles. Arp 188 sports a long, spectacular tail whose source seems to be the influence of an inconspicuous dwarf galaxy silhouetted in front of the galaxy disk itself. Arp 148 shows a collisional ring, produced when a concentrated galaxy falls almost perpendicularly through a less-concentrated disk system. The impulse of this additional, temporary inner mass produces propagating density waves and accompanying shocks, triggering the prominent rings of star formation. (Images from the Hubble Space Telescope; data for NGC 6621/2 by Keel and Borne, for Arp 148 by Aaron Evans *et al.*, and for the other two by NASA, H. Ford of JHU, G. Illingworth of UCSC/LO, M. Clampin and G. Hartig of STScI, the ACS Science Team, and ESA.)

The most extreme interactions are mergers, especially "major mergers" in which the two precursors have comparable mass. There is strong evidence that such mergers between gas-rich galaxies can (usually?) trigger such violent starbursts that the dense components of the interstellar medium are swept away in a global wind, leaving the stars alone to relax into a structure much like elliptical galaxies. Some of the best examples are found among infrared-bright galaxies, especially those with the highest infrared excess (ratio of dust-radiated emission to direct starlight) that are seen close to the epoch of merging of the nuclei. Two of the best-known were identified as such early during the IRAS mission, NGC 6240 and Arp 220. In each case, emission-line imaging reveals vast plumes of ionized gas extending tens of kiloparsecs from the main galaxy, kinematically found to be expanding at velocities up to several hundred km/s and thus likely to be unbound. A time sequence of nearby mergers, in the order suggested by numerical simulations of major mergers, indeed is also a sequence in far-infrared luminosity, stellar color, and abundance of gas tracers (Figure 5.12).

Galaxy mergers, driving massive starbursts, are sometimes considered as nearby analogs of what massive galaxies would have looked like in the early Universe. This makes sense in that they are dynamically less organized than "normal" galaxies, with collisions between gas clouds occurring at a wide range of velocities unconstrained by disk dynamics. On the other hand, a merger started out with two well-developed bound systems, each with a distinct pattern of stellar motion with position, which is something that we would not necessarily expect in the earliest stage of galaxy formation.

Galaxy mergers as a evolutionary link between disk systems and elliptical galaxies raise an obvious question—are any elliptical galaxies truly primordial? If we can still find ellipticals being made today through this route, were they all produced from mergers of disk systems? Dynamical simulations show that the resemblance between the final state of such a merger and an elliptical galaxy will be more than skin-deep, with a stellar density profile very close to what we typically find for ellipticals once the detailed structure of the merger has been smeared out by relaxation. The stars will initially include a component that is bluer than we commonly see in ellipticals, but after a few Gyr the age difference becomes more and more subtle in the integrated light of a galaxy's worth of stars.

5.7 CLUSTER PROCESSES DRIVING GALAXY EVOLUTION

The mix of galaxies in various local environments, especially in rich clusters, provides another powerful clue that external forces can drive the evolution of galaxies. The earliest surveys of the galaxy content of clusters showed that the fractions of disk and elliptical galaxies nearly reverse between the "field" and such rich clusters as Coma. More detailed work, particularly by Alan Dressler (1984), led to the formulation of a *morphology–density relation*, in which the mix of morphological types is closely set by the local space density of galaxies. Originally, this relation allowed explanations as a result either of heredity or environment. That is, such a relation could exist if conditions in what would one day become a rich cluster environment were such as to

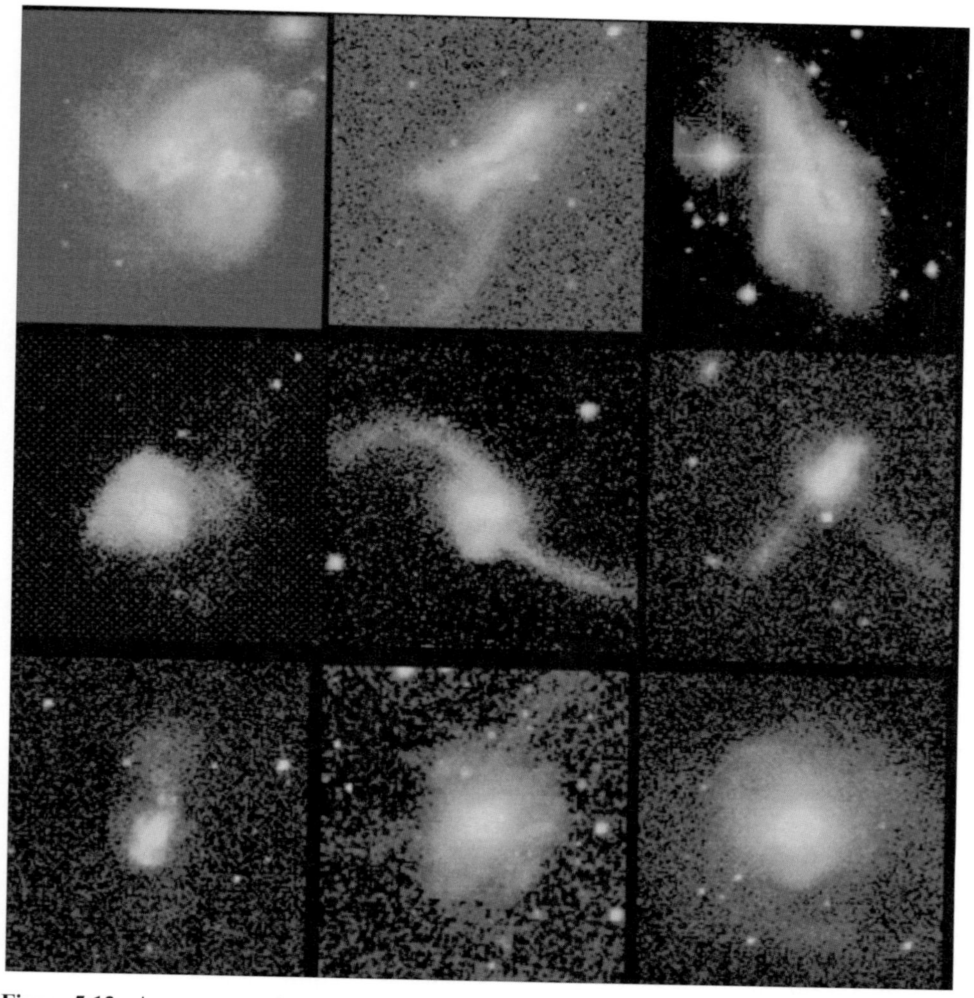

Figure 5.12. A sequence of merging galaxies, seen in visible-light images and arranged in the time order implied by numerical simulations and by their stellar populations. The sequence runs from left to right and top to bottom, comprising NGC 3690, NGC 520, NGC 6240, Arp 220, NGC 2623, IC 883, NGC 4194, NGC 7252, and NGC 7723. (Images obtained by the author at Kitt Peak National Observatory, Cerro Tololo Inter-American Observatory, Lowell Observatory, and the European Southern Observatory.)

favor the formation of some kinds of galaxies over others, or if the environment in a rich cluster brought about either the destruction of spiral galaxies or their transformation into elliptical and S0 systems.

Largely thanks to the high-resolution images from the Hubble Space Telescope, it is now clear that the morphology–density relation results at least in part from a history of galaxy transformation in clusters. This is known from the existence of hot intracluster gas, the Butcher–Oemler effect (Butcher and Oemler, 1978), and the fact that clusters are still in formation today so that their initial environments were not very different from the surroundings. There is no shortage of mechanisms for reducing the fraction of spiral galaxies in a cluster environment.

5.7.1 Intracluster gas

One of the most unexpected discoveries in the infancy of X-ray astronomy was that clusters of galaxies are strong sources of X-rays. With improvements in the achievable spatial resolution, it became clear that the X-ray emission from clusters is truly diffuse, with only a minor contribution from individual galaxies. Spectral information confirmed theoretical interpretations of this emission as being thermal radiation (bremsstrahlung) from plasma at temperatures of order 10^7 K. The mass of this plasma is comparable with that in the stars of a cluster's galaxies. The temperature is well understood, being dominated by the cluster's potential; the particles in the intracluster gas follow orbits through the overall cluster potential at nearly the same velocities as individual galaxies, which translates to the observed temperatures. More details on this material and its relation to galaxy formation are given in Chapter 6.

Cluster X-rays had more surprises in store. This gas is not primordial, chemically pristine matter which was somehow left behind in galaxy formation. Emission lines from metals show that it has significant metallicity, and thus has been processed through stars at some point before being ejected from the galaxies. Remnants of this ejection may be found in a detailed analysis of the gas temperature—for poorer clusters, the gas is slightly hotter than the cluster dynamics alone would suggest, while the cluster's internal velocities dominate for rich clusters. These can be accounted for by a fairly constant extra energy input, perhaps to be identified with the ejection energy of the gas. At this point, the X-ray data themselves don't tell much about when this massive star formation and ejection of heavy elements from supernovae happened, except that the process had been completed by the redshifts of clusters that we can study in detail (say, before $z = 0.5$). It is a reasonable supposition that the period of cluster enrichment was part of the intense period of galaxy building at early epochs, and thus that very deep X-ray observations should be able to trace the history of star formation through the emerging legacies of massive stars.

The existence of a significant intracluster medium opens the possibility that it might be responsible for the transformation of disk galaxies, most prominently through ram-pressure stripping. A disk galaxy moving through the ICM of density ρ will experience a wind at the velocity v of its movement, so that every unit area of its disk will be subject to a pressure $P = \rho v^2$. In the simplest formulation, when this pressure on a parcel of gas exceeds the restoring force from the galaxy's own gravity, the gas will be stripped. More realistically, the role of stripping should be different for different components of the interstellar medium, depending on density and distance from the galaxy's center. Even more realistically, numerical simulations show that much of the stripping would take place in a turbulent interface layer rather than in the simple dynamical scheme just described.

Evidence on the stripping of gas in cluster members long remained circumstantial. Globally, spirals in clusters are poorer in gas (most notably when the gas content is normalized to the optical luminosity) than similar non-cluster galaxies. This deficiency is most pronounced from (in fact, is almost entirely due to) the H I, which should be more vulnerable to pressure effects than H_2, which is denser and generally more centrally concentrated. The H I that does occur in cluster spirals appears truncated near cluster cores, and may be asymmetrically distributed at the outskirts, as if the inner ones have already lost their most easily-stripped gas and the outer ones are in the process of having it removed. This was beautifully shown in a massive H I study of the Virgo Cluster carried out by Cayatte *et al.* (1990) using the VLA (Figure 5.14).

It has proven much harder to find a galaxy which is clearly in the act of being stripped by a surrounding medium. There have been repeated studies of galaxies which looked promising from the standpoint of asymmetry, gas content, or surrounding X-ray structure, but confusion with tidal encounters has usually rendered the results ambiguous. This remained true for some of the asymmetric structures seen in H I; the ambiguity in their origin remained because many disks are large enough in H I to be affected by tidal perturbations more readily, and retain these disturbances much longer, than the smaller stellar disks. This difference between stellar and gaseous disks in extent can make it difficult to be sure we see the key feature of ram-pressure stripping, a difference between stellar and gaseous structures in which the stars are so dense as to be practically unaffected by the external medium (unless they are stars which manage to form in stripped gas). Perhaps the best nearby candidate for ongoing gas stripping is the Virgo cluster spiral NGC 4522 (Figure 5.15), studied by Kenney and Koopmann (1999), which shows ionized gas stretched into filaments extending more than 3 kpc beyond the star-forming inner disk, resembling a galaxy-scale bow shock, and roughly coextensive with similarly asymmetric H I emission. The outer disk shows spectroscopic evidence that star formation ceased recently, about 10^8 years before our present view, and roughly contemporaneous with the onset of stripping. Clear-cut evidence of stripping on larger scales has been found in richer clusters, which are rarer and thus seen only at greater distances; Figure 5.13 illustrates the spectacular emission-line and X-ray tail of the galaxy C153 in Abell 2125, one of the clearest examples known. The key signatures prove to be long trails of emission both in optical line emission and in soft X-rays. The origin of the X-ray emission remains ambiguous; it could result from a hot interstellar medium in the galaxy itself, being removed along with the cooler gas, or it might represent intracluster gas, cooling as the galaxy's gravitational wake drives it to higher density and therefore more efficient radiation. This sign of X-ray stripping can be seen even for elliptical galaxies with no significant cool gas.

Numerical simulations (in particular, some detailed examples carried out by Bernd Vollmer and collaborators) show that the mass loss from a spiral can be highly episodic, depending on the interplay between galaxy rotation direction and relative motion within the cluster as well as the detailed initial gas distribution. Some of the stripped gas in these simulations ends up being re-accreted by the galaxy, if its orbit carried it far enough out in the cluster before the gas becomes unbound due to either cumulative ram pressure effects or weak tidal influences.

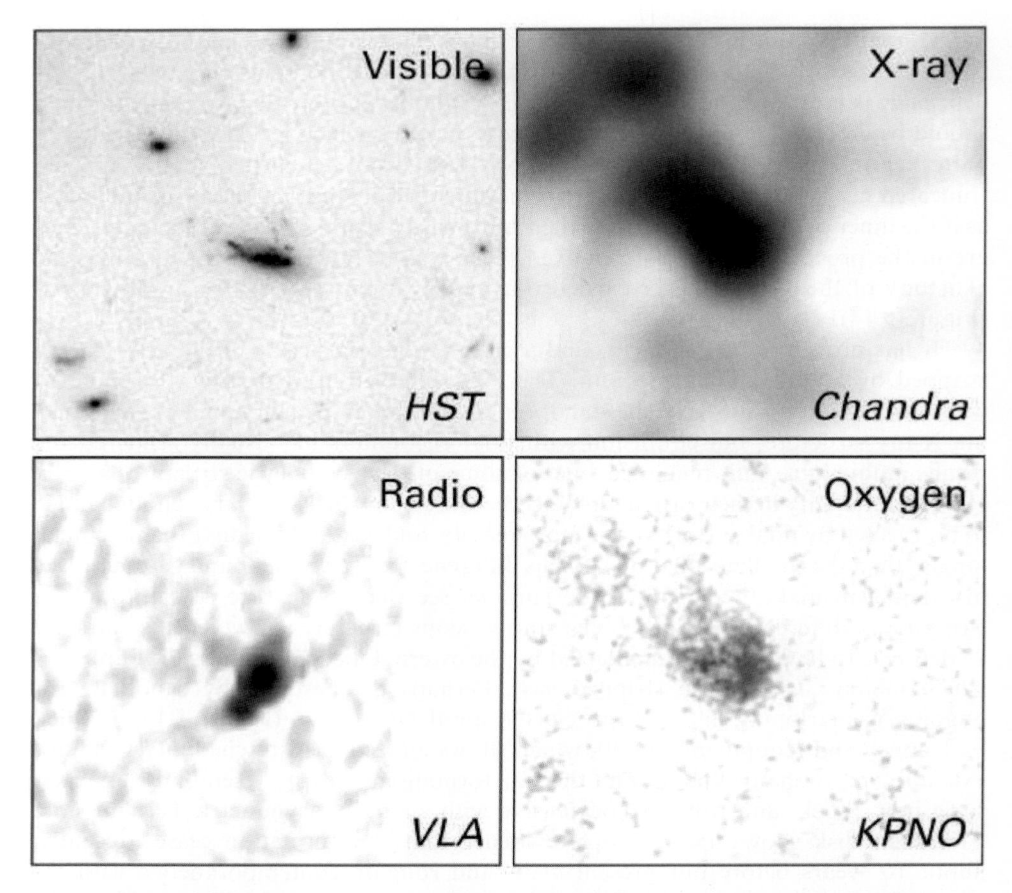

Figure 5.13. Evidence of ram-pressure stripping in the galaxy C153, part of the dynamically complex cluster Abell 2125 at $z = 0.25$. Hubble imaging shows a disturbed morphology with star-forming knots, while [O II] narrowband imaging revealed a tail nearly coinciding with a soft X-ray trail seen in Chandra data. Radio-continuum observations show a central source with bent double structure. The optical spectrum reveals multiple emission-line components and indicates that a massive starburst occurred about 10^8 years before our present view, perhaps coinciding with the initial overpressure as the galaxy was first affected by the surrounding hot gas. (W. Keel *et al.*, NASA, NRAO, NOAO, and ESA.)

5.7.2 Cooling flows, feedback, and galaxy building

In a first approximation, we treat the hot gas in clusters of galaxies as being in hydrostatic equilibrium, with the thermal pressure balancing gravity to keep the gas in a stable configuration. In this case, the gas density will increase inwards for nearly constant temperature. The cooling rate in the intracluster plasma has a quadratic dependence on density, since it radiates through a collisional process. This has interesting consequences when the central potential drives the density high

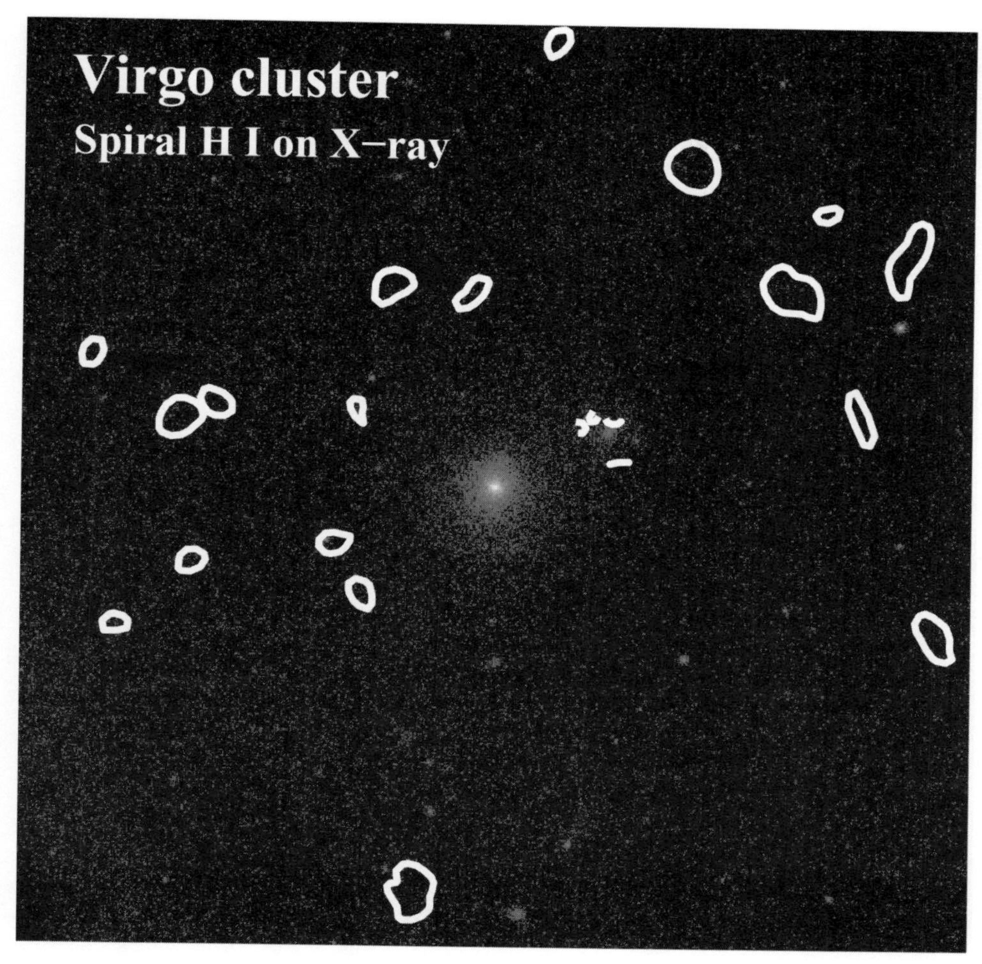

Figure 5.14. Interaction between interstellar and intracluster material. The outlines show the extent of neutral hydrogen detected in the disks of spiral galaxies in the Virgo Cluster (each enlarged by a factor of 5 for clarity), and the underlying image shows the X-ray emission from the hot intracluster gas. Spirals projected near the core have truncated H I distributions (which are also more asymmetric), while spirals in the outer regions have essentially normal properties with no evidence of environmentgal influences. These results suggest that stripping by the cluster gas is important in driving the evolution of spirals in rich environments. (The H I data were taken from V. Cayatte *et al.*, *Astronomical Journal*, **100**, 604, 1990, and the X-ray image was produced by the Skyview utility from the ROSAT All-Sky Survey data.)

enough. The cooling timescale becomes shorter, and in the cores of some clusters can be much less than a Hubble time. These clusters should then support cooling flows—inward net motions of the gas, as it cools and the pressure drops, so that the weight of overlying gas will compress it and make it cool still more rapidly. Numerous rich

Figure 5.15. Perhaps the most convincing nearby case for a galaxy being stripped by surrounding gas, NGC 4522 in the Virgo Cluster. The Hα structure shows gas being swept back on both sides of the disk, with even more extensive sweeping seen in the H I structure. Both the truncation and offset perpendicular to the disk indicate that gas is being swept from the galaxy. Its radial velocity is consistent with a rapid relative motion through the intracluster gas, which could be dense enough to affect the disk even at a projected separation of a megaparsec from the cluster core. (Images obtained by Jeffrey Kenney at the 3.5 m WIYN telescope, used by permission.)

clusters have cores which fulfill these conditions, and thus should have been feeding gas inward from the inner few hundred kpc. The computed mass flow rates past a spherical surface inside this may be hundreds of solar masses per year. This is an attractive way to build galaxies, especially since many clusters with cooling flows do have extremely luminous and large central galaxies.

However, the connection between cooling of the X-ray gas and galaxy building has proven to be elusive. Some central cluster galaxies do indeed show the ultraviolet and emission-line signatures of the appropriately high level of star formation (NGC 1275, Abell 1795, PKS 0745−191), but others show little or no trace of ongoing star formation. What happens to the gas as it cools is a puzzle. These cluster cores do not show the levels of emission in H I or CO that would accompany these amounts of star formation—and, star formation or not, the gas must be going somewhere.

One answer lies in feedback. Both starbursts and active galactic nuclei can heat surrounding material, through both radiative and mechanical processes. Global winds will slow or stop a cooling flow, at least temporarily. Around some AGN, X-ray mapping shows clear signs that jets or shocks have reshaped the X-ray gas (Figure 5.16). Feedback may occur at both the galaxy and cluster scales. If this is responsible for limiting the cooling of X-ray gas in cluster cores, it would require that most central cluster galaxies experience fairly frequent episodes of nuclear activity. Our ability to quantify the expected effect of such an episode is still very crude, because most of the work must be done by mechanical rather than radiative effects. Most straightforward observable quantities are closely related to emerging radiation; mechanical energy input requires mapping of tracers related to the local velocity field.

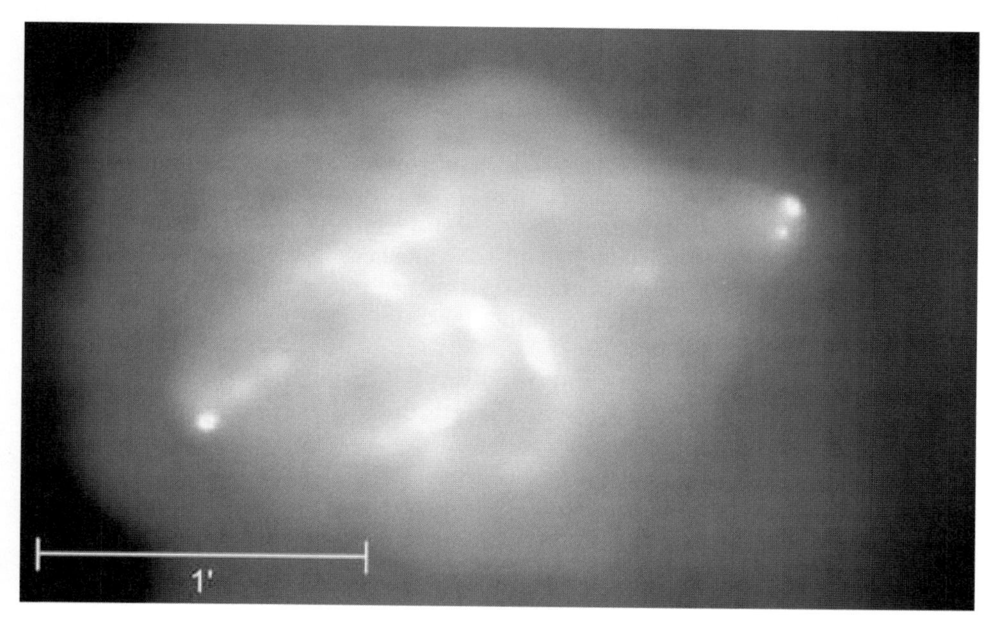

Figure 5.16. A Chandra X-ray image of the powerful radio galaxy Cygnus A. The radio jets, advancing toward the hot spots visible near the left and right edges, have opened prominent voids in the emission of cluster X-ray gas. This is among the clearest instances of feedback from an AGN shaping gas in the center of a galaxy cluster. (NASA/University of Maryland/ A. Wilson *et al.*)

5.7.3 The Butcher–Oemler effect—galaxy transformation in action

One of the first, and most powerful, empirical signs of galaxy evolution has been the Butcher–Oemler effect. This is a progressive shift in the colors of galaxies in rich clusters as one looks to greater and greater redshifts, first reported by Harvey Butcher and Augustus Oemler in 1978. The elliptical and S0 galaxies which predominate in today's rich clusters have a very narrow distribution of optical colors, reflecting their old stellar populations. In some rich clusters above $z = 0.25$, and in all but a few at $z = 0.5$, there is a significant blue tail to this color distribution, indicating that many of the cluster members are actively forming stars. Something has changed over this span of cosmic time, a change which seems to have been completed recently compared with the galaxy lifetimes. A number of subtleties have entered into our elucidation and understanding of this phenomenon. At significant redshifts, care must be taken to avoid interloper galaxies from outside the cluster, whose redshifts may be broadly similar to cluster members. We must verify that the galaxy colours do not reflect a central active nucleus, but star formation within the galaxy. Statistical controls are required to avoid biases that might be introduced from observing richer clusters at larger redshifts, both because we find such rare objects only at large distances and because richer clusters are easier to find in the presence of such large numbers of foreground and background galaxies. Comparison with galaxies at similar redshifts but outside of rich clusters is also important, to be sure that we are seeing a process actually associated with cluster members and not simply seen there because clusters are the easiest places to find large numbers of galaxies at a particular redshift.

The Butcher–Oemler effect is showing us wholesale transformation of galaxies in the cluster environment. HST images have shown that much of the transformation is in morphology, not just in stellar content. Images of several clusters from $z = 0.4$ to $z = 0.8$ have revealed that many of the "extra" blue galaxies are clearly spirals (Figure 5.17), a population which has all but vanished by the present epoch. They also let us estimate how many of these systems are interacting or merging, which is important in telling how important these processes might be in galaxy transformation.

When the morphology–density relation was first defined, there was speculation that galaxy collisions might be responsible, through the stripping of ISM to be expected in a direct high-speed collision between gas-rich disk systems. A painting by Antonio Petruccelli, done for a widely-reprinted 1954 article in *Life* shows two spiral galaxies passing through each other with their stellar patterns unscathed even as their gaseous components collide and heat to incandescence. This reflected the then-conventional wisdom, based on the fact that the mean free path between collisions for individual stars is much larger than the size of galaxies. The notion was that, since the separation between stars is so large compared with their sizes, that direct stellar collisions during such a penetration would be vanishingly rare, and the only bulk effect would be direct collisions between interstellar clouds. However, as in violent relaxation of stars in a galaxy, this is not the appropriate standard of comparison, since the interaction between the changing overall gravitational potential of the

Figure 5.17. The Butcher–Oemler effect in the rich cluster Abell 851 (Cl 0939+4713) at $z = 0.39$. This section of a red-light HST image, roughly matching the emitted B band, shows numerous disk galaxies with clear spiral structure. This stands in sharp distinction to the nearly pure elliptical and S0 galaxy mix of local clusters which are even less populous. To show structure over a wide range in surface brightness, an offset logarithmic intensity mapping has been applied. (Image from archival HST data, original PI was A. Dressler.)

combined galaxies and a star will change the star's orbit, and by extension the form of both galaxies.

As our understanding of tidal effects in galaxies grew, it was recognized that the tidally mediated *perestroika* in cases of galaxy collisions could have profound and permanent structural effects. Rich clusters, with internal velocity dispersions much greater than the speed of internal motion within individual galaxies, turn out not to be promising environments for galaxy mergers, or even very strong tidal distortions. Encounters are hyperbolic, and don't last long enough to build up an elaborate tidal response. However, a rather different kind of cumulative damage from tidal effects

Figure 5.18. Galaxy harassment in action. This pair, NGC 4435/4438 near the core of the Virgo Cluster, is undergoing a hyperbolic encounter at a relative velocity of at least 730 km/s, far above the internal velocities. The larger disk of NGC 4438 shows very asymmetric tidal damage, reflected in its H I and CO distributions even more strongly than in the starlight shown in this red-light image. The more compact disk of the SB0 galaxy NGC 4435 is escaping relatively unscathed. Further encounters, even weak ones, could result in most of the material in the disturbed disk being lost to NGC 4438. (Image by the author using the 0.9-m telescope of the SARA Observatory on Kitt Peak.)

has been identified in clusters, under the name of *galaxy harassment* (Figure 5.18). In the frantic environment of a rich cluster, hyperbolic encounters, which individually produce fairly weak tidal impulses, occur at a faster rate than a disk system can return to its previous dynamical state. The result is that the tidal tail pulled out by one encounter can be further distorted, and often become unbound, during a further encounter. In fact, a significant fraction of large disk galaxies could be completely shredded by this kind of action over a Hubble time.

The cumulative effect of such galaxy destruction is a growing population of stars floating around in a cluster free of an ordinary galaxy. In fact, a significant fraction of the starlight in some clusters comes in such a diffuse form—from stars outside the most liberal boundaries of individual galaxies. This can be seen via individual stars and planetary nebulae in the Virgo cluster. Careful baffling of cameras from stray light, and masking of starlight, allows the detection of the dim diffuse light of stars which have been stripped from individual galaxies (Figure 5.19). This intracluster light shows detailed structure attesting to the individual events which have pulled stars out of the galaxies. Several other nearby clusters show these faint wisps that may be the last shredded remnants of harassed galaxies. Structures of stars will be

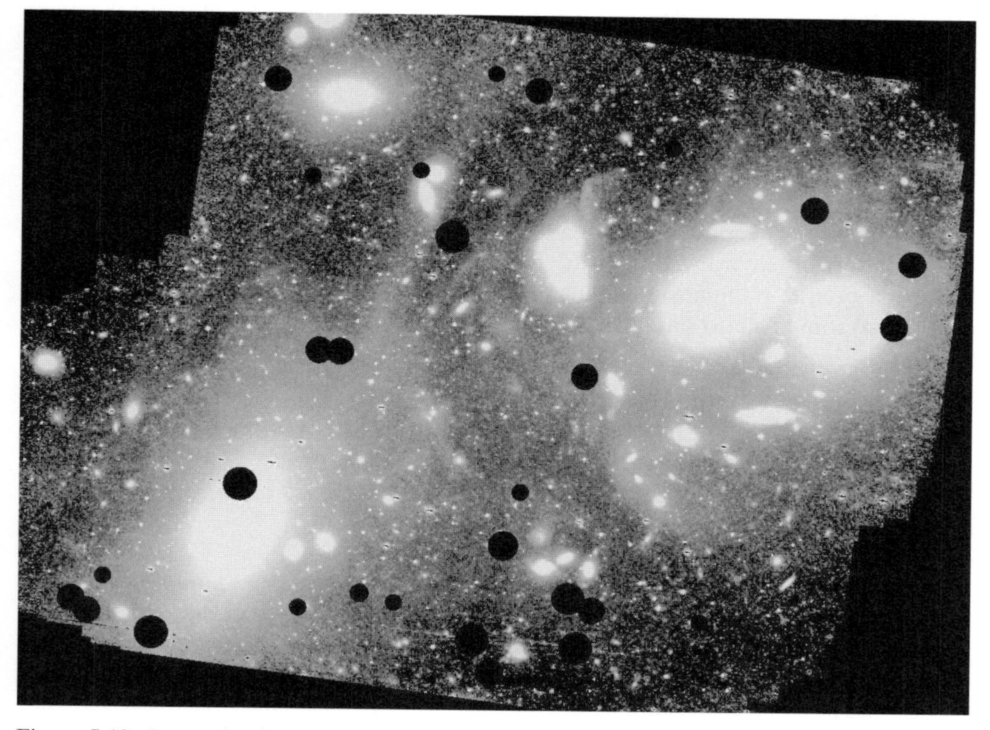

Figure 5.19. Intergalactic starlight in the Virgo Cluster. This deep wide-field image shows streams of stars outside the galaxies of the Virgo Cluster, testimony to a history of their involvement in events such as the hyperbolic encounter seen in Figure 5.18. In fact, the NGC 4435/8 system is visible here above center. Black regions are corrupted by bright stars. (This image spans a field 1.5° across, and was made possible by careful baffling of light-scattering paths within the telescope as well as extensive use of flat fields from sky observations—Christopher Mihos, Case Western Reserve University.)

recognizable longer than similar structure in gas, since the mean free path for ions in the intracluster medium is short; as in galaxies, stellar orbits respond only to gravity, and retain tails and filaments long enough to trace the recent details of a cluster's history. Some of these bits of tidal debris may eventually be recycled.

At the centers of many clusters, poor as well as rich, we find galaxies with very extended outer halos of stars, some of them the largest and most luminous galaxies in today's Universe—the cD galaxies. They may in part represent "starpiles" growing through gravitational accumulation of some of the stars stripped from smaller galaxies over cluster history.

5.7.4 Cluster growth

Rather than being relatively static structures formed in the early Universe and evolving in isolation, galaxy clusters have a rich history of continued growth, with acquisitions and mergers occurring just as we see for individual galaxies. Perhaps the key recognition here came from a study intended to measure the Hubble constant. Aaronson *et al.* (1980) showed that distances within 20 Mpc indicated systematic departures from a smooth Hubble flow, in the sense that would result if all the galaxies around the Virgo cluster are affected by its gravity and being accelerated in its direction (Virgocentric infall). Similar distortions in redshift-distance mapping are now well-documented in many clusters, and can be used to measure mass densities and the status of dynamical development of large structures. If clusters are still growing now, they cannot have been as rich or dense at early epochs, making it less likely that their environment could have dictated the formation of what would become a unique population of galaxies. Still, the idea of biased galaxy formation does allow some role to have been played, if only as a statistical constraint, by the local mass density at early times.

5.8 BIBLIOGRAPHY

Books

Adams, Fred and Laughlin, Greg (1999) *The Five Ages of the Universe: Inside the Physics of Eternity* (Free Press). A unique discussion of what happens to galaxies over such long timespans that dynamical relaxation proceeds to its conclusion. The authors then go on to describe the cosmic effects of the decay of not only protons, but black holes.

Cowie, Lenox (1988) "Protogalaxies", in *The Post-Recombination Universe* (Kluwer, pp. 1–18).

Hamilton, A.J.S. (1998) "Linear Redshift Distortions: A Review", in D. Hamilton (ed.), *The Evolving Universe* (Kluwer, pp. 185–276). A lengthy review of redshift distortions and the growth of structures with cosmic time. These mismatches between the depths of structures in transverse dimension and in redshift space are important cues to the dynamical status of clusters and superclusters, especially as particular regions turn around to drop out of the Hubble flow.

Hill, Edward (1986) *My Daughter Beatrice: A Personal Memoir of Dr. Beatrice Tinsley, Astronomer* (American Physical Society). Beatrice Tinsley's life and work are recounted by her father.

Journals

Aaronson, M.; Mould, J.; Huchra, J.; Sullivan, W.T., III; Schommer, R.A.; and Bothun, G.D. (1980) "A Distance scale from the infrared magnitude/H I velocity-width relation. III—The expansion rate outside the local supercluster", *Astrophysical Journal*, **239**, 12–37. As part of an extensive distance survey using the Tully–Fisher relation, they showed that the local pattern is much better fit if there is significant infall (including us) into the Virgo Cluster. This was the first widely accepted evidence for systematic departures from the Hubble Flow on scales beyond individual galaxy clusters.

Bloom, J.S.; Kulkarni, S.R.; and Djorgovski, S.G. (2002) "The Observed Offset Distribution of Gamma-Ray Bursts from Their Host Galaxies: A Robust Clue to the Nature of the Progenitors", *Astronomical Journal*, **123**, 1111–1148. An analysis of the available data on the host galaxies of bursts, and where the bursts come from in these galaxies. The results suggest a connection between massive star formation and gamma-ray bursts.

Butcher, H.R. and Oemler, A., Jr. (1978) "The evolution of galaxies in clusters. I—ISIT photometry of C1 0024+1654 and 3C 295", *Astrophysical Journal*, **219**, 18–30. The initial report of increased fractions of blue galaxies in clusters at substantial redshift, now known as the Butcher–Oemler effect.

Cayatte, V.; van Gorkom, J.H.; Balkowski, C.; and Kotanyi, C. (1990) "VLA observations of neutral hydrogen in Virgo Cluster galaxies. I—The Atlas", *Astronomical Journal*, **100**, 604–634. Extensive H I data on galaxies in the Virgo cluster, showing a clear connection between the neutral gas content of spiral galaxies and their distance from the cluster center. This fits with a strong role for stripping by intracluster gas in the evolution of cluster members.

Dressler, A. (1984) "The Evolution of Galaxies in Clusters", *Annual Review of Astronomy and Astrophysics*, **22**, 185–222. Defines observational constraints on ways galaxies in clusters have evolved, most of which still apply.

Franx, M.; Illingworth, G.D.; Kelson, D.D.; van Dokkum, P.G.; and Tran, K.V. (1997), "A Pair of Lensed Galaxies at $z = 4.92$ in the Field of CL 1358+62", *Astrophysical Journal Letters*, **486**, L75–L78. One of the first uses of gravitational lensing to place moderate-luminosity galaxies within reach to very high redshifts.

Holmberg, Eric (1941) "On the clustering tendencies among the nebulae. II: A study of encounters between laboratory models of stellar systems by a new integration procedure", *Astrophysical Journal*, **94**, 385–395.

Kenney, J.D. and Koopmann, R. (1999) "Ongoing Gas Stripping in the Virgo Cluster Spiral Galaxy NGC 4522", *Astronomical Journal*, **117**, 181–189. Evidence for ongoing stripping of gas from a spiral galaxy by the intracluster gas in the Virgo cluster.

Klebesadel, R.W.; Strong, I.B.; and Olson, R.A. (1973) "Observations of Gamma-Ray Bursts of Cosmic Origin", *Astrophysical Journal Letters*, **182**, L85–L88. The announcement of the existence of gamma-ray bursts.

Lamb, D.Q. and Reichart, D.E. (2000) "Gamma-Ray Bursts as a Probe of the Very High Redshift Universe", *Astrophysical Journal*, **536**, 1–18. Predictions of the properties of gamma-ray bursts produced by the hypernovae of first-generation stars.

Larson, R.B. and Tinsley, B.M. (1978) "Star formation rates in normal and peculiar galaxies", *Astrophysical Journal*, **219**, 46–59.

Loh, E.D. and Spiller, E.J. (1986) "Photometric redshifts of galaxies", *Astrophysical Journal*, **303**, 154–161.

Madau, P.; Ferguson, H.C.; Dickinson, M.E.; Giavalisco, M.; Steidel, C.C.; and Fruchter, A. (1996) "High-redshift galaxies in the Hubble Deep Field: Colour selection and star formation history to $z \sim 4$", *Monthly Notices of the Royal Astronomical Society*, **283**, 1388–1404. An estimate of the cosmic history of star formation based on Hubble Deep Field data. The subsequent argument has centered on the role of extinction in the ultraviolet, and possible selection effects due to cosmological surface-brightness dimming.

Moore, B.; Katz, N.; Lake, G.; Dressler, A.; and Oemler, A., Jr. (1996) "Galaxy harassment and the evolution of clusters of galaxies", *Nature*, **379**, 613–616. Discusses the important cumulative effects of minor tidal encounters in the cluster environment, in which disk galaxies do not return to a symmetric state before subsequent encounters unbind the material. Especially in concert with ram pressure from the hot gas, this seems to be an important driver of galaxy transformation in clusters.

Owen, F.N.; Keel, W.C.; Wang, Q.D.; Ledlow, M.J.; and Morrison, G.E. (2006) "A Deep Radio Survey of A2125 III. The Cluster Core: Merging and Stripping", *Astronomical Journal*, **131**, 1974. Observations of ongoing stripping in Abell 2125, as depicted in Figure 5.13.

Prochaska, J.X. and Wolfe, A.M. (2002) "The UCSD HIRES/Keck I Damped Ly α Abundance Database. II. The Implications", *Astrophysical Journal*, **566**, 68–92. Shows that evidence for a change in the chemical enrichment of gas in galaxy disks over time is extremely weak for $z < 3.5$, a challenge to many models for cosmic star-forming history.

Reeves, J.N.; Watson, D.; Osborne, J.P.; Pounds, K.A.; O'Brien, P.T.; Short, A.D.T.; Turner, M.J.L.; Watsin, M.G.; Mason, K.O.; Ehle, M. *et al.* (2002) "The signature of supernova ejecta in the X-ray afterglow of the gamma-ray burst 011211", *Nature*, **416**, 512–514.

Rowan-Robinson, M.; Broadhurst, T.; Lawrence, A.; McMahon, R.G.; Lonsdale, C.J.; Oliver, S.J.; Taylor, A.N.; Hacking, P.B.; Conrow, T.; Saunders, W. *et al.* (1991) "A high-redshift IRAS galaxy with huge luminosity—hidden quasar or protogalaxy", *Nature*, **351**, 719–721. The discovery of IRAS F10214+4724 at $z = 2.3$, with an implied luminosity of several times 10^{14} solar luminosities. This object has since been found to be amplified by gravitational lensing, but remains an example of the kind of dusty system that should be common in the early Universe.

Spergel, D.N. and Hernquist, L. (1992) "Statistical mechanics of violent relaxation", *Astrophysical Journal Letters*, **397**, L75–L78. An exploration of the physical processes involved in violent relaxation.

Steidel, C.C. and Hamilton, D. (1992) "Deep imaging of high redshift QSO fields below the Lyman limit. I—The field of Q0000−263 and galaxies at $z = 3.4$", *Astronomical Journal*, **104**, 919–949. The introduction of the Lyman-break technique for identifying galaxies at high redshift, based on the recognition by multiple authors around 1990 that galaxies should have a very strong, perhaps complete, spectral break at their Lyman limits.

Tinsley, B.M. (1968) "Evolution of the Stars and Gas in Galaxies", *Astrophysical Journal*, **151**, 547–566. Tinsley's classic paper on the effects of galaxy evolution. This one was selected by the editor of *Astrophysical Journal* for a centennial commemorative volume.

Toomre, A. and Toomre, J. (1972) "Galactic Bridges and Tails", *Astrophysical Journal*, **178**, 623–666. Numerical simulations, primitive by today's standards, but adequate to show that many of the long tails and plumes seen in interacting galaxies could be understood in detail as tidal distortions.

Tyson, J.A. (1988) "Deep CCD survey—Galaxy luminosity and color evolution", *Astronomical Journal*, **96**, 1–23. This paper presents the first very deep CCD observations of faint-galaxy fields, and constitutes the first clear statement of the "faint blue galaxy" excess, which proved important in understanding the luminosity function as well as evolution of galaxies.

Vorontsov-Velyaminov, B.A. (1977) "Atlas of interacting galaxies, Part II and the concept of fragmentation of galaxies", *Astronomy and Astrophysics Supplement*, **28**, 1–117. A catalog of interacting galaxies based on systematic examination of the Palomar Sky Survey. This was the most extensive sample available in its time, although the Atlas plates were almost all from the Palomar Survey and much less informative than in the Arp atlas. Numbers 1–355 appeared in part I of the atlas, published photographically in a very small printing. The arrangement and discussion here are very idiosyncratic.

Internet

http://cossc.gsfc.nasa.gov/ The mission and results of the Compton Gamma-Ray Observatory (CGRO).

6

The intergalactic medium

6.1 INTRODUCTION

Galaxies do not exist in isolation, either from one another or from their gaseous environments. Unless galaxy formation was implausibly efficient, we might expect some ordinary matter to be left over, still filling the vast intergalactic spaces. Something closely analogous happens when stars are formed. Outflows, whether collimated into narrow jets or in expansive winds, are important to the process, and it appears that the mass of material ejected from a star-forming cloud is roughly equal to that in the final star. A similar rough balance may hold for galaxies and gas which has been expelled from them, though the physics must be quite different. The hot gas bound in clusters of galaxies has shown that not only was there leftover matter, but that this matter was enriched by material expelled from galaxies after being processed through massive stars. Rather than the empty void we might once have thought of, intergalactic space, like interstellar space, turns out to be a lively and important region. C.S. Lewis might almost have been expressing this shift in viewpoint when he wrote in *Out of the Silent Planet* that "... the very name Space seemed a blasphemous libel for this empyrean ocean of radiance ...".

The intergalactic medium (IGM) can act as both a source and sink for galactic matter. The whole issue of intergalactic material is bound up with the formation and evolution of galaxies through its mass, chemistry, and ionization. The relative baryonic mass in galaxies and the IGM tell us how efficient galaxy formation was, and has cosmological implications. Chemical abundances in the IGM reflect both the fossil abundances from the early Universe and subsequent injection of processed material from star-forming galaxies. Such injection happens from global winds lofted by the energy of massive stars and supernova explosions, and probably also from stripping of galaxies as they move through surrounding gas, via ram pressure and turbulent stripping at the boundary between interstellar and intergalactic media.

The ionization of the intergalactic medium tells us about energy input from imbedded sources—particularly the ultraviolet and soft X-ray ionizing radiation from star-forming galaxies and active galactic nuclei. The ionization level is important both in tracing the energetic history of the ionizing agents and in understanding how we should look for direct evidence of intergalactic gas.

The amount of ionization in the IGM, and by inference the intensity of ionizing radiation averaged through intergalactic space, can be inferred from the ionization states of gas seen in absorption against distant quasars. This can be done at various redshifts, and compared with estimates of the amount of ionizing ultraviolet radiation available from quasars and star-forming galaxies. An approach to this problem which removes some of the uncertainty in modeling the state of the gas is to use the *proximity effect*. This is a statistical tendency for fewer Lyman α absorption lines from intergalactic gas to appear close in redshift to the background quasar. The interpretation is that the quasar's radiation ionizes suriounding gas so completely (in what might be viewed as an extremely large, if very faint, gaseous nebula) that we cannot detect any remaining neutral hydrogen. The radius of this zone, as measured in redshift, and the quasar's ultraviolet luminosity, together tell us how far from the quasar its radiation balances the surrounding mean radiation field, and then how strong that ambient radiation must be. The intergalactic ionizing radiation has declined strongly, by a factor of 5 or more, since $z = 2$, and remained fairly constant at epochs from $z = 2$–4.5. At low redshifts and recent cosmic times, quasars and other active nuclei are sufficient to produce this radiation. Some additional contribution, such as from star-forming galaxies, is likely to have been important at early times. In the dense, early Universe, with density scaling with observed redshift as $(1 + z)^3$ and recombination time scaling as the inverse square of the density, the radiation requirement to maintain the same ionization levels grows dramatically. Meanwhile, the number of quasars peaked around $z = 2.2$, and even the recent dramatic findings of quasars to $z > 6$ does not alter the conclusion that their numbers were growing rapidly with cosmic time until $z \approx 2$. Thus, early on, there were almost certainly not enough quasars to ionize the intergalactic medium, and they may not have been the dominant players in reionizing the intergalactic medium.

The classic observational test for intergalactic gas is the *Gunn–Peterson effect*. Cold, neutral hydrogen will show absorption in Lyman α, and each volume element along a given line of sight will show this absorption at the appropriate redshift. The integrated signature of smoothly distributed gas would therefore be a continuous absorption trough from the observed wavelength of Lyman α in the background radiation source toward zero redshift. Once QSOs had been identified with large enough redshifts to observe shortward of Lyman α with ground-based telescopes, it became clear that there is no important fraction of the cosmic mass density that could reside in neutral hydrogen; there was no detectable "step" in the continuum radiation across Lyman α. By itself this result had two possible interpretations—that there is no significant IGM (a situation which is both implausible and astrophysically uninteresting) or that it does not consist of neutral hydrogen (e.g., because it is largely ionized). An ionized IGM, produced in a Universe which had already become neutral at the epoch of recombination, would imply that a huge energy input occurred at

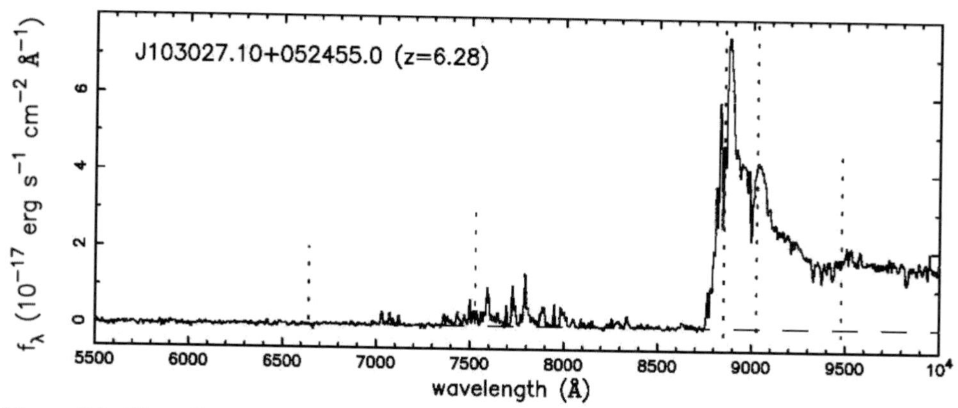

Figure 6.1. The Gunn–Peterson effect seen in the spectrum of the SDSS quasar 10327.10+052455.0 at $z = 6.28$, as observed with one of the 10-m Keck telescopes. Vertical dashed lines show the redshifted locations of typical QSO emission lines. Absorption by a widespread and significantly neutral intergalactic medium appears as a lack of any detected flux just shortward of Lyman α, at observed wavelengths roughly 8100–8750 Å. (From Becker *et al.*, *Astronomical Journal*, **122**, 2850, 2001; courtesy R. Becker, used by permission of the American Astronomical Society.)

some time between then and the earliest epochs we can see with galaxies, an event known as *reionization*.

Recently, observations of the Gunn–Peterson effect in the highest-redshift quasars, from redshifts $z = 5.8$ to 6.4, have finally shown a convincing absorption trough increasing rapidly in opacity with redshift over this range (Figure 6.1). All such quasars known to date have been found from the Sloan Digital Sky Survey (SDSS), using color selection to winnow efficiently through enormous samples of starlike images. These data indicate that we may be seeing the final stage of the ionization of the IGM, although only the very final stages. Only a small fraction of the material need be neutral to give a strong Gunn–Peterson signal, so the fact that we see a change in absorption with redshift may be more important than the exact derived optical depth.

If a more local intergalactic medium exists but its hydrogen is largely ionized, one might find traces of elements with higher ionization potentials. The obvious candidate is helium, which is abundant throughout the Universe, and whose second ionization potential (i.e., from photoionization of He^+ to He^{++}) is four times higher than hydrogen's ionization energy (54.4 versus 13.6 eV), with an ionization edge at $912/4 = 228$ Å. For sufficiently high-redshift quasars, this is redshifted past our own galaxy's hydrogen absorption, and eventually into the wavelength range for observations with the spectrographs on HST (whose primary mirror has useful reflectivity for wavelengths greater than about 1150 Å). Only a handful of bright, high-redshift quasars lie on lines of sight free of intervening galaxies, whose own hydrogen would absorb the deep ultraviolet, and wide searches were made using the "snapshot" mode of HST to locate such quasars. It was also a high priority to look for He II

Gunn–Peterson absorption at somewhat lower redshifts using specialized telescopes, both to trace its possible evolution and because the confusing effects of large numbers of narrow absorption systems (as seen at Lyman α) would be much reduced. This was seen as a more sensitive way to look for the most diffuse intergalactic gas, because the least dense regions would have the longest recombination times and thus highest ionized-gas fractions. He II Gunn–Peterson observations were priority items for both flights of the shuttle-based Hopkins Ultraviolet Telescope (HUT) and for the Far-Ultraviolet Spectroscopic Explorer (FUSE). Unlike the traditional H I Gunn–Peterson effect at high redshifts, this test has been limited to the small number of QSOs that are not only unabsorbed but quite bright, since the requisite space instruments have much smaller apertures than are available on the ground.

In each case, it turns out that high-resolution spectroscopy of the hydrogen Lyman α forest (see below) in the same redshift range is crucial to untangling the effects of He II lines associated with the same discrete features and a truly diffuse absorption. There are dense regions in the intergalactic medium which produce narrow absorption lines, whose effects must be treated separately in order to isolate any smooth absorption. For both hydrogen and singly-ionized helium, the most prominent tracer is absorption in their respective Lyman α lines, which arise from the transition between ground and first excited states of the atom's single electron. This line occurs at wavelengths of 1216 Å (in the object's frame) for hydrogen and 304 Å for helium.

The HUT observations by Davidsen *et al.* (1996) targeted the QSO H1700+64 at $z = 2.743$. They found that half of the continuum depression seen at low spectral resolution comes from the He II Lyman α forest, with the true continuum feature having an optical depth $\tau = 1.00 \pm 0.07$ around $z = 2.4$ (where the fraction of residual light after passing through optical depth τ is, by definition, $e^{-\tau}$). Working at higher redshifts, for which the He II limit is accessible at longer wavelengths at the price of the Lyman α forest making a greater contaminating contribution, HST observations have concentrated on the QSO 0302−003 at $z = 3.286$. A low-dispersion prism observation by Jakobsen *et al.* (1994) showed a flux discontinuity or complete cutoff at the He II limit, and a higher-dispersion spectrum using the Faint-Object Spectrograph (FOS) by Hogan *et al.* (1997) which gave a wavelength-averaged He II opacity of $\tau = 2.0^{+1}_{-0.5}$ at $z = 3.15$. Significantly improved results came from use of the higher throughput of STIS, as reported by Sally Heap and collaborators. They found a high continuum opacity $\tau > 4.8$ at $z = 3.15$ after modeling the contribution of discrete lines. Their data also show a rather abrupt step in opacity across the range $z = 2.9$–3.0, dropping to $\tau = 1.88$ at $z < 2.87$. This indicates an increasing ionization with time, as would be expected if the intergalactic radiation field became harder (as more QSOs appeared, for example). They also found that a region near the QSO in redshift space has decreased opacity, with all the absorption accounted for in discrete clouds. This is in the same sense found for the proximity effect in hydrogen Lyman α absorbers, indicating a signature of the ionization provided by the QSO itself.

A key further step in understanding the ionization structure and history of the intergalactic medium came with FUSE observations of the QSO HE2347−4342 at

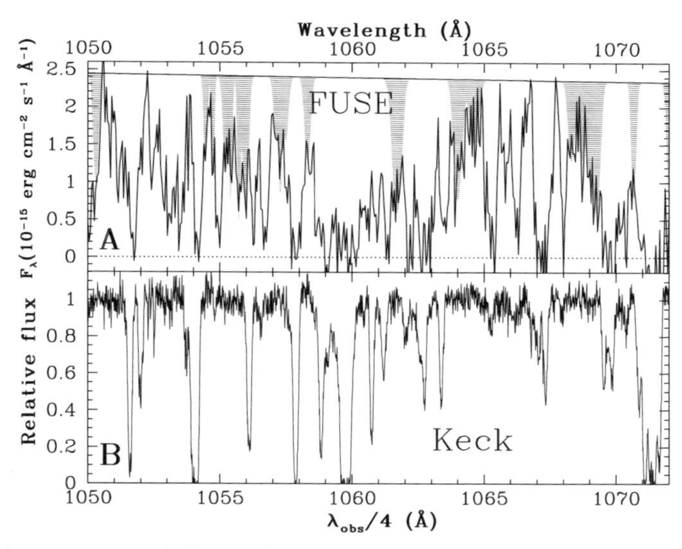

Figure 6.2. A comparison of He II and H I absorption from intergalactic gas along the line of sight to the QSO HE2347−4342. The upper panel shows a portion of the FUSE far-ultraviolet spectrum, with the lower panel showing a spectrum of H I absorption in the Lyman α forest properly aligned in redshift (which gives a factor 4 wavelength difference between H I and He II). Not only do H I lines have He II counterparts, but there is additional absorption (shaded) from highly ionized gas in regions with no detected H I absorption. The thick curve at the top of the He II spectrum represents the extrapolated continuum level from which the absorption is measured. (Figure courtesy G. Kriss of STScI.)

$z = 2.885$, at 16th magnitude the only known QSO suitable for such observations. As reported by FUSE Data Processing Scientists Gerard Kriss *et al.* (2001), the He II absorption from $z = 2.3$ to 2.7 is resolved into a discrete forest (Figure 6.2), with more lines than the matching H I forest. Incorporating observations of the hydrogen lines, the ratio of He II to H I column density ranges widely from 1 to values above 1000, with the mean value favoring He II by a factor of 80. It is this large typical value that makes accounting for the narrow lines so important in finding the truly diffuse IGM from He II observations; in a sense the He II Lyman α forest is deeper than the corresponding hydrogen lines. The ionization is highest in low-density regions, and the fact that discrete lines are present at these levels shows that structure in the IGM exists in low-density regions as well as the clumps that are most prominent in the hydrogen line. Comparing the redshift structure of these data with models of structure formation when dominated by cold dark matter, they find the IGM to comprise about 10% of the overall mass density.

There seems to be a distinct cutoff to the occurrence of He II absorption from an essentially diffuse medium, around $z = 2.7$. This may mark the final reionization of the intergalactic medium, since He II at lower redshifts is found entirely in a discrete forest analogous to, but somewhat richer than, the Lyman α forest. It is interesting that this ionization comes at lower redshift than the new H I detection, since He II should remain ionized longer, coming preferentially from lower-density regions with longer recombination times. Any change in the spectrum of ionizing radiation will also enter; a harder spectrum, such as might come from AGN with substantial absorption from local gas, further favors He II over H I in detectability.

6.2 THE LYMAN α FOREST AS PART OF THE IGM

While the Gunn–Peterson effect has not shown significant hydrogen absorption spread throughout the Universe since the epoch of recombination, we do observe numerous more or less distinct absorbing clouds, appearing as the Lyman α forest. We treat this now as a positive item for observation, rather than as a hindrance to observing high-redshift objects (Chapter 9). Every quasar at high enough redshift to measure its spectrum below Lyman α with ground-based instruments shows a large number of very narrow absorption features starting at its own Lyman α peak and running shortward, as would be produced by Lyman α absorption from clouds of intervening material (Figure 6.3).

For two decades, it appeared that these features arose from discrete, bound clouds of gas. Their measured neutral-hydrogen densities fall so far short of what

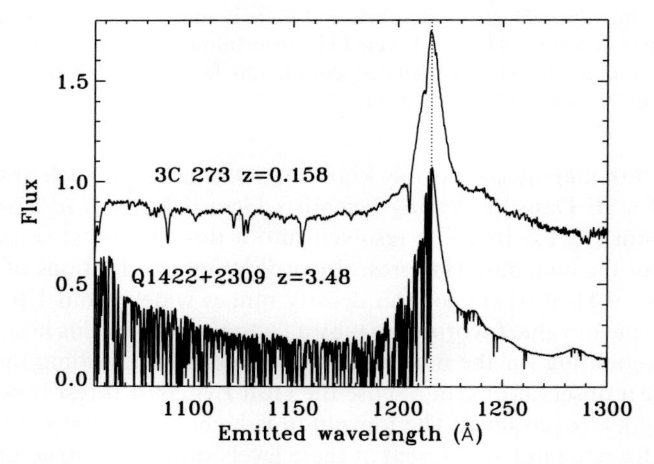

Figure 6.3. The Lyman α forest in the spectra of nearby and distant quasars. The nearest luminous QSO, 3C 273, shows only a handful of Lyman α absorption lines, which appear shortward of its own Lyman α wavelength (dashed line) when plotted in its own frame. The high-redshift object 1422+2309 shows a well-developed forest, with more distinct lines than this page can show. This comparison, in which the spectrum of 3C 273 has been shifted upward by 0.4 units, shows the evolution of the intergalactic medium with redshift. The 3C 273 data are from archival Hubble Space Telescope data with the Goddard High-Resolution Spectrograph and Faint-Object Spectrograph, originally obtained by R. Weymann. (The data for 1422+2309 are from the Keck I telescope and High-Resolution Spectrograph, kindly provided by Michael Rauch.)

is needed to hold them together gravitationally that they could be long-lasting features only if they were highly ionized. In fact, the proximity effect for absorbers close to the redshifts of the quasars suggests that these features are indeed highly ionized and can become completely ionized. Evidently a bright quasar can ionize gas within several Mpc so completely that there isn't enough neutral gas to detect.

These absorbers are quite numerous, so that their evolution with redshift could be tightly constrained from ground-based data for $z > 1.6$. Over this interval, they are disappearing with cosmic time—the implied space density increases with redshift, and stronger lines, from clouds of higher column density, become relatively more numerous. Since this was happening at high redshifts, it was a natural question to ask whether the disappearance of these gaseous objects could be related to galaxy formation. With the evidence that many galaxies continue to accrete more or less pristine gas for long times after their initial formation, the clouds of the Lyman α forest were a promising source for such material. On the other hand, ionization arguments made it clear that the observed absorption was coming from a tiny fraction of the gas by mass, so that it could be that a change in ionization state with redshift was being seen.

The complexion of our understanding of the Lyman α forest has changed twice within the last decade. The first good ultraviolet spectra of nearby quasars were obtained by HST (a "Key Project" which could be done even in its initial aberrated state for such bright objects), showing that the Lyman α forest isn't completely gone. A few low-redshift absorbers can be detected along each line of sight. These are now at such low redshift that it makes sense to see just how they are related to the galaxy distribution, since ordinary galaxies can be enumerated at these redshifts (e.g., there are absorbers within the Virgo Cluster seen in front of 3C 273). These generally follow the galaxy distribution, but there is no detailed one-to-one match with halos of particular galaxies. However, these features relate to galaxies, some of them are still around; the gas hasn't been entirely incorporated into galaxies today.

The second revolution came with a series of numerical simulations—particularly by Renyue Cen *et al.* (1994) at Princeton—tracking the behavior of intergalactic gas in an expanding Universe, starting with a realistically inhomogeneous distribution and watching how clusters and other structures develop. These simulations show the now-familiar texture of clusters, filaments, sheets, and voids. What was really new was the ability to simulate the observation of gas along particular lines of sight (Figure 6.4). Two effects conspire in these numerical realizations to make the gas distribution as traced by Lyman α appear to be much clumpier and more concentrated into individual clouds than is in fact the case. One is that there is a bias to observe denser regions, since the fraction of gas which is neutral in a mostly ionized plasma is set by the recombination rate. This rate varies as the square of the particle density (more pedantically, the product of electron and proton densities), since recombination is a collisional process. On top of this, matter in clumps and filaments departs from the smooth Hubble expansion to a degree depending on the density in the structure. This is a manifestation of the same redshift distortion effects seen in the early history of clusters and groups of galaxies. If we are looking perpendicularly through a sheet of material which is decoupling from the Hubble expansion, and thus

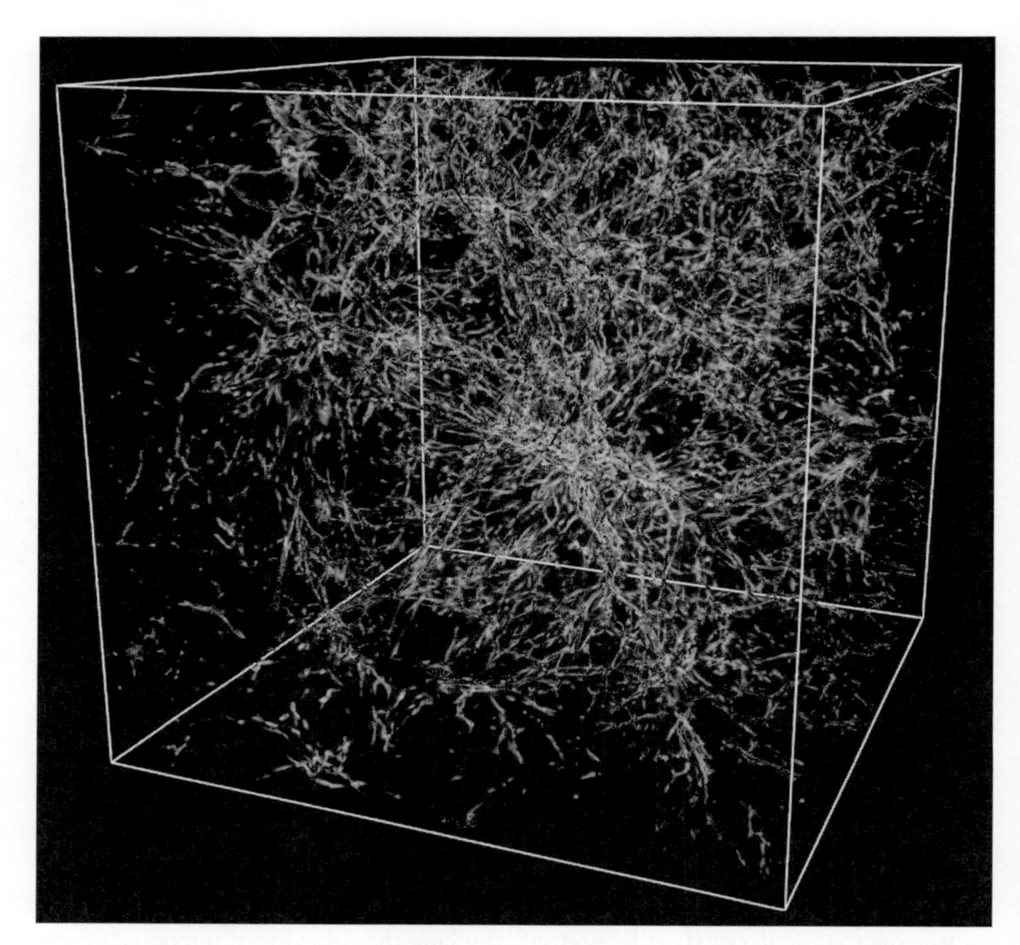

Figure 6.4. A numerical simulation of the growth of structure of the intergalactic medium. This particular visualization shows a perspective view of the density of H I, as seen in the Lyman α forest, within a 25-Mpc (comoving) cubic region at $z = 4$. The combined effects of density and radiation field make the structure in H I appear more tightly concentrated into filaments as seen in absorption than it actually is in mass. Most of the gas is ionized and only those regions with the most rapid recombination (highest density) remain neutral. See also color section. (Simulation courtesy of Dr. Renyue Cen of Princeton University.)

expanding more slowly along our line of sight than its size alone would predict, redshifts of particles in this sheet will be more closely bunched than would be otherwise predicted. In a simple redshift-position diagram, this sheet would appear thinner than its true physical thickness, because it is contracting *relative to the mean cosmic expansion*. The combination means that Lyman α clouds need not be distinct, separate objects, but are quite consistent with being part of a network of gas filling the Universe and more or less vaguely following the kinds of structures seen in the galaxy distribution.

6.3 FILLING SPACE WITH GALACTIC WINDS

As we have seen, there exists a Lyman α forest not only for hydrogen, but for ionized helium, shifted upward in energy by a factor of 4, and sampling a larger fraction of space. Somewhat surprisingly, there also exists an O VI forest, detected from its pair of strong resonant absorption lines with laboratory wavelengths near 1034 Å. This forest was found using HST spectra by Todd Tripp *et al.* and reported in 2000. Extending the trend from H I to He II, there are also more O VI than H I lines along a given line of sight. This means that the O VI absorption comes preferentially from the least dense parts of the IGM, which are generally most remote from luminous galaxies. The presence of significant oxygen in the IGM shows that there was interaction between galaxies and their surroundings very early in cosmic history, in order to have the products of stellar processing spread so widely beyond them. Just as with stars and the interstellar medium, the interplay between galaxies and their environment is unexpectedly rich.

Perhaps the most probable source for these metals in the lowest-density parts of the Universe is dwarf galaxies, since they are very numerous, and their shallow potential wells mean that most supernova ejecta will be lost to the individual galaxies, and for galaxies outside cluster potentials, to their immediate environment as well. From measurements of winds in star-forming dwarf galaxies today, Crystal Martin (1999) finds that the global winds in these galaxies have enough energy to escape for systems with rotational velocities less than about 130 km/s, which corresponds to systems with slightly less than one-tenth of the Milky Way's current luminosity. Somewhat surprisingly, the mass flow in escaping gas can be several times the rate of star formation.

Bursts of star formation, and indeed star-forming regions in general, impart energy to surrounding gas in several ways. When enough of this is converted to kinetic energy, a wind can be set up, sometimes fast enough to escape the galaxy's potential well. Supernova explosions are a demonstrated driver of winds, beginning with individual supernova remnants. For most galaxies, a wind can be started only if the supernova rate heats the interstellar medium faster than the supernova remnants can cool. Most of the material involved was not part of the supernova, but has been swept up during the expansion of the remnant in a "snowplow" phase. In realistic situations, multiple supernova explosions will produce expanding shells or chimneys of hot gas.

There is ample observational evidence of these global winds from local starburst galaxies, particularly dwarf systems. Their effects would have been even more important at early epochs, when the typical star-formation rate for most galaxy types was higher. The prototypical starburst wind is seen in M82, in our own backyard only about 3 Mpc distant. This starburst system exhibits an extensive network of filaments in optical emission lines, vaguely following a bipolar pattern perpendicular to the galaxy's edge-on disk (Figure 6.5). Spectroscopy shows that these filaments are flowing outward in a configuration concentrated along the surfaces of twin cones, at velocities sufficient to overcome the galaxy's potential and escape completely.

Figure 6.5. The nearby starburst galaxy M82 (NGC 3034), with its prominent wind of ionized gas. The wind appears red in this composite image from the Hubble Space telescope, due to strong Hα emission. It escapes most readily perpendicular to the galaxy disk. Most of the wind material, including metals synthesized in the massive stars that drive it, is at temperatures above 10^7 K; the optical line emission and soft X-ray emission trace only interface regions at the contact between the winds and cooler surrounding material. See also color section. (NASA, ESA, and the Hubble Heritage Team, released in honor of the observatory's 16th anniversary in orbit.)

Similar winds have now been identified in numerous starburst galaxies, using several observational tracers. Emission-line imaging can be especially effective for edge-on systems, or when the outflow is large compared with the galaxy's stellar distribution. This approach shows global winds from dwarf galaxies up to the intense starbursts found in IR-bright galaxies, with prime examples being the enormous pair of emission-line bubbles surrounding Arp 220 and the minor-axis plumes from NGC 6240. As outflows, we can find these features producing blueshifted emission or absorption features when the geometry is favorable. The gas in these outflows is hot enough to be easily detected in soft X-rays (Figure 6.6), which have been used to find many examples as well. There is, in addition, a nontrivial content of energetic particles, so that some outflows are seen by their radio-frequency synchrotron emission. The relation among the outflow structure, the optical emission-line gas,

Figure 6.6. The starburst wind of M82 as seen in X-rays by the Chandra X-Ray Observatory. The wind is largely channeled along the galaxy's minor axis by the dense interstellar medium. While the dominant contribution to this X-ray image is from soft X-rays, much of the wind material is likely to be at such high temperatures that its own radiation is quite inefficient, so the observed emission comes largely from interface regions with cooler entrained material. The numerous pointlike sources include supernova remnants and massive X-ray binary stars. (NASA, Chandra X-Ray Center, Johns Hopkins University, and D. Strickland.)

a signature of plasma near 10^4 K, and the soft X-ray emission with a temperature of a few 10^6 K, have been clarified in a few cases by detailed geometric studies, as well as theoretical modeling of physical processes in a galactic wind.

Entrainment of the ambient ISM by the hot outflow is important to kinematics, and indicates that material far from the loci of instantaneous star formation can leave the galaxy. The optical emission indeed peaks on roughly biconical surfaces, consistent with being ionized by shocks at the interface between the wind and ambient (including outswept) denser material. The soft X-ray emission is also strongest in these "surface" regions, indicating that it too arises from the interaction between the hot wind itself and cooler clouds or clumps of ambient interstellar gas, so that even in passbands sensitive to material at 10^7 K, most of the wind material (and outflowing metal production) remains unseen directly. The dynamics that we can trace show radial velocities of as much as 600 km/s and net outflow rates of 0.5 solar mass per year for modest starbursts such as the ones in M82 and NGC 253. The metallicity of the winds, and thus enrichment rate to the surrounding IGM, are still unknown from X-ray emission; we must look to absorption-line studies for possible clues to these. It is likely on very general grounds that luminous starbursts can eject a few billion solar masses into the ISM, of which a few percent is in the form of metals (i.e., enriched by a few times over solar abundances, without necessarily having the same fractions of all metallic elements).

The situation becomes more complicated for more massive galaxies, with correspondingly higher escape velocities, and for cases in which the star-formation rate does not rise to the level of "starburst". Development and stability of a galactic wind can be very sensitive to these quantities, which accounts for a long debate in the literature about whether elliptical galaxies did or did not generate winds which swept them clean of dense gas. Some solutions gave winds, some gave only episodic winds, some winds would stall and never sweep the galaxy. Episodic winds have recently been invoked to explain multiple episodes of globular-cluster formation in elliptical galaxies, when the cluster systems are too well correlated with the galaxy's overall properties to be plausible products of a late merger. For ellipticals, the outflow history must be linked to the hot intracluster gas, which has been enriched by the products of early outflows (although not necessarily including all such material, since clusters are evidently somewhat leaky).

Observations now indicate that global outflows are common at high redshifts. They can be detected most easily upon comparison of the relative redshifts of lines such as [O III] and Hβ which do not suffer optical-depth effects and hence reflect the systemic redshift, with UV absorption lines and Lyman α emission. Most strong UV absorption lines in galaxies have substantial contributions from interstellar absorption, so they will be blueshifted in the presence of an outflow. Lyman α emission shows the opposite effect. In an outflowing shell, the blueshifted foreground emission will be strongly suppressed by optical depth effects, so that the net emission is dominated by the redshifted farside of the line, giving a net redshift with respect to the systemic value. This effect is seen in nearby star-forming systems, where the gas structure can be spatially resolved.

Seeing these outflow effects at large redshifts has required high-quality spectra in both the optical and near-IR windows, since emitted UV and optical lines need to be compared. Using near-infrared spectra from the 8-meter VLT Unit Telescopes, Max Pettini *et al.* (2001) have recently shown that Lyman α and stellar absorption lines show these systematic offsets in the preponderance of 15 Lyman-break galaxies observed at $z \approx 3$.

Galaxy-scale outflows might also be driven by active nuclei. A few examples have been found in the local Universe, sometimes in galaxies which simultaneously host AGN and starbursts, so the proximate cause of the wind is not immediately clear. Some quasars and luminous Seyfert nuclei can drive very rapid outflows, which can be seen as *broad absorption-line systems*. These show absorption blueshifted with respect to the overall systemic velocity, often spanning enormous ranges in velocity greater than 10,000 km/s. There is wide variety in these systems—some are smooth in

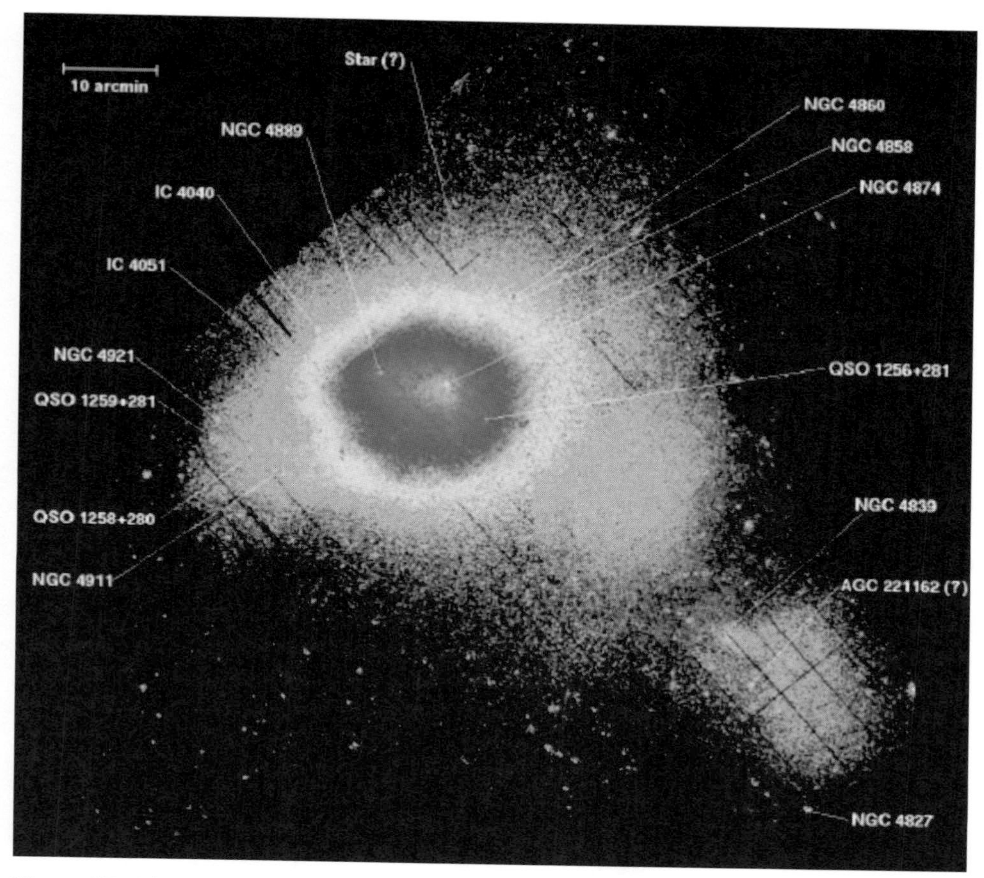

Figure 6.7. The Coma Cluster of galaxies as seen in X-rays, by the European Space Agency's XMM–Newton satellite. This view is a composite of multiple observations, some of whose circular outlines can be seen. Some individual galaxies and background quasars are marked, but the bulk of the X-rays come from a smooth, hot gaseous medium between the galaxies. The distinct lump to the southwest (*lower right*) coincides with an interesting grouping of galaxies, and has been interpreted as a smaller cluster which is now falling in to the Coma Cluster core. The field of view spans about 1.6 degrees on the sky, or about 1.6 Mpc at the cluster distance. See also color section. (European Space Agency and U. Briel.)

velocity distribution, some have distinct and separate components. These systems may be widespread but require a particular viewing angle to intercept the core's light and become visible in absorption. The high ionization of some of this absorbing gas suggests that it lies quite close to the central source. While at high velocities, this material (which we don't completely know whether to call a wind or clouds) does not entail a great deal of mass. At this point, there are no unambiguous examples of galactic winds driven purely by an active nucleus.

6.4 THE INTRACLUSTER MEDIUM

Because rich galaxy clusters are a rather special environment, with a rich set of processes going on involving the constituent galaxies, we generally distinguish between the general intergalactic medium (IGM) and the intracluster medium (ICM). The existence of a substantial ICM was one of the major surprises from the early days of X-ray astronomy. There were several surprises actually, starting with the finding that clusters of galaxies are strong X-ray sources whose spectra are consistent with thermal emission from bremsstrahlung in a low-density gas at $1\text{–}2 \times 10^7$ K. Imaging X-ray telescopes (starting with *Einstein*, launched in 1978) confirmed that the X-rays come from the cluster overall, not individual galaxies (Figure 6.7). Finally, the X-ray spectra of the ICM show emission features from highly ionized metals (most notably Fe, though lines of Ne, Mg, Si, S, Ar, and Ni can be detected as well—Figure 6.8). The overall metal abundances are roughly solar, so the ICM is not simply leftover gas that never made it into galaxies. Early in its history, it was enriched by matter which had been through massive stars and then ejected from galaxies.

Energetics and abundance patterns in clusters have led to the suspicion that the clusters need not have retained all the stellar ejecta, and that early star formation may therefore have enriched some of the genuine intergalactic medium as well. This makes sense with the detection of intergalactic O VI absorption from low-density regions, and would imply that star formation and metal production were even more brisk than the cluster gas chemistry already indicates.

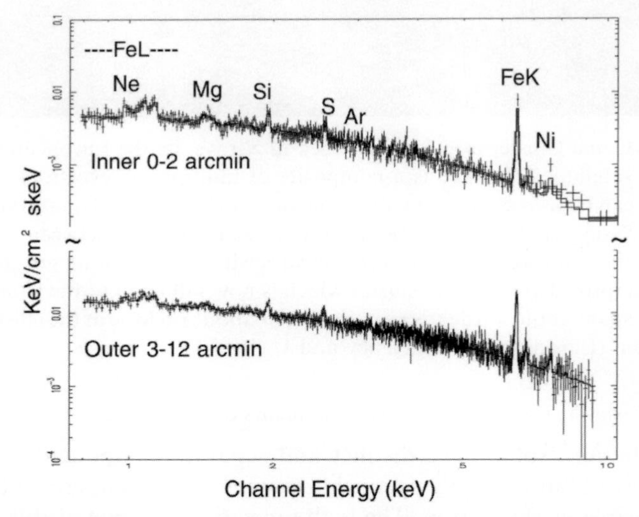

Figure 6.8. The X-ray spectrum of the galaxy cluster Abell 496, with emission features from metal ions marked. The cluster is divided into inner and outer regions, showing the radial gradient in some abundances. (The data were taken from ASCA observations analyzed by R. Dupke and R. White, III, reproduced from *Astrophysical Journal* by permission from the American Astronomical Society.)

The high temperature of the ICM has a straightforward explanation. For massive clusters of galaxies, it scales closely with the velocity dispersion of the galaxies, and both measure the depth of the cluster's gravitational potential. The mean free path of individual electrons or nuclei is so large that they are essentially pursuing orbits within the cluster, like galaxies, so that both naturally have the same characteristic velocities.

The chemical makeup of the ICM has turned into an active field, with observations now hinting at the detailed history of chemical enrichment and leakage beyond the cluster potential.

This is one more hint that "clusters of galaxies", entrenched though it is, isn't a particularly exact name for these objects. We should always beware of letting names condition how we think about the Universe.

Observations in just the last few years are now allowing us to piece together an empirical picture of the intergalactic medium. As we have seen, this turns out to be an important, if stealthy, player in the history of galaxies.

6.5 BIBLIOGRAPHY

Becker, R.H.; Fan, X.; White, R.L.; Strauss, M.A.; Narayanan, V.K.; Lupton, R.H.; Gunn, J.E.; Annis, J.; Bahcall, N.A.; Brinkmann, J. *et al.* (2001) "Evidence for Reionization at $z \approx 6$: Detection of a Gunn–Peterson Trough in a $z = 6.28$ Quasar", *Astronomical Journal*, **122**, 2850–2857. A clear Gunn–Peterson trough from hydrogen, seen in the spectrum of a QSO at $z = 6.28$ found by the Sloan Digital Sky Survey.

Cen, R.; Miralda-Escude, J.; Ostriker, J.P.; and Rauch, M. (1994) "Gravitational collapse of small-scale structure as the origin of the Lyman α forest", *Astrophysical Journal Letters*, **437**, L9–L12. One of the simulations which completely changed our view of how the Lyman α forest maps into the structure of intergalactic gas.

Davidsen, A.; Kriss, G.A.; and Zheng, W. (1996) "Measurement of the opacity of ionized helium in the intergalactic medium", *Nature*, **380**, 47–49. HUT observations targeting the QSO H1700+64 at $z = 2.743$ for the He II Gunn–Peterson effect, with less interference from lower-redshift Lyman α than in the HST studies. These authors described their modelling techniques in detail, in Zheng, W.; Davidsen, A.F.; and Kriss, G.A. (1998) "The He II opacity of the Ly-alpha forest and the intergalactic medium", *Astronomical Journal*, **115**, 391–396.

Finoguenov, A.; David, L.P.; and Ponman, T.J. (2000) "An ASCA Study of the Heavy-Element Distribution on Clusters of Galaxies", *Astrophysical Journal*, **544**, 188–203; and Fukuzawa, Y.; Makashima, K.; Tamura, T.; Ezawa, H.; Xu, H.; Ikebe, Y.; Kikuchi, K.; and Okashi, T. (1999) "ASCA Measurements of Silicon and Iron Abundances in the Intracluster Medium", *Publications of the Astronomical Society of Japan*, **50**, 187–193. These studies use X-ray spectroscopy to show that much of the supernova-synthesized matter has been lost in many clusters, escaping to the intergalactic medium. Most of the ejecta of later, largely type Ia supernovae is bound in the clusters.

Gunn, J.E. and Peterson, B.A. (1965) "On the Density of Neutral Hydrogen in Intergalactic Space", *Astrophysical Journal*, **142**, 1633–1636. Derivation of the relation between density of the intergalactic medium and continuum absorption from neutral gas. This has been the basis of the Gunn–Peterson tests for the properties of the IGM.

Heap, S.; Williger, G.M.; Smette, A.; Hubeny, I.; Sahu, M.S.; Jenkins, E.B.; Tripp, T.M.; and Winkler, J.N. (2000) "STIS Observations of He II Gunn–Peterson Absorption toward Q0302−003, *Astrophysical Journal*, **534**, 69–89.

Heckman, T.M.; Armus, L.; and Miley, G.K. (1990) "On the nature and implications of starburst-driven galactic superwinds", *Astrophysical Journal Supplement*, **74**, 833–868. Estimates of the mass flow rate in powerful starburst winds.

Hogan, C.J.; Anderson, S.F.; and Rugers, M.H. (1997) "Resolving the Helium Lyman-alpha Forest: Mapping Intergalactic Gas and Ionizing Radiation at $z \sim 3$", *Astronomical Journal*, **113**, 1495–1504; and Heap, S.R.; Williger, G.M.; Smette, A.; Hubeny, I.; Sahu, M.S.; Jenkins, E.B.; Tripp, T.M.; and Winkler, J.N. (2000) "STIS Observations of He II Gunn–Peterson Absorption toward Q0302−003", *Astrophysical Journal*, **534**, 69–89. Data at high spectral resolution are used to separate diffuse from clumpy contributions to the He II Gunn–Peterson effect in the QSO 0302−003.

Jakobsen, P.; Boksenberg, A.; Deharveng, J.M.; Greenfield, P.; Jedrkejewski, R.; and Paresce, F. (1994) "Detection of intergalactic ionized helium absorption in a high-redshift quasar", *Nature*, **370**, 35–39. They report absorption in the He II Gunn–Peterson trough from material along the line of sight to the QSO 0302-003 at $z = 3.286$.

Kim, T.-S.; Hu, E.M.; Cowie, L.L.; and Songaila, A. (1997) "The Redshift Evolution of the Ly α Forest", *Astronomical Journal*, **114**, 1–13 (1997).

Kriss, G.A.; Shull, J.M.; Oegerle, W.; Zheng, W.; Davidsen, A.F.; Songaila, A.; Tumlinson, J.; Cowie, L.L.; Deharveng, J.-M.; Friedman, S.D. *et al.* (2001) "Resolving the Structure of Ionized Helium in the Intergalactic Medium with the Far Ultraviolet Spectroscopic Explorer", *Science*, **293**, 1112–1116. The Far-Ultraviolet Spectroscopic Explorer doing one of the things it was specifically designed for, measuring the He II Lyman α forest and tracing the structure of the intergalactic medium in a redshift range where foreground structure is much less confusing than the range available to Hubble.

Lynds, R. (1971) "The Absorption-Line Spectrum of 4C 05.34", *Astrophysical Journal Letters*, **164**, L73–L78. Discovery of the Lyman α forest.

Martin, C. (1999) "Properties of Galactic Outflows: Measurements of the Feedback from Star Formation", *Astrophysical Journal*, **513**, 156–160. Galactic outflows can carry away gas at several times the rate of the star formation which powers them.

Pettini, M.; Shapley, A.E.; Steidel, C.C.; Cuby, J.-G.; Dickinson, M.; Moorwood, A.F.M.; Adelberger, K.L.; and Giavalisco, M. (2001) "The Rest-Frame Optical Spectra of Lyman Break Galaxies: Star Formation, Extinction, Abundances, and Kinematics", *Astrophysical Journal*, **554**, 981–1000. Some of the first hints at how the Lyman-break galaxies at $z > 3$ compare with nearby objects when observed in a familiar emitted passband.

Shopbell, P.L. and Bland-Hawthorn, J. (1998) "The Asymmetric Wind in M82", *Astrophysical Journal*, **493**, 129–153. A detailed study of the starburst wind in M82, tracing the bipolar outflow, and sources of ionization.

Strickland, D.K.; Heckman, T.M.; Weaver, K.A.; and Dahlem, M. (2000) "Chandra Observations of NGC 253: New Insights into the Nature of Starburst-driven Superwinds", *Astronomical Journal*, **120**, 2965–2974. Demonstration that even the X-ray emission does not actually come from the hot wind itself, but from cooling interfaces. This implies that much of the enriched gas escaping the galaxy is still not directly detected.

Tripp, T.M.; Savage, B.D.; and Jenkins, E.B. (2000) "Intervening O VI Quasar Absorption Systems at Low Redshift: A Significant Baryon Reservoir", *Astrophysical Journal Letters*, **534**, L1–L5. Discovery of the O VI forest from the intergalactic medium, and a first estimate of its relative importance.

White, R.E., III (1991) "The Metal Abundance and Specific Energy of Intracluser Gas", *Astrophysical Journal*, **367**, 69–77. Argues that intracluster gas is hotter than can be accounted for from gravitational processes alone, and that the initial kinetic energy from supernovae and winds is important (especially in lower-mass clusters, in which some of the enriched gas has been lost).

Figure 2.1. The contrasting properties of elliptical and spiral galaxies are illustrated by the color-composite image of the galaxy pair NGC 4647/9 in the Virgo Cluster. The spiral NGC 4647 shows a blue disk with bright clusters, the loci of current star formation, and a redder central bulge. The elliptical NGC 4649 (Messier 60) shows a much redder overall color and smooth texture, testifying to its lack of current star formation and old stellar population. Some members of its rich globular-cluster population appear as faint starlike objects superimposed on its outer regions. (This image was made from two-band CCD observations by the author and R.E. White, III, with the 2.1-meter telescope of Kitt Peak National Observatory.)

Figure 3.2. The galactic globular cluster 47 Tucanae. This color-composite picture shows the color and luminosity contrast between red giants and the brightest main-sequence stars, characteristic of old Population II stars. (This image was produced from two-band CCD data obtained by the author, R.E. White, III, and C. Conselice at the 1.5-meter telescope of Cerro Tololo Inter-American Observatory.)

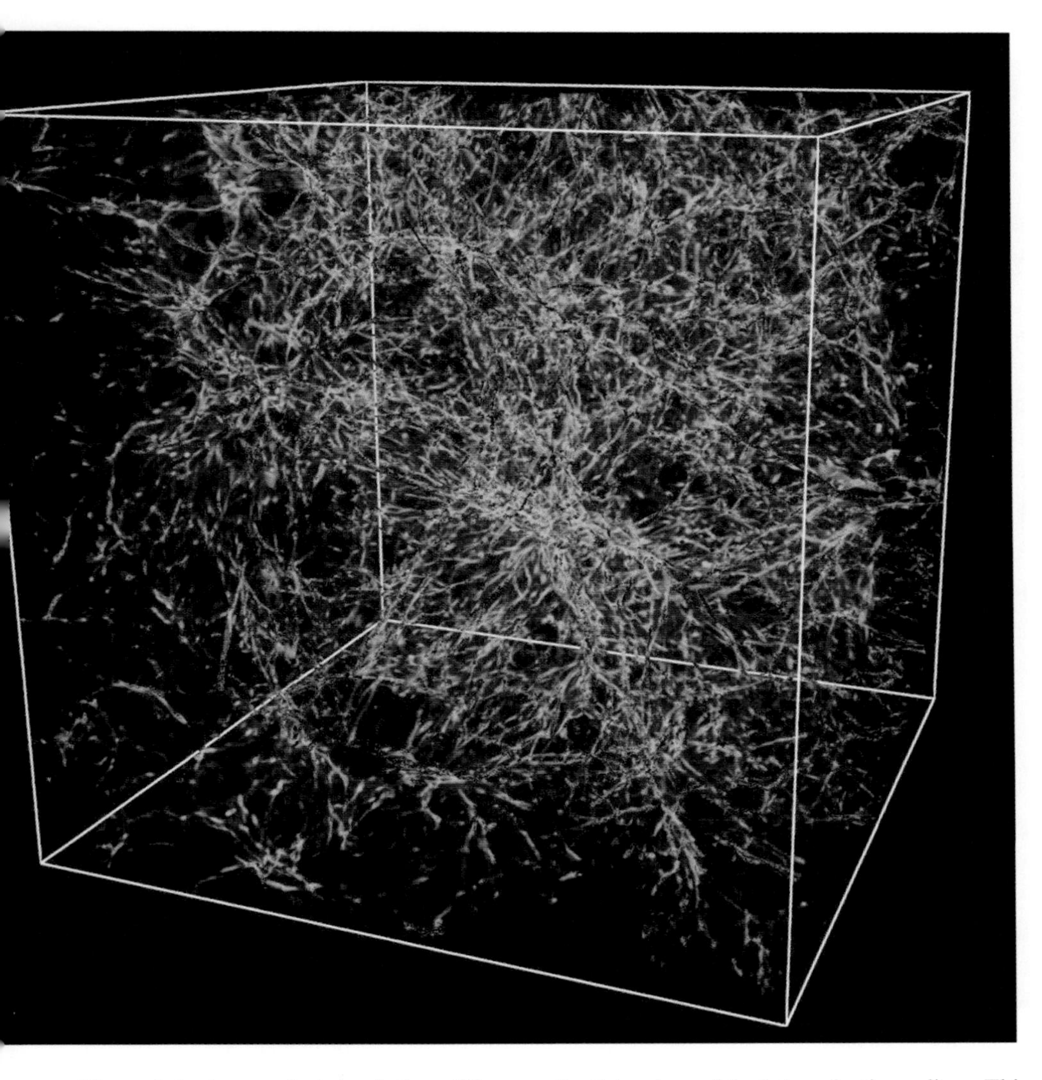

Figure 6.4. A numerical simulation of the growth of structure of the intergalactic medium. This particular visualization shows a perspective view of the density of H I, as seen in the Lyman α forest, within a 25-Mpc (comoving) cubic region at $z = 4$. The combined effects of density and radiation field make the structure in H I appear more tightly concentrated into filaments as seen in absorption than it actually is in mass. Most of the gas is ionized and only those regions with the most rapid recombination (highest density) remain neutral. (Simulation courtesy of Dr. Renyue Cen of Princeton University.)

Figure 6.5. The nearby starburst galaxy M82 (NGC 3034), with its prominent wind of ionized gas. The wind appears red in this composite image from the Hubble Space telescope, due to strong Hα emission. It escapes most readily perpendicular to the galaxy disk. Most of the wind material, including metals synthesized in the massive stars that drive it, is at temperatures above 10^7 K; the optical line emission and soft X-ray emission trace only interface regions at the contact between the winds and cooler surrounding material. (NASA, ESA, and the Hubble Heritage Team, released in honor of the observatory's 16th anniversary in orbit.)

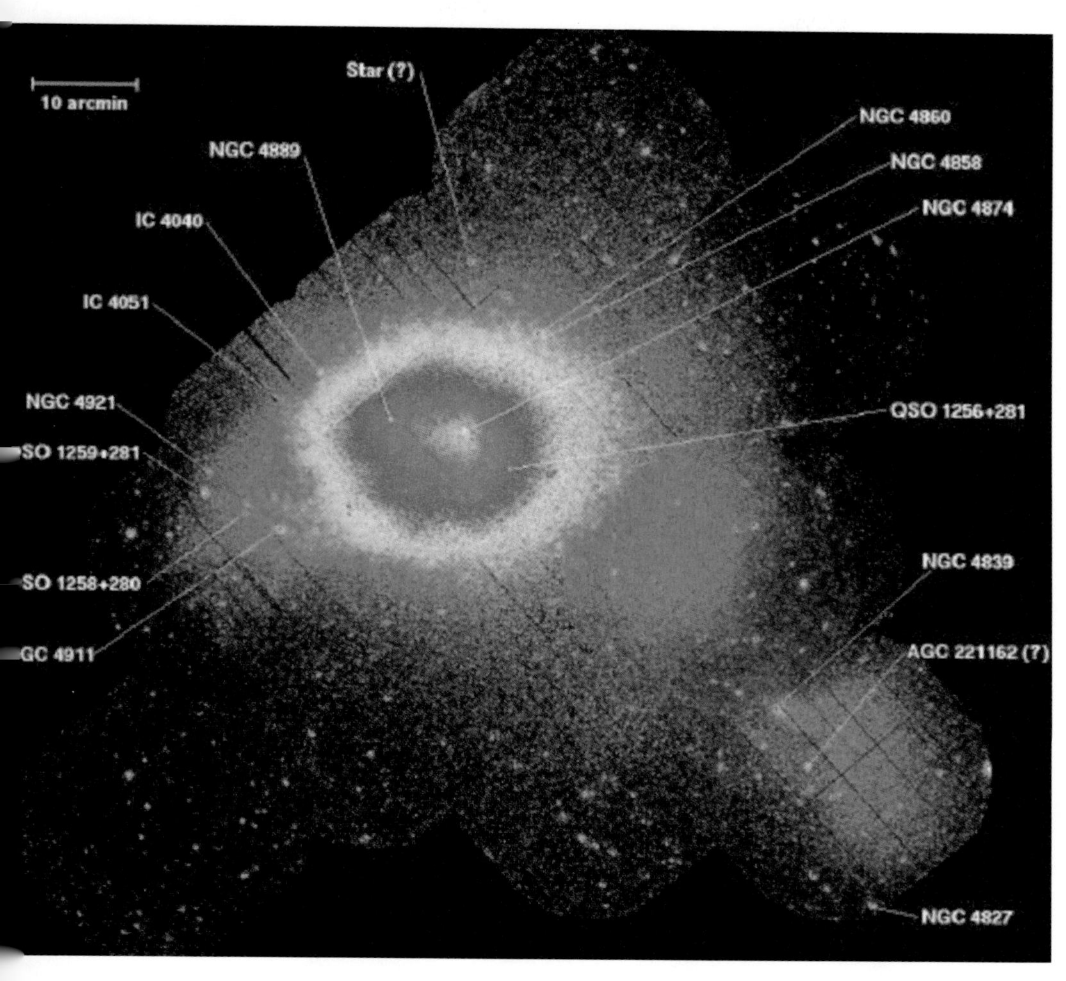

Figure 6.7. The Coma Cluster of galaxies as seen in X-rays, by the European Space Agency's XMM–Newton satellite. This view is a composite of multiple observations, some of whose circular outlines can be seen. Some individual galaxies and background quasars are marked, but the bulk of the X-rays come from a smooth, hot gaseous medium between the galaxies. The distinct lump to the southwest (*lower right*) coincides with an interesting grouping of galaxies, and has been interpreted as a smaller cluster which is now falling in to the Coma Cluster core. The field of view spans about 1.6 degrees on the sky, or about 1.6 Mpc at the cluster distance. (European Space Agency and U. Briel.)

Figure 7.3. All-sky map of the CMB from the first three years of WMAP data, processed to remove local foreground emission. Colors represent amplitudes of fluctuations about the mean brightness; the largest amplitudes have $\Delta T/T \sim 10^{-5}$. (NASA and the WMAP Team.)

Figure 9.11. Part of the original Hubble Deep Field, seen in a color-composite spanning about one-quarter of the WFPC2 field. Spiral galaxies are distinguishable to redshifts above $z = 1$, and some of the faint red images have been identified as Lyman-break galaxies at redshifts as large as $z = 5.6$. (Image by R. Williams of STScI, the HDF Team, and NASA.)

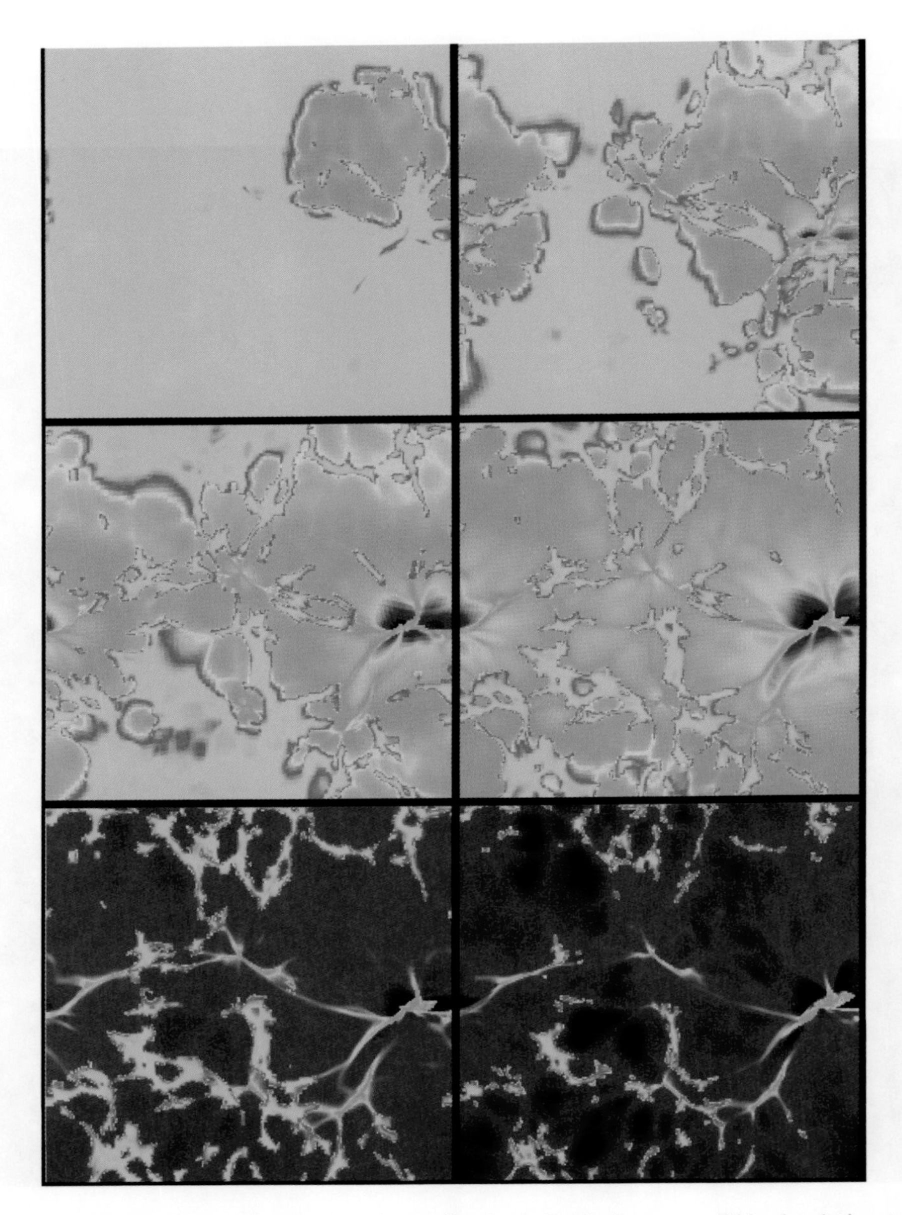

Figure 10.3. The history of cosmic reionization by individual sources. This simulation traces the ionized fraction of the intergalactic medium once star formation is widespread enough to begin ionizing the surrounding gas. The ionized regions are small, expanding around the young galaxies, and eventually merge, finally encompassing the densest regions which have yet to begin star formation. The color scheme runs from black for fully ionized material, through blue, green, yellow, and red for neutral gas. These snapshot slices cover a region 4 Mpc on a side on comoving coordinates, beginning at a redshift $z = 11.5$ and continuing to $z = 6$, where we now have our first direct glimpses of the end of reionization. (Courtesy of Nickolay Gnedin.)

7

The initial conditions before galaxy formation

Here is where cosmology really meets astronomy—in setting out the initial conditions for galaxy formation. The microwave background radiation provides our clearest picture of the fluctuations in mass density, seen just when they became free to collapse under the gravitational influence of ordinary and dark matter alike. The entire geometrical model from cosmology specifies how the competition between gravitation and cosmic expansion played out during the era of galaxy formation. The nature and clumping of dark matter are the yet-invisible field whose form controlled everything that gravity could do in galaxy formation. And primordial nucleosynthesis had already set the chemistry available to the first stars, whose effects may well predate the formation of the galaxies we see today.

7.1 BACKGROUND RADIATION

Diffuse background radiation is important in understanding the formation and evolution of galaxies, providing measurements integrated over redshift and wavelength. In optical astronomy the approach has generally been divided into counts of individually detected objects as a function of flux at various wavelengths, and measurement of the integrated extragalactic light, generally an upper limit, which then constrains the extrapolation of the counts to lower fluxes. The history of discovery in other wavebands has often driven the opposite approach, with measurement of the mean surface brightness of the sky preceding resolution of discrete sources and the ability to study their statistics. In each case the possibility of a genuinely diffuse component, not coming from individual galaxies and QSOs, has interesting physical implications. As observations have improved, so has our ability to interpret the various kinds of extragalactc background radiation. The far-infrared background constrains early star formation in galaxies, the X-ray background shows the history of active galactic nuclei, and the microwave background tells us of the combined early histories of matter and energy.

The existence of a *far-infrared* background has been claimed and dismissed several times, initially in the form of a distortion to the microwave background spectrum. Such distortions would represent modifications to the thermal history of matter, perhaps from reheating of the cooling material after recombination. The COBE results made it clear that no significant distortion exists for wavelengths longer than 0.5 mm, so attention could focus on shorter wavelengths, at which we would be detecting the thermal dust emission from galaxies integrated over redshift. This is both technically and scientifically daunting. We need to eliminate all more local sources of diffuse emission, beginning with dust in the Solar System and the Milky Way, which severely restricts the regions of sky we can use; in essence, one finds the dimmest sizable patch and takes that to mean that galactic dust emission is minimized there, if not zero. The residual contribution from the microwave background can be calculated precisely based on its longer-wavelength peak and temperature. What is left, after all these additional sources have been removed based on these kinds of external evidence, is an upper limit to the far-IR background. We can, strictly speaking, measure only an upper limit since it is not completely demonstrable that we have properly accounted for all other diffuse far-IR sources. Good agreement in the background intensity and spectrum for different assumptions about other contaminating sources suggests that the far-IR background has in fact been isolated, peaking just longward of 200 microns in F_ν. This peak is sufficiently different (by about a factor of 6) from that of the microwave background to confirm that we are not dealing with some modification of the microwave structure, but a distinct process and epoch. The fitted temperature of this background is about 18 K, and it follows the same sort of modified blackbody as found for grains in galaxies, which is expected if this radiation in fact comes from galaxies at a substantial redshift. Interstellar grains do not emit precise blackbody radiation, since the grains are small compared with the wavelength of most of their radiation. This makes their efficiency for radiation small and wavelength-dependent (with the efficiency estimated from data on different objects to follow a dependence between $1/\lambda$ and $1/\lambda^2$). This far-IR background intensity provides an integral constraint (in a redshift-weighted manner) on the total amount of star formation over cosmic time. In initial estimates from reprocessed COBE data, it accounted for about 20% of the Cowie limit on star formation over cosmic time. This comparison may well be revised, both because of improvements in far-IR measurements and data treatment, and a more accurate inventory of the total metallicity (and thus cumulative amount of star formation) in today's Universe. Detection of heavy elements (particularly oxygen and neon) in the intergalactic medium has made this a more open question than once thought, with the possibility that the total mass of metals outside galaxies may exceed that inside them.

The *X-ray background* was found early in the history of X-ray astronomy, partly as a result of the large field of view and poor point-source sensitivity of early detectors. Its integrated spectrum is close to a broken power-law form, with a break in the index at about 30 keV (which corresponds to a peak in the flux per decade νF_ν). This break has figured prominently in attempts to account for its origin, along with a gradual flattening of the spectrum across the 1–3 keV range. One goal of successively more capable X-ray imaging systems has been to see what fraction of the overall X-ray flux from the high-latitude sky can be shown to come from individual sources as we reach fainter and fainter, to tell whether there is a significant role for a diffuse component which might be hot intergalactic gas. Theoretically, there has been no shortage of candidates for what kinds of sources might be important. At various times, normal galaxies, clusters of galaxies, QSOs, and low-luminosity AGN have each been calculated to overpredict the X-ray background by themselves.

Recent deep surveys by the Chandra and XMM-Newton Observatories have made great progress in (quite literally) resolving the issue of the X-ray background. Various kinds of galaxies and AGN dominate the background at different energy levels, with the fainter sources more important at high energies. Observations toward deep fields already observed in the optical (such as the Hubble Deep Fields and GOODS fields), where substantial redshift information was already available, show that most of the X-ray background arises from galaxies at redshifts $z < 1$. The spectrum of the X-ray background is harder—more strongly weighted to higher energies—than is typical for the best-known active nuclei. This suggests an important role for AGN obscured by a significant column density of surrounding gas, whose emergent spectrum will be harder than the intrinsic one and which will contribute at higher energies where the radiation can penetrate their surroundings. Thus, we find that the X-ray background is, to a large extent, a cumulative tracer of the history of active galactic nuclei, and indirectly of the growth of their central masses (Chapter 8). The greatest scope for diffuse matter remains at rather low redshifts, perhaps gas which has escaped from low-mass clusters and groups of galaxies. Oddly enough, the tenuous gas between galaxies can be so hot that it is very inefficient at radiating even via X-rays, and can be detected most readily through absorption lines of highly ionized species in the deep ultraviolet and soft X-ray ranges. As we now understand it, the X-ray background gives a useful constraint on the progress of the evolution of galaxies and AGN, but does not retain a memory of the initial conditions for galaxy formation.

7.2 THE COSMIC MICROWAVE BACKGROUND

Among the assorted kinds of background radiation, the cosmic microwave background (CMB) plays a unique role. It comes from the earliest time we can sample anywhere in the electromagnetic spectrum, and its properties tell us much about the pregalactic Universe. Indeed, analysis of the CMB in detail has become a major cosmological industry. Like so many of the seminal discoveries in radio astronomy, the uncovering of the CMB owed much to serendipity. It began with Arno Penzias and Robert Wilson at Bell Labs in Crawford Hill, New Jersey, trying to establish the source of unexpected levels of "noise" which appeared in all their radio-astronomical observations. After careful elimination of more local possibilities, including repeated efforts to chase away homing pigeons which roosted in the antenna and deposited a "white dielectric substance", they announced the detection of a uniform bath of radiation from everywhere in the sky, whose effective temperature was near 3 K as derived from their observations at wavelengths near 7 cm. The interpretation of this radiation as a cosmologically important relic was hastened by the work of a group at nearby Princeton University, who were already searching for such radiation. A cosmic background was a robust prediction of the Big Bang cosmology, one that had been derived in a somewhat mysterious way by George Gamow in 1948. The observations and their interpretation were announced in a back-to-back set of papers in 1965, leading to the 1978 Nobel Prize in Physics for Penzias and Wilson. For many cosmologists, this was the key piece of information in favor of a broad Big Bang picture, rather than the unchanging Steady State model which had considerable philosophical appeal. As described by cosmologist Michael Turner, speaking recently at Bell Labs (now Lucent Technologies), "the discovery of the cosmic microwave background by Penzias and Wilson transformed cosmology from being the realm of a handful of astronomers to a 'respectable' branch of physics almost overnight."

In retrospect, the first hint of the CMB had come from routine measurements of absorption lines produced by interstellar gas in our own Galaxy, seen against the light of distant, luminous stars several thousand light-years away. Spectral features from the lowest-lying energy states of the CN molecule at 3874 Å, states which can be excited by radiation at wavelengths of 1.3 and 2.6 mm, indicated a minimum temperature close to 3 K, showing that there exists some minimum excitation level for these molecules (already mentioned by Walter Adams in a 1941 paper). Through the 1980s, observations of these molecular absorption features, which can be done easily from the ground, continued to provide useful limits on possible departures from a perfect blackbody form for the CMB. In a more philosophical sense, these CN measurements immediately rule out any very local origin for the CMB, since it must have very similar properties across much of our Galaxy. A similar role in demonstrating the truly universal nature of the CMB is now taken by absorption lines of low-excitation states of carbon ions seen against distant quasars (as detailed below).

A blackbody spectrum is an idealization, the distribution of energy produced by a hypothetical object which is a perfectly efficient radiator and absorber of energy. Opaque solids and gases can provide good approximations, but the CMB is the closest approach to a pure blackbody observed in astrophysics. A perfect blackbody has a spectral shape (in energy per unit frequency)

$$B_\nu(T) = \frac{2h\nu^3}{c^2} \frac{1}{e^{h\nu/kT} - 1} (\text{erg cm}^{-2}\ \text{s}^{-1}\ \text{Hz}^{-1}\ \text{sr}^{-1})$$

or equivalently expressed in flux per unit wavelength, by using $\nu\lambda = c$,

$$B_\lambda(T) = \frac{2hc}{\lambda^5} \frac{1}{e^{hc/kT\lambda} - 1} (\text{erg cm}^{-3}\ \text{s}^{-1}\ \text{sr}^{-1})$$

(the Planck curve). These equations specify not only the shape but the intensity normalization of surface brightness for a blackbody. In these expressions the physical constants have their usual symbols: c for the speed of light, h for Planck's constant, k for Boltzmann's constant, T for the temperature, λ for the wavelength of observation, and ν for the frequency of observation. Units in the c.g.s. system, although not

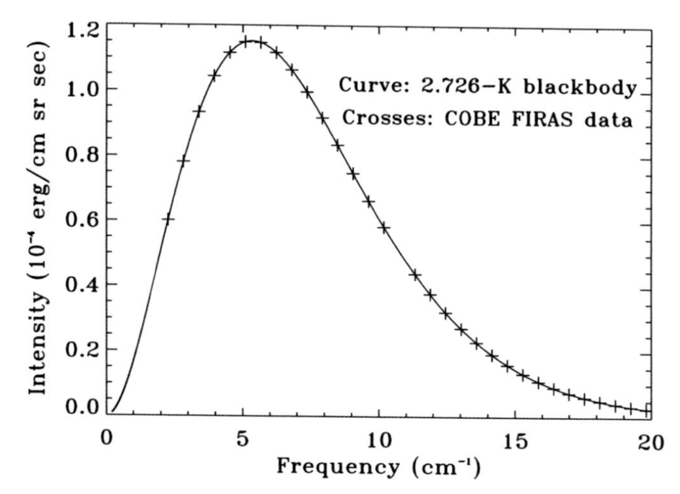

Figure 7.1. The final COBE measurement of the spectral shape of the microwave background radiation, compared with a perfect blackbody at the fitted temperature. The crosses mark measurements from the Far-Infrared Absolute Spectrometer; the error bars are too small to be shown, at about 0.03% r.m.s. The CMBR is as exact a blackbody as our data can show. (These data are from J.C. Mather *et al.*, *Astrophysical Journal*, **420**, 439, 1994.)

actually carried without units for the physical quantities, are included to illustrate the distinction in surface brightness as a function of frequency or wavelength.

The CMB temperature has been measured with great precision by the Far-Infrared Absolute Spectrophotometer (FIRAS) experiment on the Cosmic Background Explorer (COBE) satellite, at $T = 2.728 \pm 0.004$ K. The plot (Figure 7.1) from these data shows an excellent fit to a single-temperature blackbody, varying only in a single parameter T. The form of the Planck equation shows that blackbody radiation has the interesting property of being invariant to redshift—if the blackbody spectrum is redshifted, it is still exactly a blackbody spectrum of different temperature T, with a new normalization appropriate to this temperature. Specifically, a blackbody radiation field emitted at temperature T and redshift z will be observed as a blackbody field at temperature $T/(1+z)$.

The classic cosmological interpretation of the CMB is that it is the observable remnant of the oldest unscattered radiation, radiation streaming since the Universe first became transparent. This happened at the *epoch of recombination*. The name is a slight misnomer, since it was more accurately the epoch of combination, but the relevant processes are known as recombination from their occurrence in nearby nebulae and we seem to be stuck with the name. The radiation field in an expanding Universe will cool, eventually reaching a level such that neutral atoms can form permanently when the density of ionizing photons becomes low enough. This happens at a much lower temperature than given by the $kT = 13.6$ eV level corresponding to the ionization potential of hydrogen, because the number of photons in the radiation field per proton or neutron (the photon-to-baryon ratio) would have been so high that even the small fraction of radiation in the highest-energy blackbody tail could maintain substantial ionization. In a cosmologically short time the Universe at this point changes from being filled with a dilute ionized gas, whose continuum opacity leads to absorption or scattering of radiation after traveling distances very short compared with the cosmological horizon, to being filled with a neutral gas (composition ratio about 3:1 hydrogen:helium by mass), which has essentially zero opacity outside well-defined absorption lines and thus allowing radiation to stream freely to arbitrary distances. With the time–redshift mapping in an expanding Universe, the CMB forms a background "wall" to the part of the Universe we can see via electromagnetic radiation.

This epoch marks a crucial point for the formation of structure in the early Universe, as well as for our ability to observe it. Before the epoch of recombination at a redshift near $z = 1080$ (about 280,000 years into the expansion), matter and radiation were very tightly coupled. Any overdense clumps of ordinary baryonic matter would have been blown apart by the pressure of associated radiation, and underdense regions would similarly have been filled in from their surroundings. Thus, as far as the "ordinary" matter of nucleons and electrons is concerned, collapse to the precursor clumps of galaxies, clusters, or larger-scale structures we see today could begin only after recombination, when matter was left free to follow the dictates of gravity. Therefore, any fluctuations in cosmic gas density at this epoch are the initial conditions for galaxy formation. To form galaxies within a Hubble time (or many Hubble times), their seeds must have existed at recombination. Finding these seeds has been the major impetus for CMB research for two decades.

This transition also marks a major change in the thermal cooling history of the Universe. At earlier times, matter and radiation were tightly coupled, simply by the fact that the young Universe was so dense that radiation was constantly being emitted and re-absorbed by the matter particles, without being free to travel very far. In this regime the temperature behavior of both matter and radiation was the same, so that the energy density would follow a law which maps into the redshift we would observe as (density proportional to $(1+z)^3$). After matter and radiation decouple, the radiation follows the behavior of pure radiation, with an energy density dropping as $(1+z)^4$. Eventually, the density becomes matter-dominated, with the interesting result we see as today's Universe.

To the approximation that recombination was instantaneous and that the CMB represents a step-function wall in observability, fluctuations in density at recombination map into fluctuations in apparent temperature today according to $\Delta T/T = 1/3(\Delta\rho/\rho)$. Denser regions recombined later (requiring more photons, and a later time to reach the same state as less-dense regions), so we see them recombine at a smaller redshift. This overrides the slightly lower temperature at their recombination, giving higher-density regions a warmer-than-average observed temperature today. Fluctuations in the surface brightness (temperature) of the CMB tell us directly about the scale and amplitudes of fluctuations in density at recombination. Such fluctuations have now been securely detected over a wide range of angular scale. These were initially put on a solid footing by the Differential Microwave Radiometer (DMR) experiment on COBE, and have been further probed by several Earth-based experiments and the Wilkinson Microwave Anisotropy Probe (WMAP). The fluctuation spectrum and even individual hot and cool regions agree between the COBE results, the recent Tenerife experiment, and the BOOMERANG balloon flight, while the angular resolution and precision of the measurements has improved with time. The WMAP results have been particularly important; not only do they offer much better angular resolution than COBE, but they probe a large range of angular size scales with a single experiment, providing confidence that systematic uncertainties in combining the results of measurements taken by different instruments (at different times and analyzed in different ways) are not playing a role in the apparent structure of the CMB.

There are formidable technical requirements in detecting fluctuations in the cosmic microwave background. From the Earth's surface, measuring the background is nontrivial, since it represents a tiny signal against the much stronger, and varying, radiation from our own atmosphere. Since it comes from everywhere, the traditional on/off source-chopping that serves most radio and infrared astronomy very well is of no use, so a more sophisticated scheme—ultimately referred to as a cold calibration source—must be employed. Atmospheric effects, which depend most strongly on the viewing angle above the horizon, can be canceled by the use of multiple receivers pointing at the same altitude but different azimuths, so that taking the difference of their signal will largely cancel atmospheric emission but leave true structure in the astrophysical background as a differential measurement. Even for space-based measurements which never suffer from atmospheric emission, there are strong sources of cosmic foreground radiation which must be removed to yield the cosmologically interesting signal. The low-energy tail of emission from cold dust grains in the Milky Way, and the high-energy extremes of radio synchrotron emission from particles spiraling in the galactic magnetic field, both make substantial contributions near the wavelength of the CMB, so that a small error in accounting for either could mimic background structure. What makes the problem at all tractable is that these components are much brighter at other wavelengths, where the CMB is negligible, so that observations at additional frequencies can allow very accurate estimates of their individual contributions in the crucial millimeter regime.

These subtle features in the CMB sky encode a rich variety of information about conditions in the early Universe, some of which apply today as well. For large-scale structures, the CMB represents a cross-section in density. However, for smaller fluctuations we must deal with the fact that recombination was not instantaneous. Detailed tracing of the populations of H and He atoms in all relevant states yields a history of ionization such that the overall rate of recombination follows approximately a Gaussian form, peaking at $z = 1088$ with a Gaussian standard deviation $\sigma(z) = 81$, or a full width at half-maximum of about 10% in redshift. Much of this behavior could be understood with the simplified analytical model presented as early as 1968 by Peebles and by Zel'dovich and collaborators. This finite depth to the CMB "wall" means that structures smaller than this depth will appear to us washed out, since their intensity structures will blend with foreground and background regions. We expect to see the CMB appear smooth on the smallest angular scales for this reason, so it is only on larger scales, corresponding to a supergalactic structure, that we can recover the initial perturbations directly. This damping of small-scale fluctuations is the only important and observable consequence of the time (and redshift) needed for recombination. Contrary to intuition, this span hides its temperature signature. At first glance, it seems that such a range in redshifts, and necessarily range in temperatures, must alter the blackbody nature of the spectrum we see. Our initial expectations are conditioned by the behavior of radiation temperature in an expanding Universe or of matter in such a Universe. However, as long as the two were radiatively coupled, the temperature of radiation at any cosmic epoch was constrained to match that of matter, with the result that we will see no significant distortion of the pure blackbody spectrum from the finite time that was required for recombination to be completed. The blackbody form of the spectrum is not a result of the recombination process (which can give striking emission-line spectra in such objects as H II regions like the Orion Nebula), but is itself a fossil of earlier epochs when the Universe was in thermodynamic equilibrium, preserved through the period of recombination. This is interesting, because it says that any additional energy input between this period of equilibrium and recombination must have been minor. Furthermore, any energy input after recombination, and before the time when we can start to see discrete sources, can be limited observationally by the fact that we do not see distortion of an exact blackbody. Such distortions, which might have been produced by such things as extremely massive first generations of star formation, were widely discussed until the COBE spectral results ruled them out.

Most of the direct information we can glean on the start of galaxy formation comes from the minute angular fluctuations that show us that the early Universe was not exactly homogeneous. It is convenient, and conventional, to treat the fluctuations through their power spectrum, the relative amounts of power in fluctuations of different wavenumbers around the sky. This amounts to a decomposition into a set of spherical harmonic functions, the analog of Fourier transforms on the surface of a sphere (as illustrated in Figure 7.4 for the analogous case of solar oscillations, where we view the sphere from without instead of within). This representation is especially convenient for observations taken with interferometer systems involving multiple antennas, since the combined output of an antenna pair with a particular separation measures a corresponding Fourier component of the brightness pattern. This connection was exploited, for example, by Caltech's Cosmic Background Imager (CBI), located in northern Chile at an altitude of 5000 meters, to map subdegree-scale structures in the CMB.

By now, the gold standard for CMB fluctuation measurements comes from the NASA Wilkinson Microwave Anisotropy Probe (WMAP), a highly successful follow-up mission to COBE. Since COBE data had determined the mean spectral shape of the background radiation, subsequent experiments can take this as given and concentrate on purely differential measurements of the spatial changes in its intensity. WMAP (Figure 7.2) uses two off-axis 1.5-meter optical systems on opposite sides of the spacecraft's axis of rotation, simultaneously comparing the intensity in regions 140° apart as the spacecraft rotates about once every 2.2 minutes and precesses over longer timescales to cover the entire sky. For greater freedom in planning a scanning strategy to cover the whole sky without interference from the

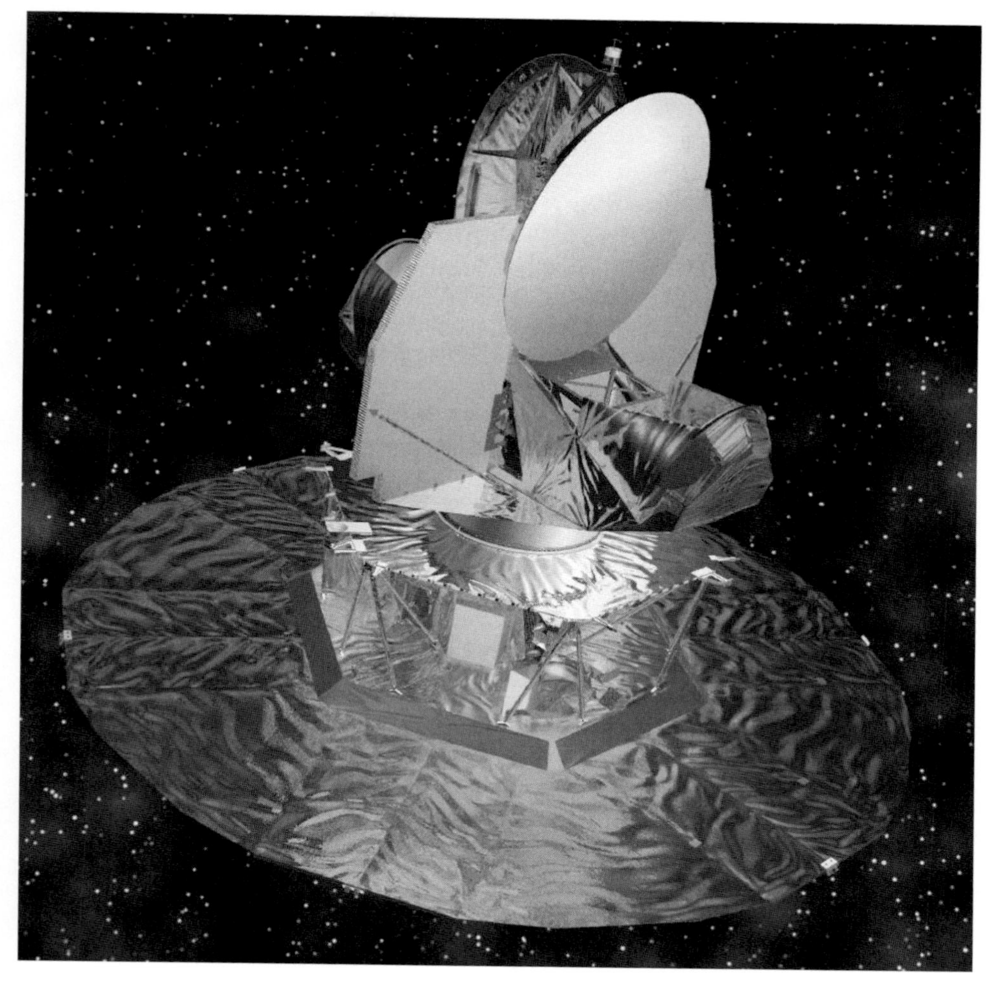

Figure 7.2. Digital rendering of the WMAP spacecraft at work. The two back-to-back Gregorian optical systems are seen at the top, delivering radiation in an off-axis beam to the receivers. The solar arrays are at the bottom, on the unseen side of the sunshade. During operation, the spacecraft rotated about an axis which is vertical in this view to scan the sky with two beams 140° apart. (NASA and the WMAP Team.)

Earth, WMAP was launched into a path oscillating about the outer Lagrange point (L2) of the Earth–Sun system. This point lies about 1.5 km outward from the Earth, where the combined gravitational effects of Sun and Earth keep an object aligned with the two. After launch in June 2001, WMAP spent three months reaching its operational orbit, including a lunar swingby for gravitational assist. Data collection continues in an extended-mission phase; the final results for each year's sky mapping require a global analysis of the entire data stream.

Figure 7.3. All-sky map of the CMB from the first three years of WMAP data, processed to remove local foreground emission. Colors represent amplitudes of fluctuations about the mean brightness; the largest amplitudes have $\Delta T/T \sim 10^{-5}$. See also color section. (NASA and the WMAP Team.)

The first year's WMAP results appeared in 2003, and determined a number of cosmological parameters with unprecedented precision. These fit well with the "concordance model" previously driven by *Hubble* distance measurements and high-redshift supernova photometry: a Hubble constant of $72 \pm 5\,\mathrm{km\,s^{-1}\,Mpc^{-1}}$, baryon density $\Omega_b = 0.046$, total matter density $\Omega_m = 0.27$, and geometry as nearly flat as could be measured (which implies that the remaining fraction of $\Omega = 1$ must reflect dark energy). The constraints tightened somewhat when additional kinds of data were fit simultaneously. One puzzle was that the optical depth to scattering implied by the available temperature–polarization correlation data was uncomfortably high without invoking a longer time to reionize the Universe than suggested by either models of the first stars or data on the highest-redshift galaxies; multiple reionization epochs even seemed possible.

The WMAP team continued data collection (Figure 7.3), not only to refine these estimates but in hopes of detecting ever-more-minute effects involving the polarization of the background. These would be detectable if even a small fraction of CMB radiation had ever been scattered since the epoch of recombination. When the year-to-year differences in their sky maps exceeded what was expected from statistical errors, the team spent over two years on an end-to-end reworking of the data system, identifying a number of subtle issues to improve (some as simple as roundoff issues in a calculation, others more subtle involving tests for changes in detector response with details of the spacecraft state). The three-year combined data provided yet tighter constraints on the major cosmological parameters, and cleared up some of the curious issues raised by the one-year results. In particular, the implied amount of electron scattering since the epoch of recombination is substantially smaller than found from the initial data release, no longer suggesting a more complex reionization

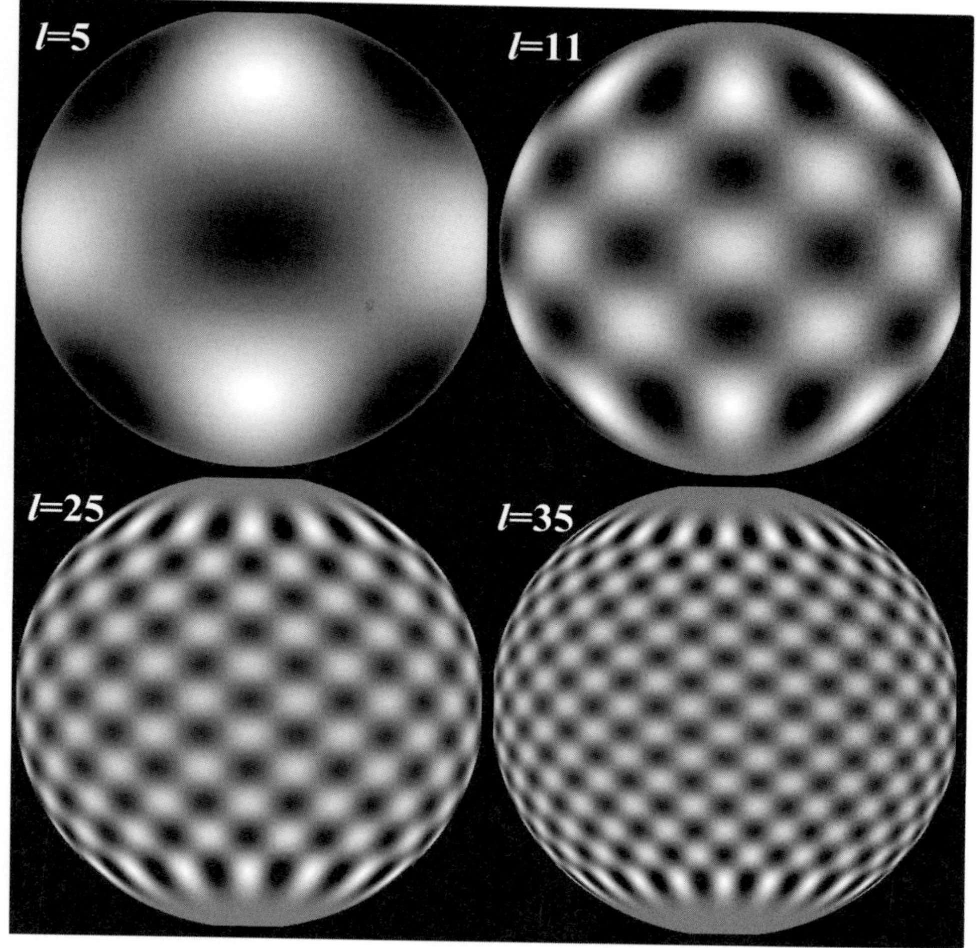

Figure 7.4. Visualization of spherical harmonics. Individual sets of the spherical harmonic functions, as used to decompose the power in fluctuations of the microwave background on various angular scales, have been projected onto a sphere as viewed from outside. The l-values give the number of positive and negative nodes of that mode around the sphere. A node number in the polar direction, m, must be specified for a visualization, although this has no meaning in analysis of the microwave background since there is no preferred axis in the problem. This differs from another important astrophysical application, solar oscillation, in which the rotation axis does play a special role. (Visualizations performed with software kindly provided by Rachel Howe of the National Solar Observatory.)

history. However, it remains curious that the largest-scale structures in the microwave sky do not match the fits to higher-order features very well.

We have no independent information on any intrinsic dipole structure ($l = 1$) in the CMB, because this would be exactly masked by the need to remove the signature

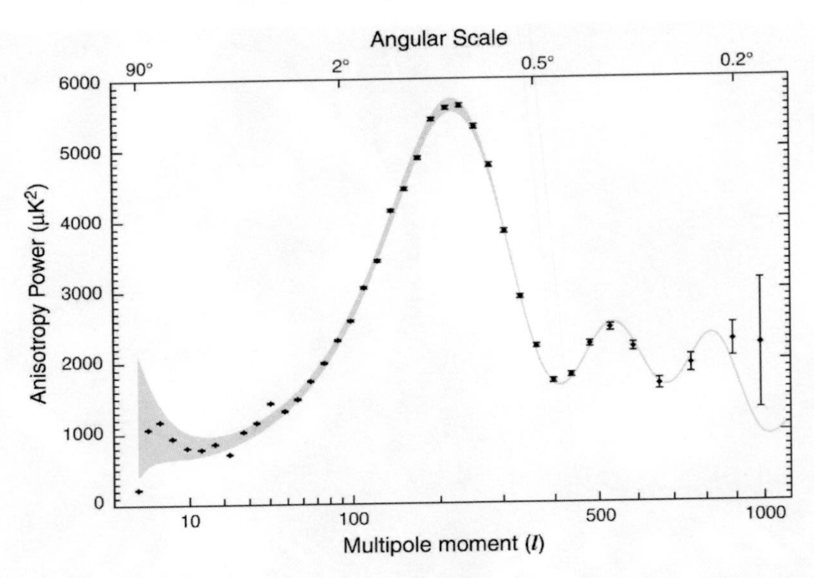

Figure 7.5. Comparison of measurements of fluctuations in the cosmic microwave background with sample predictions in an inflationary cosmology. Symbols show the results of the 3-year WMAP analysis of fluctuations on various angular scales (set by the order of the harmonics, l). The curve is a fit to the data set; the relative amplitudes and locations of the Doppler or acoustic peaks serve as tests of inflation and the properties of dark matter. The gray band shows the expected amount of cosmic variation, the unavoidable sampling error brought about by our having only a single sample of the Universe to observe in this way. (NASA and the WMAP Team.)

of our own motion relative to the mean frame of the CMB material at recombination. The next few low-order amplitudes ($l = 2$–4) are all much lower than the fit to the rest of the data would predict (Figure 7.5). More puzzling, the global quadrupole and octupole components are aligned with each other along the so-called "Axis of Evil", pointing in a direction roughly in the ecliptic plane. This effect has persisted with the reprocessing of all WMAP data using improved techniques, and seems to be a genuine feature of the sky brightness distribution. It is possible that these low-order structures appear anomalous only because of the luck of the draw of our particular viewing location, but a few workers have expressed concern that they might result from some inadequacy in our ability to remove the foreground contamination from local emission.

Contemporary CMB experiments, such as WMAP, often center on the measurement of the power spectrum of fluctuations in the CMB intensity with angle on the sky. Various schemes for the nature of fluctuations make distinct predictions about the power spectrum. Given the other information available from the local Universe, such as the mass distribution as inferred from peculiar motions, there is a robust expectation that the power spectrum will rise from small scales (large wavenumber) to a Doppler peak (sometimes called acoustic peak, though there is dispute about the applicability of the term). There will be a drop to larger scales (small wavenumber), with a set of harmonic peaks whose location and amplitude depend on the nature and origin of fluctuations in mass density (or velocity field). These peaks and valleys have to do with what perturbation sizes will have been more or less strongly damped in the early Universe, for a given cosmological model. The angular frequency l of the first (strongest) peak depends on $\Omega_{\rm tot}$, measuring the relative scale factors of the Universe between now and recombination. The baryon density Ω_b affects the ratios of height of the even and odd peaks (through dissipation acting differently for compression and rarefaction phases of the oscillations). Both Ω_b and the Hubble constant h affect the height of all the peaks, and a cosmological constant Λ affects both the amplitude and location of the first peak. These combined effects illustrate why it is so important to have uniform data spanning a wide range in l.

Several major projects, particularly WMAP, have constrained our knowledge of CMB fluctuations so tightly that some might consider the major constants of cosmology to be "solved". As a further step, the European Space Agency is constructing the *Planck*, designed to yield an even more precise measure of the fluctuation spectrum, with a 1.5-meter telescope yielding 0.1° angular resolution. Planck is planned for a 2007 launch, together with the Herschel far-infrared observatory, into the same kind of L2 halo orbit as WMAP.

The redshift behavior of the microwave background, with the temperature at a time corresponding to what we see at redshift z being given by $T = 2.728(1 + z)$, may become important in understanding how the first generation of stars formed, against a much warmer background of radiation than would allow gas to cool as far as in present-day regions of star formation. This behavior has been observed directly by using clouds of gas seen in absorption against background quasars, which set limits on T at redshift values as large (to date) as $z = 2.34$ which are in accord with this behavior. The measurements rely on observing a pair of absorption lines from neutral or singly-ionized carbon atoms in the emitted ultraviolet (at wavelengths near 1657 and 1334 Å, respectively, so that high quasar redshifts put them in the atmospheric window for ground-based detection using large telescopes). These pairs of lines are special in that one of each pair arises in the lowest-energy (ground) state of the atom, and the other comes from a slightly excited (fine-structure) state, whose excitation energy is low enough to be fulfilled by the CMB radiation. The results of these measurements are given as upper limits since in any individual gas cloud we do not know *a priori* that there is no other heating source, but taken together they strikingly support the expected history of the microwave background.

Recent statistics of the local galaxy distribution, from the Sloan Digital Sky Survey, have revealed a predicted connection from today's distribution of galaxies to structures at the epoch of recombination—baryon oscillations, a spatial signature of the same fluctuations seen in the CMB. In the early Universe, matter overdensities were the force behind acoustic waves, driven by the powerful restoring force of the coupled radiation field. These wave fronts (spherical in the idealized case of a single dense region in a uniform background) propagate outward at the sound speed as the Universe expands. However, after recombination, the sound speed falls rapidly and the mass associated with the wave peak is effectively frozen in comoving coordinates, after the wave has traveled about 150 Mpc in today's coordinates. This density enhancement will be reflected in the availability of more matter to form galaxies in these locations, and hence in the number of galaxies found in the fossils of these slightly overdense regions. We can detect this length scale only statistically, as an effect of amplitude $\sim 1\%$ superimposed on the complex fluctuation pattern of cosmic structure. Sufficiently large galaxy surveys can show this feature: the only distinct length scale defined by galaxy correlations, directly connecting the scale of the Universe today to that at the epoch of recombination. In the words of Daniel Eisenstein *et al.* (2005), the spectrum of the Universe has a single emission line. This length scale is so well defined that it may prove a competitive way of tracing the history of cosmic expansion as large galaxy surveys penetrate deeper in redshift.

These ripples date to a time when the Universe wasn't so close to a vacuum as it is today, so that acoustic processes played a crucial role. Rather than shadows of forgotten ancestors, we see their echoes. For galaxies, if not literally the Word, in the beginning was the Sound.

7.3 DARK MATTER

Structure in the CMB shows us how baryons themselves were behaving at recombination. This is only a small part of the story for beginning galaxy formation, since the density of dark matter was a dominant factor. For nonbaryonic dark matter, we have little constraint on its behavior before recombination. If it coupled much more weakly to the radiation field than baryons, the dark matter density field could have been quite distinct from baryonic matter. Only when the coupling between radiation and baryons was broken would the baryons begin to trace the potential defined by the dominant mass of dark matter.

Our understanding of the importance of dark matter utterly changed the way we view galaxy formation. Malcolm Longair, in his book *Galaxy Formation*, devotes an entire chapter to this transition. Before dark matter (BDM), this appeared to be a classical Jeans-mass problem, with a role for cooling in the gas accelerating gravitational collapse. Afterward (ADM), the visible matter does little more than follow potential wells defined by the dark matter until its own internal interactions allow it to cool and clump more strongly, on the small scales of galaxies, than dark matter.

Considerations of the behavior needed to produce galaxy-sized masses were important in favoring cold dark matter over possible forms of hot dark matter. Broadly, we speak of hot dark matter as having particle velocities much greater than the present-day velocities of galaxies in groups or clusters, including relativistic velocities, although the more formal definition involves the particle velocities around the epoch of recombination. The most direct approach we have to the dark-matter distribution at high redshift involves using the present-epoch density field as a ground-truth comparison with simulations of a cosmologically evolving density field. At this point, this can be done on large scales, but the numerical resolution to trace the mass distribution within a numerical "galaxy" very accurately in a simulation that includes a cosmologically interesting volume is still a problem for the future.

7.4 PRIMORDIAL NUCLEOSYNTHESIS

The initial chemistry for galaxies, which defined how material could cool and what the raw material for the first stars would be, was set long before recombination, starting after the first three minutes. In the first seconds of cosmic history, the temperature and density (of both matter and radiation) were high enough to maintain the population of protons and neutrons at the number ratio set by their masses

$$(n/p) = e^{(m_p - m_n)c^2/kT}$$

as long as the rates for the three $n \leftrightarrow p$ reactions are fast compared with the expansion timescale. When the temperature is low enough that particle energies are below about 1 MeV, these processes slow dramatically as the density drops as well. Nucleosynthesis begins in earnest once the temperature drops so low that photodissociation of D is rare (below about 8×10^8 K). The final isotopic distribution in the early Universe is set by the competition among nucleosynthetic processes, decay of free neutrons, and declining density of available particles as the Universe expands. The relevant reactions are:

$$p + n \rightarrow \mathrm{D}$$

$$\mathrm{D} + \mathrm{D} \rightarrow {}^3\mathrm{He} + n$$

$$\mathrm{D} + \mathrm{D} \rightarrow {}^3\mathrm{H} + p$$

$${}^4\mathrm{He} + {}^3\mathrm{H} \rightarrow {}^7\mathrm{Li} + \gamma$$

$${}^7\mathrm{Li} + p \rightarrow {}^4\mathrm{He} + {}^4\mathrm{He}$$

$${}^7\mathrm{Be} + n \rightarrow {}^7\mathrm{Li} + p.$$

The large initial fraction of free neutrons differs from the usual stellar nucleosynthesis; there is no weak reaction needed to make D, which takes a very long time in a stellar interior. Lithium is the most massive species we need to consider; the gap at atomic mass number 8 is bridged in negligibly small numbers, a situation which differs from stellar interiors in which the triple-α process operates, since the density was much lower during primordial nucleosynthesis. Once the neutron half-life (932 seconds) has been specified by contemporary experiments, the eventual frozen-out ratios for the light isotopes are set by the initial baryon density (since only baryons partake of nucleosynthesis). We can seek consistency among our cosmological model and the primordial chemistry by asking whether there is a single value of early expansion rate which fits the ratios of all these species, as observed in pristine objects.

What environments count as pristine enough to see pregalactic chemical abundances, uncontaminated by any star formation? To use the relatively straightforward tools of emission-line spectroscopy, we face the contradictory requirements of seeing gas which has not been enriched by massive stars close to just such stars, whose ionizing radiation excites the same gas. Accordingly, attention has focused on star-forming regions in low-mass galaxies, and those in which all the heavy-element abundances are the lowest known, suggesting minimal chemical influences from stellar sources. The blue compact galaxy I Zw 18 is especially important in this regard, having held the record for lowest known emission-line abundances for two decades. Its oxygen abundance is only 1/50 of the value in the solar neighborhood, so that contamination by stellar products should be minimized for helium.

To guard against stellar production of He, one common approach has been to consider He/H as a function of O/H, extrapolating to zero oxygen abundance as an estimate of the initial cosmological value. At the level of accuracy needed for cosmological conclusions, many subtleties enter into analyzing the spectroscopic data. Optical depth for the helium emission lines themselves, and the amount of transfer between helium in various excitation states, can be important. However, for understanding the first stars and galaxies, our current accuracy of primordial helium constituting 22% by mass of the initial gas composition is almost certainly close enough.

The deuterium abundance can be approached in a different vein. Helium can be produced inside stars, while essentially all stellar processes only destroy deuterium (usually through the net effect of transmuting it into helium). Similarly, interstellar processes such as spallation from energetic cosmic rays can dissociate deuterium, but no known contemporary process produces deuterium without also destroying it at a faster rate. Thus, a lower limit to the primordial deuterium abundance can be found in a wide variety of environments. Terrestrial oceans, the Jovian atmosphere, the local interstellar medium between the Sun and α Centauri, and QSO absorption-line systems at high redshift have all been considered for this purpose. The spectroscopic approach uses the isotopic shift between Lyman α for ordinary hydrogen and deuterium to measure their relative abundances. Atomic parameters can then translate the ratio of line strength to a number ratio, although the measurement itself is tricky to measure because the ordinary line is strongly saturated whenever the D line is detected (at 1215.32 Å, only 0.35 Å shortward of the ordinary Lyman α line). The best chance to detect the D line requires that we look through a single gaseous component with small velocity dispersion, to keep the hydrogen line as narrow as possible at a given column density. Thus, the best local determinations will come for lines of sight toward nearby stars, and the best distant measures will come from the rare QSO lines of sight which pass through galaxy halos producing intrinsically narrow but strong (damped) Lyman α absorption. However, any single QSO absorption system leaves the risk of confusion with an unrelated Lyman α forest line, so that multiple objects are needed for a reliable value. Indeed, the difference between the first ones successfully measured using Keck spectra is likely traceable to chance contamination by such an absorber. Three of the four secure QSO data sets give a low D/H value, near $3x \times 10^{-5}$. From the recent review by David Tytler *et al.* (2000), the current results for ^{4}He, D, and ^{7}Li overlap for a baryon density range 3–4×10^{-31} g cm^{-3}, or a baryon mass fraction 0.04–0.05 of critical density (for $H_0 = 65$).

A simultaneous fit to the various ratios of light elements can indeed, in principle, constrain the cosmological picture, since various ratios have quite different behavior with Ω_b. The ^{4}He/H ratio increases with density, while the abundances of ^{3}He and D decline. ^{7}Li has its greatest abundance for both low and high densities (Figure 7.6).

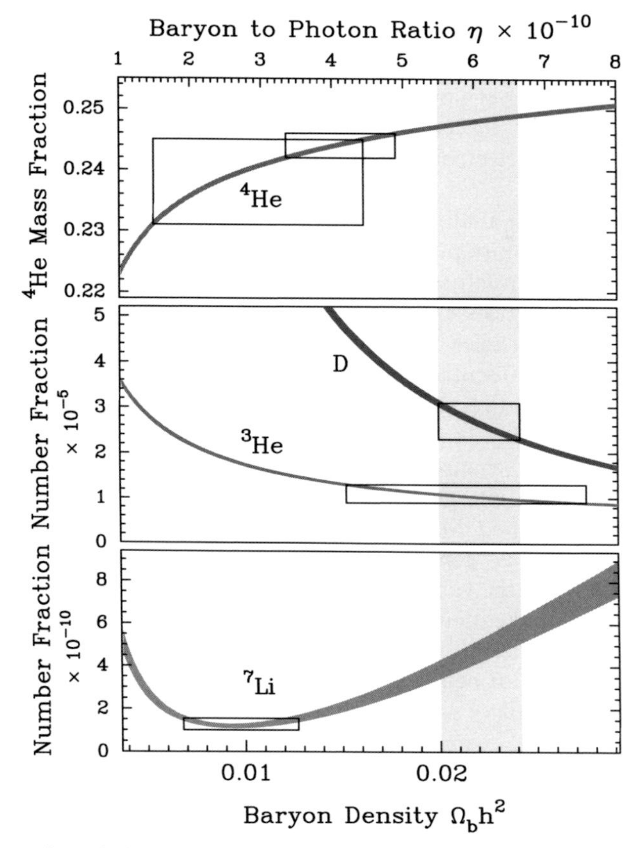

Figure 7.6. Isotopic and elemental abundances compared with predictions for nucleosynthesis in the Big Bang. The horizontal scale shows changing values of the density of baryons today, which predicts the various curves for amounts of light elements relative to hydrogen. The boxes show the ranges implied by data for various species. In a homogeneous Universe, and if all errors are properly accounted for, their bounds would overlap in a region including the actual baryon density. Further work centers on refining the deuterium abundance (especially in QSO absorption-line systems and the galactic interstellar medium), and understanding possible depletion of lithium in the observed environments. In this scale the favored region in gray is centered near a baryon density about 5% of the closure density. (Courtesy David Tytler and Nao Suzuki.)

Adopting these values as universal assumes a homogeneous Big Bang. Inhomogeneous conditions would give somewhat different bulk results, since the production of some species would be wieghted toward the densest regions, as well as possibly changing the ratios in various environments if they have not been mixed over cosmic time. While these possibilities have been explored, the observations to date are not compelling that inhomogeneity was sufficient to matter for present purposes.

7.5 DEEPER BACKGROUNDS

There are two additional, yet-unobserved backgrounds which have great theoretical potential to tell us about the early Universe. In keeping with the statement that the CMB is the earliest electromagnetic radiation we can see, neither fall in the electromagnetic spectrum.

In a process somewhat analogous to the freezing out of atomic nuclei in the early Universe, we expect a relic population of neutrinos. A bath of neutrinos and antineutrinos would have been left when the Universe dropped below a temperature of about 10^{10} K, when no processes energetic enough to keep them in thermodynamic equilibrium with other particles could proceed. This background is substantial, with over 100 neutrinos and antineutrinos per cubic centimeter of today's Universe. These neutrinos are far beyond direct detection today, with characteristic energies about 2 eV (whereas contemporary experiments focus on solar and extrasolar neutrinos at MeV energies). However, because they are so numerous, these neutrinos contribute a significant fraction of the energy density of the Universe—they exceed the energy density of photons from the CMB, for example. The potential contribution of these neutrinos to the overall mass density of the Universe has driven particular interest in the value of the masses of the various neutrino flavors. Average masses of only a few electron volts would make neutrinos important contributors, and at one point massive neutrinos were thought to be an excellent dark-matter candidate. However, current experiments rule out neutrino masses large enough to make them both cold enough to condense on galaxy scales or to be a significant contributor to the cosmic mass density.

Moving masses create gravitational radiation, if the description in general relativity continues to hold on all scales. Mass certainly moved during galaxy formation, but the motions were so slow that we don't expect significant gravitational waves from this cause. However, if inflationary cosmologies are correct, we do eventually expect to see a particular flat spectrum of power versus frequency in a cosmological background of gravitational radiation. This would be found at a power level about 10^7 times lower than limits from the constancy of measured pulsar periods, but great sensitivity increases are possible with, first, laser interferometry on the ground, and eventually interferometry between spacecraft. As John Peacock notes, finding this kind of gravitational-wave background would do for inflationary cosmology what the cosmic microwave background did for the Big Bang.

7.6 BIBLIOGRAPHY

Books

Alpher, R.A. and Herman, R. (2001) *Genesis of the Big Bang* (Oxford University Press); *Proceedings of the National Academy of Sciences*, **58**, 2179. This memoir sets out the once-forgotten early history of predictions of a cosmic thermal background in a Big Bang cosmology. George Gamow's initial prediction of $T = 5$ K appeared in such a terse (perhaps even obtuse) form that Russian astrophysicist Artur Chernin has devoted an article to unraveling it: "George Gamow and the Big Bang", *Space Science Reviews*, **74**, 447–454 (1995).

Longair, Malcolm (1998) *Galaxy Formation* (Springer).

Mather, John C. and Boslough, J. (1996) *The Very First Light: The True Inside Story of the Scientific Journey Back to the Dawn of the Universe* (Basic Books); and Smoot, G. and Davidson, K. (1993) *Wrinkles in Time* (W. Morrow). Two of the principal investigators for the COBE mission, who went on to share the 2006 Nobel Prize in Physics, have given their individual views on its conduct and results in these books.

Partridge, R.B. (1995) *3 K: the Cosmic Microwave Background Radiation* (Cambridge University Press). A detailed description of studies of the microwave background is provided.

Peacock, John (1999) *Cosmological Physics* (Cambridge University Press); and Longair, Malcolm (1998) *Galaxy Formation* (Springer). Considerable detail on the epoch of recombination and how the density fluctuations might be traced to an inflationary stage are given in these works.

Weinberg, Steven (1984) *The First Three Minutes*. Summarizes the standard Big Bang picture up to the time of nucleosynthesis.

Journals

Adams, W.S. (1941) "Some Results with the Coudé Spectrograph at Mt. Wilson", *Astrophysical Journal*, **93**, 11–23. This paper includes the report of CN absorption from interstellar gas, observed following a suggestion by Andrew McKellar. Adams notes the detection of lines from both the ground state and a low-lying excited state, which proved in retrospect to be a first detection of the cosmic microwave background by its excitation of molecular gas. On this basis, McKellar noted that the rotational temperature of interstellar CN was about 2.2 K, probably the first unknowing step towards measuring the 2.7-K temperature of the Universe.

Alexander, D.M.; Brandt, W.N.; Hornschemeier, A.E.; Garmire, G.P.; Schneider, D.P.; Bauer, F.E.; and Griffiths, R.E. (2001) "The Chandra Deep Field North Survey. VI. The Nature of the Optically Faint X-Ray Source Population", *Astronomical Journal*, **122**, 2156–2176. Counts of active nuclei identified in very deep X-ray observations. Depending on the energy range, 14–21% of the background has now been identified from individually resolved active nuclei.

Chernin, Artur (1995) "George Gamow and the Big Bang", *Space Science Reviews*, **74**, 447–454.

Davidson, K. and Kinman, T.D. (1985) "Primordial helium, spectrophotometric technique, and I Zwicky 18", *Astrophysical Journal Supplement*, **58**, 321–340. This report of abundance measurements includes an extensive list of caveats to be observed in measuring the He/H ratio to high precision. Radiative transfer effects, and line absorption in both interstellar space and the terrestrial atmosphere, can become important, though more typical observing programs can often ignore them, because the accuracy needed to derive cosmologial implications for He/H is quite high.

Eisenstein, D. *et al.* (2005) "Detection of the Baryon Acoustic Peak in the Large-Scale Correlation Function of SDSS Luminous Red Galaxies", *Astrophysical Journal*, **633**, 560. Reports the detection of baryon oscillations in the galaxy distribution.

Fixsen, D.J.; Dwek, E.; Mather, J.C.; Bennett, C.L.; and Shafer, R.A. (1998) "The Spectrum of the Extragalactic Far-Infrared Background from the COBE FIRAS Observations",

Astrophysical Journal, **508**, 123–128. A measurement of the far-infared background from processing COBE data. This background comes mostly from the integrated dust emission of galaxies at redshifts $z = 1$–3.

Hasinger, G. (1996) "The extragalactic X-ray and gamma-ray background", *Astronomy and Astrophysics Supplement*, **120**, 607–614. A brief introduction to the known contributors to high-energy backgrounds.

Molaro, P.; Levshakov, S.A.; Dessauges-Zavadsky, M.; and D'Odorico, S. (2002) "The cosmic microwave background radiation temperature at z(abs) = 3.025 toward QSO 0347−3819", *Astronomy and Astrophysics*, **381**, L64–L67. Uses absorption from an excited state of C II in QSO absorption-line systems to limit the temperature of the CMB at $z = 3.02$, showing consistency with the predicted proportionality between T and $(1 + z)$. This is the latest in a series of observations by several groups, gradually increasing the redshift range over which we can test the thermal history of the CMB.

Peebles, P.J.E. (1968) "Recombination of the primeval plasma", *Astrophysical Journal*, **153**, 1–11.

Penzias, A. and Wilson, R. (1965) "A Measurement of Excess Antenna Temperature at 4080 Mc/s", *Astrophysical Journal*, **142**, 419–421. The marvellously pedestrian title conceals the discovery announcement of the cosmic microwave background. The paper by Dicke, Peebles, Roll, and Wilkinson, presenting the cosmological interpretation, appeared on the immediately preceding pages 414–419, and marked perhaps the first appearance in print of a diagram for the entire thermal history of the Universe.

Silk, J. (1967) "Fluctuations in the cosmic fireball", *Nature*, **215**, 115–116. A first introduction to cosmic microwave background fluctuations and galaxy formation.

Spergel, D.N.; Verde, L.; Peiris, H.V.; Komatsu, E.; Nolta, M.R.; Bennett, C.L.; Halpern, M.; Hinshaw, G.; Jarosik, N.; Kogut, A. *et al.* (2003) "First-Year Wilkinson Microwave Anisotropy Probe (WMAP) Observations: Determination of Cosmological Parameters", *Astrophysical Journal Supplements*, **148**, 175–194. Presents the first year's WMAP data and their implications for cosmology. Year 3 data remain in press, with the paper on cosmological parameters accessible already in electronic form from *http://www.arxiv.org/abs/astro-ph/0603449*

Tytler, D.; O'Meara, J.M.; Suzuki, N.; and Lubin, D. (2000) "Review of Big Bang nucleosynthesis and primordial abundances", *Physica Scripta*, **T85**, 12–31.

Zel'dovich, Ya.B.; Kurt, V.G.; and Sunyaev, R.A. (1968) "Recombination of hydrogen in the hot model of the universe", *Soviet Physics-JETP*, **28**, 146 (in Russian, **55**, 278–286).

Internet

Information on some current projects to measure the fluctuations in the microwave background can be found at the following WWW sites:

Cosmic Background Imager *http://www.astro.caltech.edu/~tjp/CBI/*
Planck *http://astro.estec.esa.nl/SA-general/Projects/Planck/*

8

Active galactic nuclei in the early universe

Hints of cosmic reionization, and most of what we know about the intergalactic medium, come from QSO absorption lines. Being able to reach such conclusions requires the presence of QSOs as background sources for the absorbing material. Their high luminosity has long made quasars attractive probes of the distant Universe, not only through absorption of their light by foreground material, but as background sources for gravitational lensing. The role of active galactic nuclei in our picture of galaxy formation and evolution extends well beyond these indirect applications. These objects, especially the luminous quasars, directly pose profound issues about events in the early universe. Their existence at high redshifts indicates that some galaxies (perhaps the most concentrated or most massive) had formed by then, and the chemical makeup of their emitting gas shows that their immediate environments must have undergone rapid and intense star formation.

8.1 ACTIVE NUCLEI TODAY

There are several kinds of objects lumped together as "active galactic nuclei" or AGN, distinguished in various flavors by their optical spectra or radio-frequency properties. They all share the defining characteristic of having an intense radiation source arising in a very compact nucleus, one too powerful to be accounted for from the normal processes of star formation and evolution. The varieties are considered in the following subsections.

8.1.1 Seyfert nuclei

Seyfert galaxies were recognized as a class by Carl Seyfert in a 1943 paper. They are defined spectroscopically as central regions of galaxies whose optical spectra show emission lines which are *strong*, *broad* by the standards of galaxy kinematics (500–5000 km/s), and of *higher ionization* level than seen in star-forming regions. Seyfert nuclei were further divided into two types based on whether all kinds of emission lines have similar widths. In type 1 Seyferts, certain emission lines—so-called permitted lines, which can arise in gas at high densities—are much broader than the rest, the so-called forbidden lines which arise only in material with densities lower than a few thousand ions per cubic centimeter (Figure 8.1). In type 2 Seyferts, all these spectral features have comparable widths. Seyfert nuclei, especially those of type 1, appear as starlike cores against their surrounding galaxies. Seyfert nuclei occur mostly in spiral galaxies of early Hubble types (Figure 8.2), Sa–Sb, and S0 systems, occurring in about 5% of luminous spirals. Their defining optical spectra are quite similar to those of quasars and many radio galaxies.

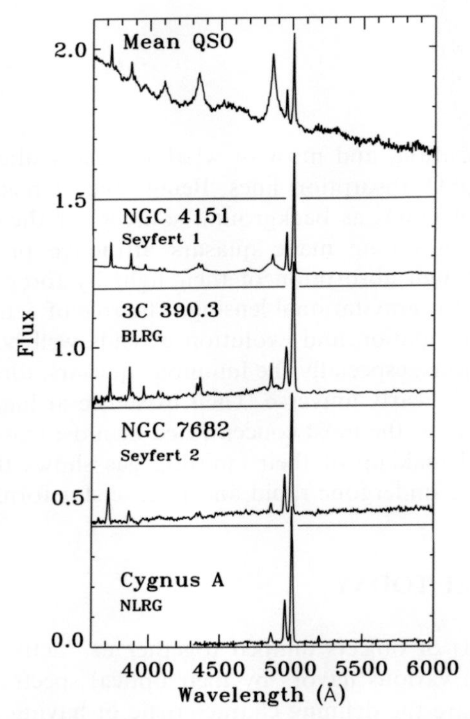

Figure 8.1. The optical spectra of several kinds of active galactic nuclei, all compared in the emitted frame. They have been offset vertically for clarity, with the zero level indicated for each. The prominent narrow emission lines near 5000 Å are the [O III] forbidden lines, with the Hβ recombination line just blueward. These lines furnish good examples of broad and narrow features—for example, between the narrow- and broad-line radio galaxies and the two kinds of Seyfert nuclei. The very blue nonstellar continuum light in QSOs and type 1 Seyferts is also in evidence. (Data from a compilation by Paul Francis *et al.* for the QSO spectrum, from Charles Lawrence for 3C 390.3, and by the author for the others.) The blue end of the Cygnus A spectrum was not observed.

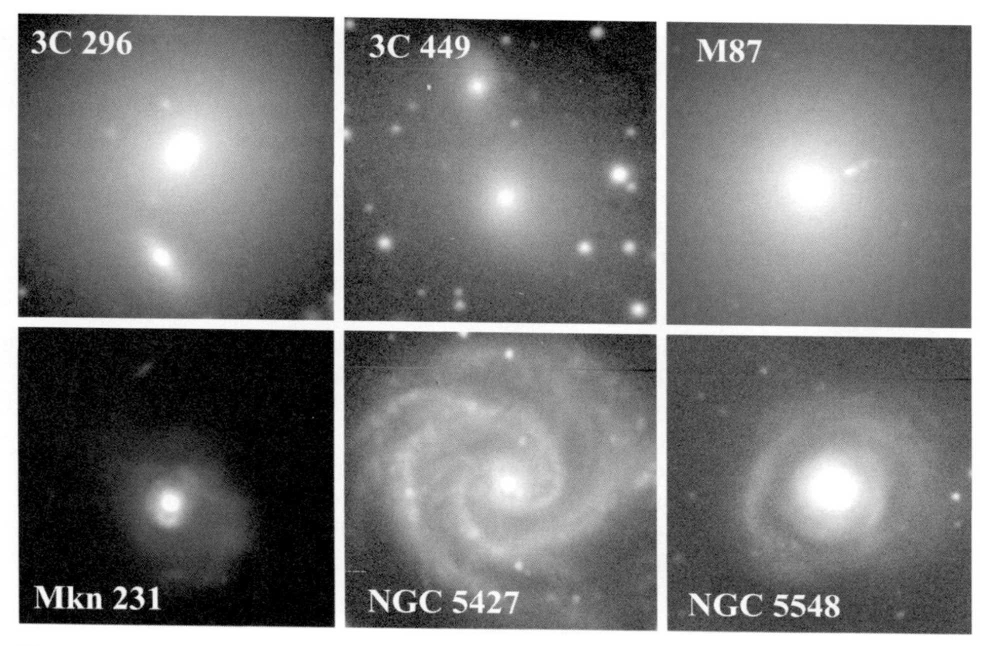

Figure 8.2. Overall views of radio galaxies and galaxies with Seyfert nuclei. The radio galaxies in the upper row show the typical elliptical morphology, often with interacting companions. M87 shows a well-known optical counterpart of the radio jet as well. The Seyferts, in the lower row, are generally disk systems, in some cases with strong tidal disturbance as in Mkn 231. (Images by the author, with the 2.1-meter telescope of Kitt Peak National Observatory.)

8.1.2 Radio galaxies

Radio galaxies were originally recognized from their strong radio emission. This often occurs in a pair of *radio lobes*, emitting components symmetrically located on either side of the galaxy. High-quality radio mapping eventually showed that these lobes are often connected to the galaxy's core by jets of emitting matter. Many radio galaxies show no unusual features in their optical spectra. Of those which do show emission lines (Figure 8.1), we find examples which have very broad permitted lines (broad-line radio galaxies) and in which all lines have similar, narrower widths (narrow-line radio galaxies). These spectra are very similar to those of type 1 and 2 Seyfert galaxies. The galaxies themselves are generally either elliptical galaxies or merger remnants (Figure 8.2), in contrast with the spiral hosts that are common for Seyfert nuclei.

8.1.3 Quasars

Quasars (originally QSRS, a contraction for quasistellar radio source) and quasi-stellar objects (QSOs) trace their discovery to identification of a handful of radio sources from early surveys, which the modest positional accuracy of those surveys could not associate with any peculiar or otherwise interesting optical counterpart. Improved techniques, such as timing the disappearance of radio sources when occulted by the Moon, and above all the rise of radio interferometry, led to the identification of some of these radio sources with starlike optical objects. Their optical spectra, enigmatic at first, proved to show familiar emission lines at unprecedented redshifts. These spectra (Figure 8.1) are in fact quite similar to those of Seyfert nuclei. As sky surveys proceeded, a more numerous population of similar objects without detectable radio emission was found; when distinction is important, quasars are radio-loud and QSOs may include both radio-loud and radio-quiet types. Major lessons from quasars were the high luminosity and small size possible for AGN. By definition, a quasar does not show a surrounding galaxy through a typical ground-based telescope (otherwise we would call it a Seyfert galaxy), and to be so bright at their redshifts some quasars must be extraordinarily luminous, beyond a thousand times the optical output of an L_* galaxy. This makes the brightest quasars so powerful that they can be found at high redshifts more easily than any other kind of object. The strong variability of quasars, at many wavelengths, was also important in establishing the small size of the emitting region, an important factor in trying to explain their enormous energy production.

8.1.4 BL Lacertae objects

BL Lacertae objects were originally catalogued, in several cases, as variable "stars" found in photographic surveys. Their optical spectra are nearly featureless (Figure 8.1), a defining characteristic which frustrated the initial attempts to figure out just where, let alone what, they are. Eventually, it was possible to measure the spectra of the surrounding galaxies, and of the weak residual spectral lines in a few BL Lac objects themselves, showing them to reside in the cores of galaxies and share common features with quasars. A related class of quasars exists, known as optically violently variable (OVV) QSOs. BL Lacertae objects are intense X-ray sources, and some of them are the highest-energy sources of extragalactic gamma rays known, accounting for numerous detections by the Compton Gamma-Ray Observatory and several measurements of radiation at TeV energies (10^{12} electron volts) using ground-based detection of Čerenkov radiation from atmospheric interactions.

8.2 ACCRETION IN AGN

Despite the different ways in which their activity is manifested, all these manifestations of activity have important common aspects. The central engine is uniformly a source of high-energy radiation, bright when seen in X-rays, and with a spectrum extending to much higher energies than any kind of star can manage. Variability shows that radiation arises in a tiny volume, the size of the Solar System or even less. We often observe evidence of a preferred axis, seen from the jets in radio galaxies and quasars, and from signs of radiation emerging in opposite pairs of cones in some Seyfert and radio galaxies. These considerations have led to the common picture of active nuclei as sites of accretion into massive black holes, from millions of solar masses upwards. Accretion at a high enough rate from material with a characteristic orbital plane, or into a black hole with large intrinsic angular momentum, will settle rapidly into a thin disk, and under a wide range of conditions magnetic effects can launch jets of material at relativistic speeds from near such a disk. The high temperatures of matter in such a disk, its small size, and the flywheel stability of a massive accretion disk, make this a popular picture for the innermost regions of an active nucleus. The gas we see in emission lines would come from around and outside the disk, illuminated by the X-rays and ultraviolet radiation produced closer in. It does remain a puzzle, however, that we have not been able to find definitive evidence of the accretion disks proper. We can see disk-like arrangements of material extending many light-years outward from some nearby AGN, but too large and cold to be the accretion disk proper. Even when we can measure X-ray emission lines from very hot material, we do not see the symmetric double-peaked profile modified by relativistic effects which are a clear prediction for radiation from an accretion disk. Either the accretion disk produces only continuum radiation, or we have yet to find its signature at all.

Several key observations have led to *unification schemes*, in which some kinds of AGN are in fact the same class of objects observed at different orientations. At least some narrow-lined (type 2) Seyfert nuclei are in fact broad-line type 1 objects seen behind dense clouds of obscuring material often considered to form a dense torus close to the core and perhaps aligned with the smaller accretion disk. Similarly, some kinds of radio galaxy may be quasars seen from a similar direction. A more extreme orientation works well to explain BL Lacertae objects—AGN seen almost exactly along the axis of a relativistic jet so that the jet's emission, dramatically amplified by relativistic effects, overwhelms the other features of the core. This same set of effects also compresses the object's timescale as we see it, accounting for the dramatic variability shown by these objects.

8.3 QUASAR EVOLUTION

It has been well established for nearly two decades that the quasar population evolves on a cosmological timescale. (I use quasar and the more formal QSO interchangeably here, since the strength of radio emission is not important for these arguments.) Statistically, quasar evolution is described by changes in their luminosity function with redshift. At the powerful luminosities which can be traced to high redshifts, the space density increases with redshift out to about $z = 2.2$, reaching a peak several orders of magnitude greater than the local density that we can measure in our own neighborhood. At still higher redshifts, the density drops rapidly (the so-called QSO cutoff). Radio galaxies do something similar up to a point, but the relative ease of getting redshifts for large QSO samples in comparison has made QSOs the easier tracer with much more statistically secure results.

Establishing how the luminosity distribution of QSOs has changed with redshift does not give a unique physical picture for just what is evolving. The luminosity function is close to a power-law form, lacking any characteristic reference feature such as L_* in the galaxy luminosity function, which engendered years of debate about whether the evolution was expressed in luminosity or space density. For such a featureless distribution, measures of the luminosity function itself could not distinguish changes that corresponded to vertical (space density) or horizontal (luminosity) displacements of the curve. Improved data sets, and ways of accounting for the inevitable selection biases at various redshifts, eventually showed changes in the form of the luminosity function which made it clear that the amount of density evolution depends on luminosity; or, more transparently, the shape of the luminosity function becomes flatter at earlier epochs for the most luminous objects. In more physical terms, the relative number of very luminous QSOs was larger at early times, above and beyond the entire population having a greater space density (in comoving coordinates, so that the space density as measured at that time would have been still greater by a factor $(1+z)^3$).

Evolution in the QSO luminosity function, while an adequate description of changes in the population with redshift, doesn't tell us the whole story. For AGN in the local Universe, there are hints that the duty cycle of powerful activity is much less than unity, as we might expect if special conditions are needed to feed the central engine and produce the radiating byproducts that make AGN so bright. At low luminosity, Seyfert galaxies comprise about 5% of the population of luminous galaxies. This might in principle mean that these 5% are special and are always Seyferts, or it might mean that all similar galaxies spend 5% of their time as Seyfert galaxies, with the central mass quiescent and invisible the rest of the time. There are indeed also physical reasons to suspect that quasars are "on" for only a fraction of cosmic time. There is a simple and powerful argument based on the luminosities of luminous QSOs, since we believe these to be fed by accretion of matter around a central black hole. Such accretion can convert up to half the rest mass of this matter into energy, though the efficiency could be much less depending on the details of accretion. The nearest really powerful quasar, 3C 273, has an overall energy output of about 10^{46} ergs/s. This would require one-third of a solar mass per year at the highest possible efficiency, meaning that keeping it so bright over a Hubble time would have required the consumption of over 5 billion solar masses' worth of stars and gas. We can still see the host galaxy around 3C 273, so that it hasn't been consumed (and this probably didn't happen for the even more powerful QSOs we can see at higher redshifts). More circumstantially, HST images of the galaxies around QSOs at moderate redshifts ($z = 0.3$) show a remarkable fraction interacting with very compact companion galaxies, a high enough fraction that these unusual kinds of interaction seem to have something to do with the occurrence of QSO activity. If we see a statistical connection between, say, powerful nuclear activity and some kind of galaxy interaction, the episode of activity cannot last much longer than the interaction does, or the correlation would be weakened. This suggests that, at least for this kind of trigger phenomenon, an episode of luminous QSO symptoms lasts for a time comparable with a galaxy interaction or merger, typically a few times 10^8 years.

Recently, Amy Barger *et al.* (2001) have used a census of X-ray sources in deep exposures taken by the Chandra X-ray Observatory to address this problem in a different way, tallying the total amount of X-rays from AGN at various flux levels. Since X-rays are an energetically important part of the total output of AGN, and insensitive to absorption by ordinary amounts of gas in galaxies, this should be a powerful and unbiased survey. The X-ray intensity can be linked to the accretion rate, using the kinds of efficiency arguments above. Taking the estimated present-day statistics of massive black holes in galaxies, they find that a typical AGN has spent a total of about 0.5 Gyr "on", a duty cycle of about 4%. While there may be wide variations from galaxy to galaxy about this characteristic value, and in how the 0.5-Gyr span was occupied in one or many episodes, this is interestingly close to the local demographics as to the fraction of galaxies that are "on" at a given time. The history of accretion into massive black holes may also be bounded from above, as an important contributor to the X-ray background. The integrated, redshifted X-ray emission from all AGN cannot exceed the mean X-ray intensity from "blank sky" at any energy. As imaging techniques allow resolution of ever-larger fractions of this background emission into individual sources, it has become clear that we must recognize an important contribution from objects that are heavily absorbed at low energies by surrounding gas, so that the mean background spectrum is "harder" than we would expect from nearby, well-observed AGN.

In this light, QSO evolution may have consisted of changes in the typical luminosity of an episode of activity, its duration, or both. The link between stellar and central masses (below) suggests that, if the growth of the two was linked, the central engine was being fed constantly if not necessarily continuously by accretion. The overall evolution of QSOs may then indicate that they were "on" a larger fraction of the time earlier in the Universe. In fact, when they were most numerous at epochs around $z = 2.2$, they accounted for a substantial fraction of all luminous galaxies.

8.4 THE MYSTERIOUS CENTRAL ENGINE

The most widely held picture of active galactic nuclei involves a massive black hole (10^6–10^9 solar masses) surrounded by an accretion structure, with jets, a torus of absorbing matter, and ionization cones as common byproducts. Direct evidence of the black hole, and in fact of accretion disks, has proven elusive. The picture rests largely on the fact that it seems more plausible, and requires fewer assumptions about new kinds of object or new physical laws, than do various alternatives. As Martin Rees pointed out in 1978, all the good candidates for the central engine either already *are* black holes or will *become* black holes in short order. There are three major driving factors for this interpretation:

(1) The radiation in QSOs comes from very small volumes, as shown most famously by their variability. Unless relativistic beaming plays a role (which it can't for the whole population of QSOs), no source variation can be observed on a timescale faster than the light-travel time across the source. QSOs commonly vary in the

optical and UV ranges over timescales from weeks to months, and X-ray variations may be faster. One category of Seyfert galaxies (known as narrow-line Seyfert 1 nuclei) shows rapid X-ray flickering on timescales of a few hours, and the power spectrum of X-ray variability has a detectable signal down to times of order 100 seconds in some active nuclei. Detailed study of how emission lines respond to changes in the central ionizing flux (reverberation mapping) shows that the emission lines come from regions several light-days across (with the lowest-ionization lines extending out to a light-year). For comparison, the Schwarzschild radius for a black hole of 10^8 solar masses is nearly 1000 light-seconds.

(2) The emission lines show very large Doppler widths, so the potential well must be very deep to keep this gas bound to the nucleus. Characteristic velocities of 5000 km/s are common, and 15,000 km/s are known (especially among broad-line radio galaxies). A fair estimate of the central mass comes from a simple Keplerian calculation based on the radius of the broad-line region and this characteristic velocity. A characteristic radius of 5 light-days (typical from reverberation studies) and transverse velocity of 5000 km/s, using Kepler's third law, give a binding mass Rv^2/G of 2.5×10^7 solar masses. While this number must be uncertain by a factor of a few depending on how the gaseous emission is distributed and what kinds of orbits the individual emitting clouds follow, it already indicates that no combination of stars or other known objects can account for so much mass within this small volume without producing other observable signatures that we don't see. Indeed, collections of stars dense enough to satisfy this mass density are dynamically unstable to gravitational collapse, on short timescales.

(3) Accretion into a deep gravitational well is the most conservative available energy source, in the sense of requiring the least matter to fuel it. Direct matter–antimatter annihilation is more efficient (at 100%), but AGN do not show the characteristic annihilation spectral features (such as the 511-keV spectral line from mutual annihilation of electrons and positrons), quite aside from the enormous problems in how antimatter could be produced in large amounts to begin with. Accretion onto relativistic compact objects can release up to $mc^2/2$ in energy, and efficiencies of $0.1mc^2$ do not seem unlikely in typical circumstances around black holes. Accretion power seems the only known way to power a QSO for a significant time without consuming more than a galaxy's worth of material in the process.

8.5 THE G IN AGN

We can examine the galaxies surrounding nearby active nuclei, particularly Seyfert and radio galaxies, in some detail. These forms of activity show a marked preference for systems with substantial bulge components—early-type spirals for Seyferts and elliptical galaxies or their relatives for most radio galaxies. The situation for quasars has been observationally much more difficult, of course, since one initial defining characteristic of a quasar was that no surrounding galaxy light showed up on typical photographs. Eventually, ground-based imaging did show galaxy-scale "fuzz" around many QSOs at moderate redshifts, and a notable success was the measurement of stellar absorption features in the spectra of the extended light (e.g., by Boroson and Oke in 1982, an approach extended by several groups since then). It had long been expected that the Hubble Space Telescope would enable great progress on the problem of host galaxies of QSOs, and after the refurbishment of its instruments during the STS-61 mission of the space shuttle *Endeavour* this promise was indeed realized.

Images of a number of moderate-redshift QSOs ($z < 0.5$) were obtained using a variety of filter and detector configurations. Even with the vastly improved point-spread function, the limit in ability to extract information on the structure of the surrounding galaxies comes down to how accurately we can subtract the brilliant core. Small time-dependent changes in the point-spread function, as measured using field stars, and in the structure of the diffraction spikes, limit how close to the core the galaxy's profile can be measured. Even so, a remarkable variety of galaxy forms emerged, not always following the simple expectation based on Seyfert and radio galaxies. It had been suspected for some time that QSOs would prove to be largely more powerful versions of these other AGN varieties, with radio-quiet QSOs like scaled-up Seyfert galaxies in spiral hosts and radio-loud ones, more closely related to radio galaxies, in elliptical host systems.

Spiral and elliptical hosts are clear, as are some merging systems (Figure 8.3). There is an excess of companion galaxies compared with a random sample, which constitutes perhaps the clearest piece of evidence linking galaxy encounters to the triggering of nuclear activity. However, the expected dichotomy in types of host galaxy did not appear. Some radio-quiet QSOs are in elliptical galaxies, and some radio-loud ones are in merging systems which may have once had disks, although no radio-loud QSO has shown up in a normal spiral galaxy. For understanding the early history of galaxies, these results may indicate that certain kinds of young galaxies could be distinguished by examining the relative fractions of radio-quiet and radio-loud QSOs with redshift. The number of radio-loud QSOs may drop at high redshifts, which might suggest that the dense environments of ellipticals take longer to form. This result would be at odds with other evidence that ellipticals formed at least their dominant stellar populations very early on.

It would be very desirable to survey the host galaxies of QSOs to higher redshifts, to tell what kinds of galaxies hosted them when they were most numerous (at which time the number density suggests that most luminous galaxies hosted a QSO), and, at the highest redshifts, what the structure of these earliest dense mass concentrations was like. However, observational limitations make such observations progressively more difficult. The redshift moves the optical and near-IR bands, where older stellar populations are brightest, progressively farther into the infrared, where ground-based observations are severely restricted by atmospheric foreground radiation and space-based observations are not yet possible. Longer-wavelength observations also suffer from limited spatial resolution, simply because any telescope's diffraction limit is larger on going to longer wavelengths. On top of this, the cosmological $(1+z)^4$

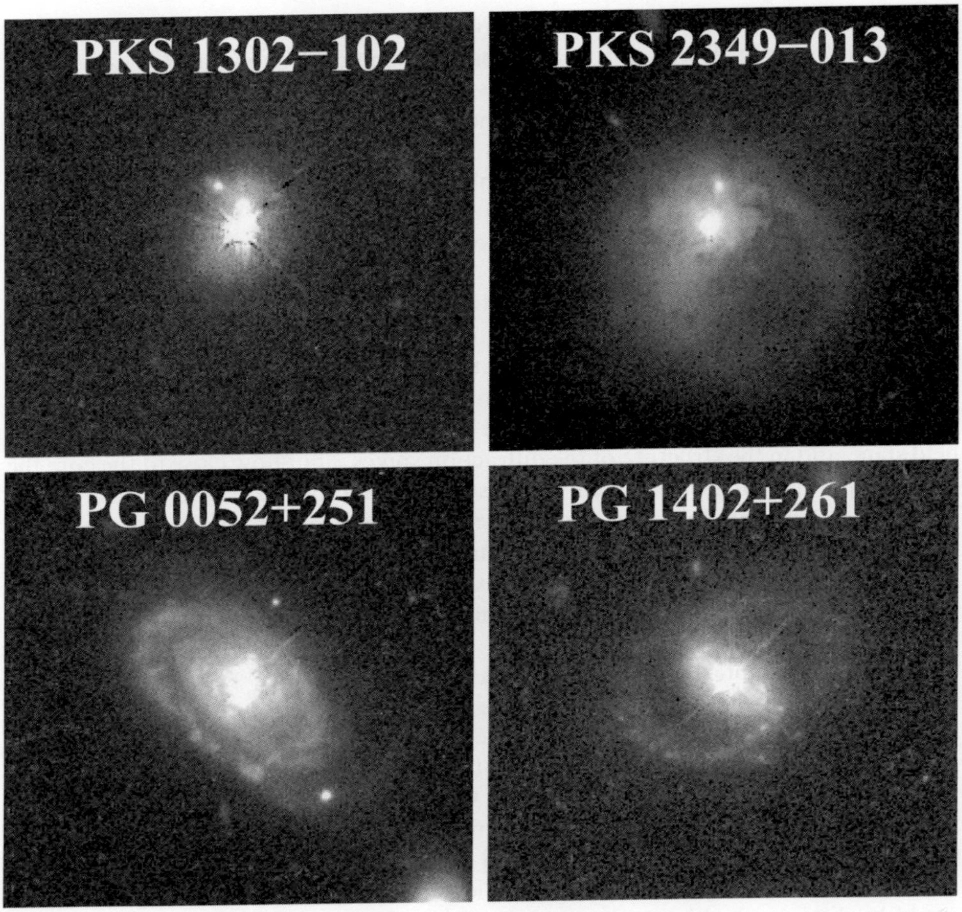

PKS 1302−102 PKS 2349−013 PG 0052+251 PG 1402+261

Figure 8.3. Quasar host galaxies. These Hubble Space Telescope images show the galaxies around quasars at redshifts $z = 0.15–0.28$. The images are 22 arcseconds across, or 75–140 kiloparsecs for these objects. The upper two objects are radio-loud, and the lower ones are radio-quiet. Elliptical, spiral, and strongly interacting host galaxies can be seen. (Data retrieved from the NASA Hubble Space Telescope archive, from observations with J. Bahcall and F. Macchetto as principal investigators.)

surface-brightness dimming means that almost no ordinary galaxy would be detectable outside the glow of the bright core of a $z > 4$ QSO. Any galaxies seen around high-redshift QSOs will have to be very unusual in surface brightness. There have been some tantalizing hints from ground-based data that a few QSOs at $z > 2$ have resolved galaxies around them, and the implied connection (below) between the onset of QSOs in the Universe and the growth of galaxy bulges gives some hope that high-redshift quasars are sometimes embedded in regions of extraordinarily intense star formation.

As noted in Chapter 5, the X-ray structure of the intracluster medium around some AGN suggests an important role for feedback, in which mechanical energy from the central galaxy slows or shuts down net inflow from surrounding gas which would otherwise cool and flow inwards. Similar feedback is likely to be important within the host galaxies of AGN as well, although here again most observational approaches can address its importance only indirectly. This mechanism has been proposed in particular to explain the cutoff in the distribution of baryonic mass of galaxies, roughly corresponding to the exponential cutoff in the Schechter function in luminosity. Simulations of structure formation dominated by cold dark matter generally predict a continuous spectrum of galaxy masses to larger values than we see; having episodes of AGN activity, especially early in galaxy history, could explain why there is a characteristic value for galaxy baryonic mass above which a power-law distribution no longer applies. In particular, models with even a heuristic allowance for feedback do better at matching the different luminosity (and implied stellar mass) of the most luminous blue versus red galaxies.

A remarkable recent finding relating the histories of AGN and their host galaxies came from the statistics of narrow-line AGN, predominantly type 2 Seyferts, from the Sloan Digital Sky Survey; these are the only species of AGN which could be identified reliably from the SDSS spectra at a wide range of signal-to-noise ratios. The distinct red and blue color sequences for the overall galaxy population allow most galaxies to undergo only a single major shutdown of star formation (Chapter 4). The fraction of galaxies showing a (type 2) AGN peaks strongly in the "valley" between the two color sequences. This may hint that fueling of the central engine is uniquely efficient during this one-time period in galaxy history.

8.6 QSO CHEMISTRY AND THE GROWTH OF GALACTIC BULGES

The strong emission lines characteristic of quasars arise from a variety of ions, with prominent lines from H, He, C, O, Si, and blends of Fe seen in various spectral ranges. These lines permit at least schematic estimates of the chemical abundances in the gas around the central engine, limited by how much density and ionization structure is present in the region giving off these lines (for high-redshift objects, most of what we observe comes from the broad-line region, containing clouds or filaments of dense gas in a region some light-days in size). High-redshift QSOs are spectroscopically very similar to local examples (Figure 8.4), with no hint of any lower metallicity that we might associate with their being in less evolved galaxies. In fact, some prominent metallic emission lines are actually stronger in QSOs at $z > 4$ than in those at low redshift, indicating that some abundance ratios were actually more weighted toward heavy elements (for once, literally metals) than we see today.

The abundance patterns among heavy elements are important: as well as showing generally supersolar metallicities, iron is enhanced relative to the so-called

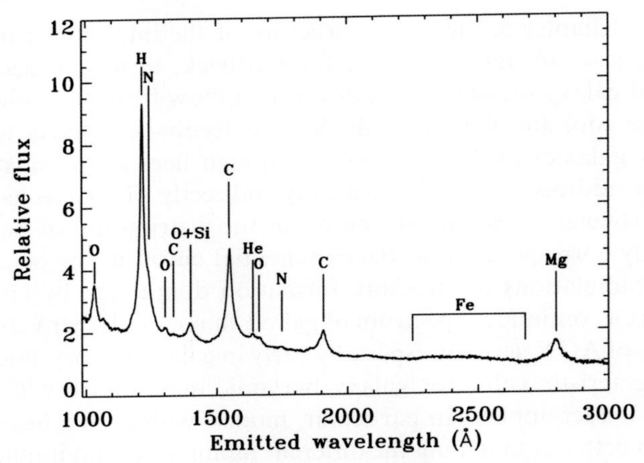

Figure 8.4. Ultraviolet quasar spectrum, showing emission features useful for tracking the metal abundance of surrounding gas. Ionization models show that the O and N features are particularly sensitive to the chemical composition. Iron is important since it arises from different stars than nitrogen and carbon. This spectrum represents a composite from many objects, averaging over intervening absorption systems; objects with narrower emission lines have the nitrogen line cleanly separated from the hydrogen Lyman α feature. (These data were provided by Wei Zheng, based on Hubble Space Telescope observations.)

α-elements[1] as compared with solar abundances. For an evolving starburst population, Fe production comes mostly from type Ia supernovae, which start to appear about 1 Gyr after the onset of star formation. Thus, the abundance patterns in high-redshift QSOs seem to favor element production in regions which have been forming stars for a time of order 1 Gyr before the epoch we observe, and in potential wells deep enough to capture all the ejecta from supernovae. This is a powerful constraint on when significant star formation began and on how quickly galaxies became dynamically isolated, strongly condensed entities. No QSOs have been seen in low-metallicity environments, even though we should still be able to detect them by their Lyman emission lines however low the metallicity becomes. Thus, QSOs appear uniquely in those environments where star formation in deep potential wells had been active for about 1 Gyr (the full span for various models is 0.3–3 Gyr) before the epoch observed at the highest QSO redshift (currently $z = 6.3$). For the current consensus cosmology, we can't take the 1-Gyr span too literally, since there had yet to be a gigayear by $z = 6.3$. Even the minimum span usually allotted for any SN Ia to appear, 300 Myr, leaves only 0.5 Gyr between recombination and the onset of star formation. These must be, not surprisingly, the (proto)galaxies least likely to allow outflows, since SN ejecta must have been bound with high efficiency, by deep gravitational potentials, to get the observed metallicities.

This correlation between quasars and high metallicity, when other high-redshift galaxies show lower metallicities than we would see today at their luminosities,

[1] The ones built up in red giants starting from a helium core (O, Mg, etc.).

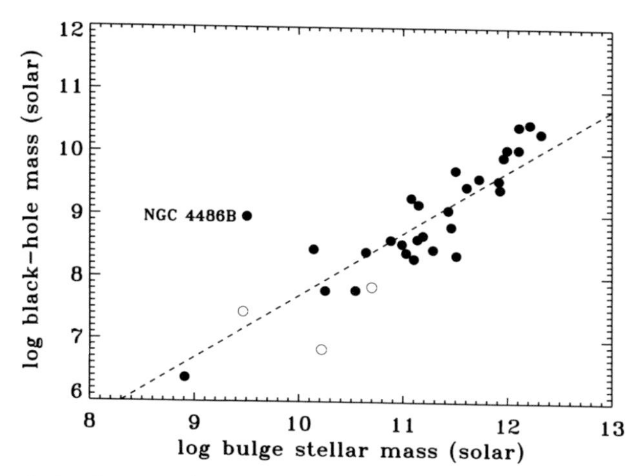

Figure 8.5. The relation between the masses of black holes in galactic nuclei and the stellar mass in the surrounding bulge, derived from kinematic measures with the Hubble Space Telescope. Both are plotted logarithmically, with solar masses as the implied unit. The dashed line marks a constant fraction 0.5% of the bulge mass, which is a fair representation over three orders of magnitude here. The outlier is NGC 4486B, a companion to M87 in the Virgo Cluster which shows evidence that many of its stars have been tidally stripped during encounters with M87 itself. The three open symbols represent upper limits in black-hole mass. (The data were taken from Magorrian *et al.*, *Astronomical Journal*, **115**, 2285, 1998.)

suggests something special about their surroundings: that there is a close link between early star formation in deep potential wells and the occurrence of a powerful AGN. It has been possible to use dynamical evidence to measure the masses of the central unseen objects in a variety of galaxies, especially using the Space Telescope Imaging Spectrograph (STIS) on the Hubble Space Telescope. The demographics of these central masses in galaxies today suggests that the link between these masses and their surroundings extends quite generally to the formation of galactic spheroids. For a wide range of galaxies, active and inactive, there is a good correlation (Figure 8.5); often called the Magorrian relation between the stellar mass in the spheroidal bulge component and the mass of the central object (presumably a supermassive black hole), such that the central object has about 1/200 the mass of the stars. This is most naturally explained if growth of the central object is a byproduct of evolution in the spheroid, since if it came first there is no obvious way to regulate the amount of surrounding material condensed into stars. In contrast, the growth of a central point mass in a stellar system can be self-regulating, and will saturate in a non-interacting system once all stars on "death orbits" have been disrupted. Scattering of additional stars into the region of phase space that is vulnerable to such deep encounters is very slow, so the central black hole will stop rapid growth once its mass becomes dynamically significant in the inner regions.

It is important that this mass relation holds for such a wide variety of present-day galaxies, spiral bulges as well as ellipticals from M32 to M87. This makes it difficult to avoid the conclusion that all these galaxies went through at least one AGN phase as their spheroidal components were formed, with the central black hole growing through accretion of gas (possibly itself liberated by tidal disruption of stars). It is very attractive to link this growth with the evolution of the QSO luminosity function, which would then trace in some sense the formation history of bulges. This complements the star-formation tracers which are sensitive to all galactic environments.

8.7 ACTIVE NUCLEI AS DRIVERS AND QUENCHERS OF STAR FORMATION?

The environments of powerful AGN are busy places. In addition to a powerful radiation field and gas clouds in rapid motion, there must often be outflows spanning large solid angles, to account for the extended emission-line regions and absorption-line structures seen in some active nuclei. This opens the possibility that AGN can trigger star formation in the surrounding galaxies, by compression of gas clouds on impact of the outflowing gas or at the interface with the rapidly emerging jets. One

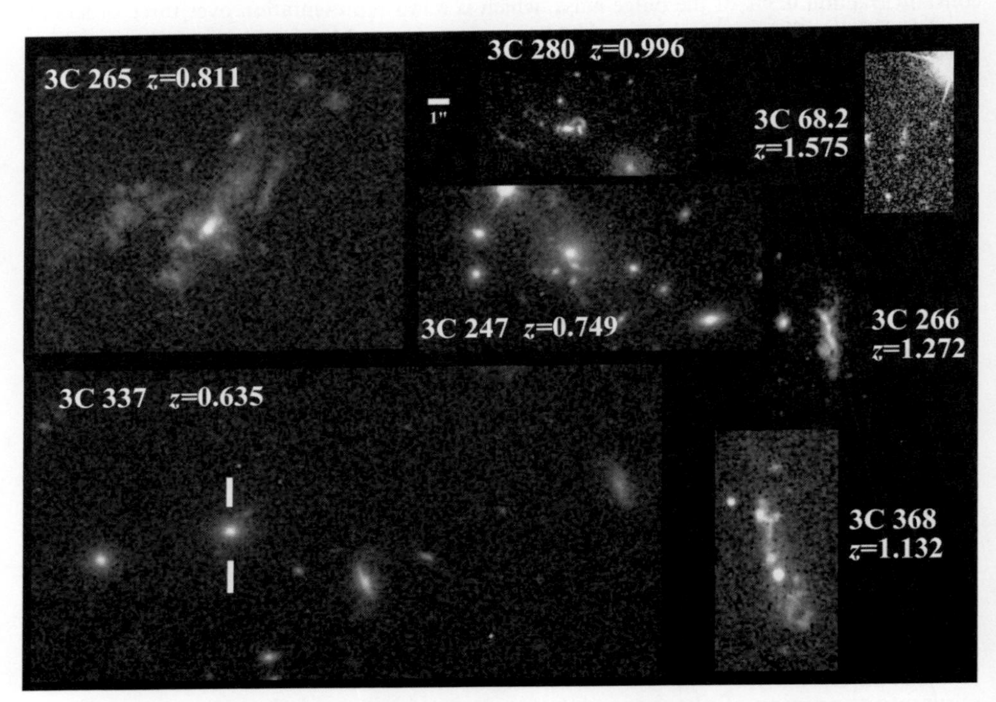

Figure 8.6. This montage of Hubble images shows radio galaxies at a range of redshift, all to the same angular scale and all in the emitted near-ultraviolet. Low-redshift objects are normal ellipticals, while high-redshift examples are less symmetric and preferentially aligned with the radio source axes. (Data from the NASA/ESA Hubble Space Telescope archive, with observations originally obtained by P. Best and M. Longair.)

might suspect this process in the early Universe, so that AGN would help start galaxy growth if not formation, and it has been proposed to explain the *alignment effect* seen in radio galaxies at high redshift.

In the local Universe we do not see any particular alignment between the direction of radio jets emerging from active galaxies and the apparent shape outlined by the galaxies' starlight. This changes as we look to high redshifts (Figure 8.6), so that for $z > 0.4$ the radio and optical axes are more often nearly parallel (hence "alignment effect"). At the highest redshifts, radio galaxies can be very elongated and lumpy as seen in the continuum, again well aligned with the radio source. This effect seems to mix several causes. One is purely observational. At high redshifts, we are often observing at shorter emitted wavelengths, into the ultraviolet, where recent star formation and scattering by dust grains are both more important than in visible light. Indeed, local radio galaxies can appear more "aligned" when seen in the ultraviolet. Still, we find that the ultraviolet light from radio galaxies is better aligned at high redshifts, even when these biases are taken into account. Possible reasons might be that the radio source induces star formation as it emerges and continues to emanate

Figure 8.7. Hubble image of Minkowski's Object (*left*), in which star formation is apparently being triggered by interaction with a jet from the radio galaxy NGC 541 (*left*). Numerous bright star clusters appear in the disrupted gas-rich system; the radio jet is deflected and loses collimation at this same point. (Data from the NASA/ESA Hubble Space Telescope archive, originally obtained by Stefi Baum.)

from the inner regions, that the radio source is most easily produced along a particular axis of a nonspherical young galaxy, or that we're not seeing starlight but light from the active nucleus, scattered from surrounding dust and seen where the radiation emerges most easily along the axis of escaping radio plasma. Triggered star formation remains a possibility in many instances of this effect.

While there has been considerable speculation on the role of AGN in triggering starbirth, there is less in the way of systematic data. There are a few local examples in which the process seems to be ongoing as we watch. In the nearest moderately powerful radio galaxy, Centaurus A or NGC 5128, we see supergiant stars and ionized gas tracing one side of the radio jet as it leaves the dense part of the galaxy, things that are not seen elsewhere so far from the center (even given that there is plenty of gas and star formation in the prominent dust lane). The association certainly suggests that the radio jet has something to do with these short-lived stars. On a grander scale, the radio source in NGC 541 (a radio galaxy in the cluster Abell 194) points directly to a bright blue galaxy known as Minkowski's Object. This object, oddly for a cluster member, shows a high rate of star formation and irregular morphology (Figure 8.7). Furthermore, the radio source seems to know about Minkowski's Object, being deflected and broadened on passing it. These observations have led several groups to suggest that star formation in this case was brought on as the radio jet reached the interstellar medium of a dwarf galaxy, compressing much of its gas into gravitational collapse.

8.8 BIBLIOGRAPHY

Books

Osterbrock, Donald E. and Ferland, Gary J. (2006) *Astrophysics of Gaseous Nebulae and Active Galactic Nuclei*, 2nd ed. (University Science Books). The classic reference for the astrophysics of emission-line plasmas and their interpretation, in a revised and extended edition.

Peterson, B.M. (1997) *An Introduction to Active Galactic Nuclei* (Cambridge University Press). A broad view of the physics of quasars and other active galactic nuclei.

Journals

Barger, A.; Cowie, L.; Bautz, M.; Brandt, W.; Garmire, G.; Hornschemeier, A.; Ivison, R.; and Owen, F. (2001) "Supermassive Black Hole Accretion History Inferred from a Large Sample of *Chandra* Hard X-Ray Sources", *Astronomical Journal*, **122**, 2177–2189. Uses X-ray fluxes and source counts to address the overall evolution of AGN and the growth rate of central black holes.

Boroson, T.A. and Oke, J.B. (1982) "Detection of the underlying galaxy in the QSO 3C48", *Nature*, **296**, 397–399. The first spectroscopic detection of starlight from the host galaxy of a classical quasar, made possible by then-new instrumentation and CCD detectors at Palomar.

Fassett, C.I. and Graham, J.A. (2000) "Age, Evolution, and Dispersion of the Loose Groups of Blue Stars in the Northeast Radio Lobe of Centaurus A", *Astrophysical Journal*, **538**,

594–607. Star formation near the radio jet in the nearest radio galaxy, so close that we can count the luminous blue stars. All the other star formation in this system is associated with the (accreted?) gas disk and dust lane, which implicates the radio plasma in triggering the events seen along the jet to the northeast. This has been proposed as a laboratory for examining processes at work, with much greater strength, in high-redshift galaxies.

Hamann, F. and Ferland, G. (1999) "Elemental Abundances in Quasistellar Objects: Star Formation and Galactic Nuclear Evolution at High Redshifts", *Annual Review of Astronomy and Astrophysics*, **37**, 487–531. The interpretation of chemical abundances in the emission-line regions of quasars at high redshift, and their implications for the early star-forming history of the surrounding galaxies.

Heckman, T.M. and Kauffmann, G. (2006) "The host galaxies of AGN in the Sloan Digital Sky Survey", *New Astronomy Reviews*, **50**, 677–684. An initial description of the relation between the red and blue galaxies sequences and the occurrence of AGN.

Magorrian, J.; Tremaine, S.; Richstone, D.; Bender, R.; Bower, G.; Dressler, A.; Faber, S.M.; Gebhardt, K.; Green, R.; Grillmair, C. *et al.* (1998) "The Demography of Massive Dark Objects in Galaxy Centers", *Astronomical Journal*, **115**, 2285–2305. This paper collected evidence that central unseen objects typically have 0.6% of the mass of the stars in the surrounding bulge, thereby promulgating the "Magorrian relation" which proved important to our view of how galaxies and AGN have evolved together.

Merritt, D. and Quinlan, G.D. (1998) "Dynamical Evolution of Elliptical Galaxies with Central Singularities", *Astrophysical Journal*, **498**, 625–639. Calculations of the simultaneous evolution of stellar orbits in a galactic growth and a central supermassive black hole. These results are broadly in accord with the observed relation between masses of black holes and surrounding bulges.

Rees, Martin (1978) "Accretion and the quasar phenomenon", *Physica Scripta*, **17**, 193–200. A broad overview of the basic issues in understanding active galactic nuclei, particularly the arguments that accretion onto massive black holes is the most likely energy source.

Seyfert, C.K. (1943) "Nuclear Emission in Spiral Nebulae", *Astrophysical Journal*, **97**, 28–40. The defining paper on the class which came to be known as Seyfert galaxies.

van Breugel, W.; Filippenko, A.V.; Heckman, T.M.; and Miley, G. (1985) "Minkowski's object—A starburst triggered by a radio jet", *Astrophysical Journal*, **293**, 83–93. Observations of a strong candidate for a radio-galaxy jet inducing star formation, in this instance a gas-rich dwarf companion to the radio galaxy PKS 0123−016A (NGC 541).

9

Approaching the Dark Ages

We can observe an early phase in cosmic history through the microwave background. Our direct view next picks up with galaxies and quasars at redshifts now exceeding $z = 6$. The timespan in between, encompassing several hundred million years, and whose exact length depends on the cosmological parameters, has been termed the *Dark Ages*. They are "dark" in two rather different ways. Observationally, we can't yet see what happens in this interesting and busy period, because for part of it there were no luminous objects to see, and for the rest of it we haven't yet found the strongly redshifted signatures that the first discrete objects would have. From a theoretical viewpoint, a shorter period would have been truly "dark" because there were as yet no sources of radiation except the almost featureless, slowly cooling gas—no stars to shine by fusion and enrich surrounding gas, no quasars to release gravitational energy through accretion. These energy sources have profound effects on the gas and its behavior, so that the "first light" sources set the stage for what was to come. Much of the work in understanding galaxy formation is thus involved with the Dark Ages, either in trying to trim its boundaries observationally or understand its history theoretically.

One of the most gratifying advances in astronomy during the last 15 years has been the ability to observe galaxies at high redshifts—beyond $z = 3$. Starting with a few radio galaxies, new techniques and instruments now deliver wholesale samples of galaxies at $z > 3$, and heroic efforts combining HST images and the largest ground-based optical telescopes can deliver a few examples beyond $z = 6$, sneaking toward the Dark Ages. This is a far cry from the situation up to the mid-1980s, when the only objects observable beyond $z = 1$ were a relative handful of luminous quasars. We can now probe cosmic evolution with objects selected in several distinct ways, allowing us to piece together the outlines of galaxy evolution and point to features dating to their formation.

9.1 QUASARS

The first probes to high redshift were active galaxies of various kinds. Quasars were known to beyond $z = 2$ by 1966, only three years after these objects were initially identified. By 1973, QSOs at $z = 3.5$ had been found, rising gradually thereafter to temporary records of $z = 3.80$ in 1986 and jumping to $z = 4.7$ in 1989, as observing strategies were honed to find higher-redshift QSOs efficiently. A change in approach was needed because, at high redshifts, quasars no longer show the blue colors which are characteristic at low redshifts, but may appear quite red as the strong Lyman α emission appears at longer wavelengths, and the Lyman α forest absorbs more of the light blueward of this line. Recent projects, including the Sloan Digital Sky Survey, have exploited this characteristic to yield large, complete samples of QSOs extending to redshifts (currently) as high as $z = 6.4$. There were a number of studies assessing the use of QSOs as cosmological tracers, making use of possible correlations between luminosity and spectral properties, since these were then the only objects we could observe at such large redshifts. However, our very imperfect knowledge of how QSOs are connected to the evolution of galaxies meant that they could tell us very little about how galaxies in general might be forming and evolving with redshift, beyond the fact that as active galactic nuclei they probably did not predate galaxies completely.

9.2 RADIO GALAXIES

The first set of galaxies observable as such to large redshifts came from radio galaxies. Some of these are spectacularly powerful objects, as shown by the fact that the venerable 3C catalog, containing the few hundred brightest radio sources in the northern sky, includes radio galaxies as distant as $z = 2.47$. Much deeper radio surveys than this were possible by the 1970s, and it was apparent that these must include galaxies at substantially higher redshifts than the shallow 3C list. Given that high-redshift galaxies are optically quite faint, especially for that time before CCDs and 10-meter telescopes, and that most faint galaxies are not at such high redshifts, there was considerable gain from using radio emission to tell which faint galaxies might be intrinsically luminous and distant, since powerful radio sources occur preferentially in optically luminous galaxies. Thus, great effort went into the identification of optical counterparts of radio sources from these deeper surveys, and spectroscopy of likely candidates. This was initially very tedious work, entailing stacking multiple nights of observation of blank patches of sky. It was speeded up when CCDs were introduced to astronomy, and further with the proliferation of very large telescopes. Still, many galaxies turned out to have the (by then) rather ordinary redshifts in the range $z = 0.5$–1. The next substantial improvement in the high-redshift population came with the realization, pushed by George Miley and collaborators, that the highest-redshift radio galaxies had unusually steep radio spectra (i.e., if the flux varies with frequency ν as $F_\nu \propto \nu^{-\alpha}$, unusually large indices α). This criterion was used to find hundreds of radio galaxies at $z > 2$, with examples at least to $z = 5.2$. These samples gave our first clear view of *some* kind of galaxy at early epochs.

Just what we see in these powerful radio galaxies has been less clear. Such faint objects may have a measured redshift only if they have extremely strong emission lines, and these are generally of such high ionization as to suggest that the gas is ionized either by the active nucleus of by rapid shocks, and therefore don't tell us much about the properties of the galaxy as a stellar system. In a few of these distant radio galaxies, spectral absorption features from hot stars have been measured, so we can be sure that we're seeing starlight. In this sense these objects let us start tracing the history of galaxies. But which galaxies? The alignment effect is strong for even low-power radio galaxies at these redshifts, so some of the structure we see may be due to the effect of the AGN, which may or may not tell us about the history of typical galaxies. Furthermore, in examining such extreme and powerful objects we are faced with a degeneracy between rarity and cosmic evolution. If we are looking at the most powerful 10 radio sources in the observable Universe, then statistics tells us that the nearest will most likely lie at a large redshift and be seen at a significant lookback time, so we would infer that they existed only in the early Universe, *regardless of whether this population evolves or not*. This has been some of the rationale for surveys which have sampled successively fainter radio sources, to trace truly comparable samples of galaxies across redshift and time.

These weaker radio sources are less afflicted with the "diseases" that make it hard to know how to interpret the most powerful ones. They give results for the characteristic star-forming history of their host galaxies (usually ellipticals) that match what we see from purely optical surveys, and still allow reasonably efficient selection of high-redshift objects. Current radio surveys are sensitive enough that, in a sense, we have reached a natural limit. We could now catalog *all* modestly powerful radio sources to beyond $z = 10$ (i.e., as long as we have much reason to believe they have existed) within a field many arcminutes across, with a few hours of observation. Going even deeper doesn't show us more distant sources, it reaches deeper into their luminosity function at modest redshifts—to what frequent radio surveyor Jim Condon has called "the dregs of the Universe". Thus, the faintest radio sources in these surveys are often spiral galaxies, especially interacting systems, at redshifts typically $z = 0.5$. At these levels, as in our own galaxy, the radio emission reflects star formation (particularly through emission from supernova remnants—for massive stars, the death rate is as informative as the birth rate) rather than a central active nucleus.

Even with these caveats, radio galaxies are clearly useful as tracers of galaxy overdensities, leading to some of the first-known groupings of objects at high redshift. A particularly interesting example, which may be hierarchical galaxy formation seen in action, is provided by MRC 1138−262, the "Spiderweb" radio galaxy (Figure 9.1), which is surrounded by low-mass star-forming clumps which are likely to merge with the central galaxy on timescales of a Gyr or less. Rich collections of Lyman α emitters are found around radio galaxies to $z = 4.1$.

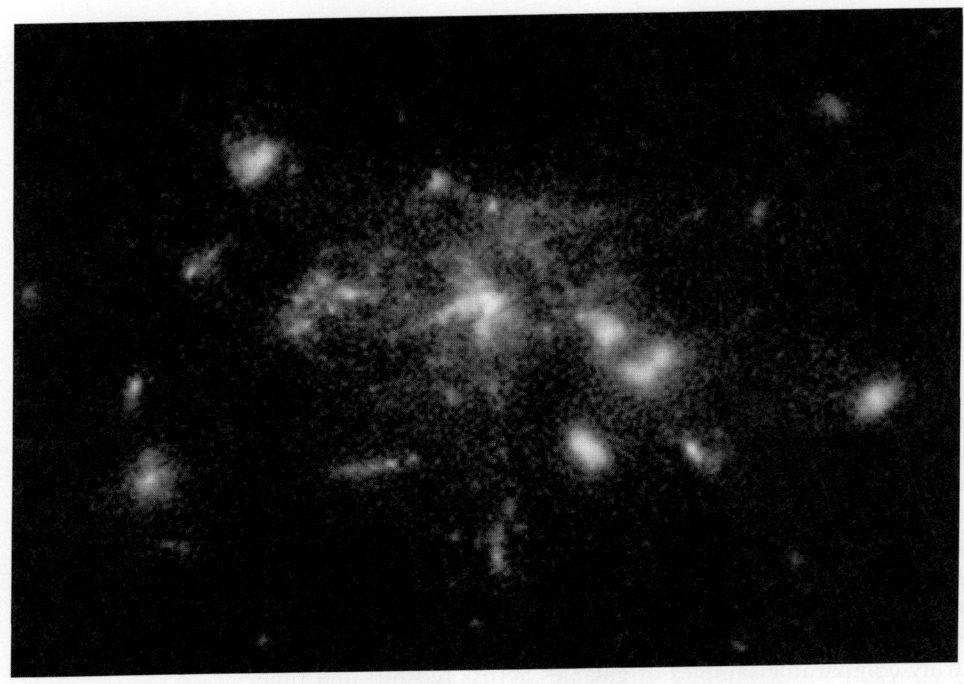

Figure 9.1. The radio galaxy MRC 1138−262, named the "Spiderweb" for its network of surrounding star-forming clumps. This system looks as one might expect from predictions of hierarchical galaxy formation, with multiple low-mass clumps joining into a more massive entity. (NASA, ESA, G. Miley and R. Overzier of Leiden Observatory, and the ACS Science Team.)

9.3 THE OVERALL GALAXY POPULATION

A safer way to probe galaxy evolution, without the potential biases introduced through using systems with powerful active nuclei, is to use the whole galaxy population. An obvious approach, albeit one very expensive in telescope time, is to spectroscopically survey large, well-defined samples of faint galaxies, preferably selected at wavelengths which minimize the role of ongoing star formation. Until recently, with large telescopes and spectrographs which can target more than 100 galaxies at once, this was such a slow endeavor that various kinds of selection still had to be applied to enrich the sample in higher-redshift systems. For example, one could target the reddest galaxies at a particular magnitude, since these will mostly consist of luminous elliptical galaxies. Donald Hamilton used this approach in the early 1980s to limit the amount of recent star formation in giant elliptical galaxies, since he found no change in the strength of the age-sensitive spectral break at 4000 Å out to redshift $z = 0.8$. In retrospect, a few objects in that sample at $z = 0.9$ would have shown evidence for evolution, as later studies involving both optically- and radio-selected galaxies showed.

These spectra of faint galaxies yield not only redshifts, but spectroscopic parameters which can be related to the age mix of the constituent stars and the amount of internal reddening (both of which are correlated with galaxy morphology in the local Universe). The sizes of samples available has risen dramatically with several kinds of technological development. The real breakthrough was multiobject spectroscopy—the ability to simultaneously measure spectra of objects scattered throughout a telescope's field of view. The most important implementation for faint objects has been multislit assemblies in the focal plane, where instead of the traditional single spectrograph slit, one incorporates a custom plate with a precision-machined set of slitlets whose locations will exactly match target galaxies when the telescope is properly pointed. The first such setups could observe perhaps 12–15 objects at once, with the spectra recorded on CCDs which were quite small by present standards. Richard Kron, David Koo, and collaborators used this approach with the 4-meter Mayall telescope on Kitt Peak to stretch the meaning of "long-term program" beyond a decade, amassing a sample of 739 galaxy redshifts within four small fields, extending fainter than a red magnitude of 20. Similarly extensive surveys have been carried out by groups at the University of Hawaii, and the large collaboration known as the Canadian Network for Observational Cosmology. More recent additions, such as physically larger detectors with more pixels, multiple detectors, and setups with the spectra shorter than a detector side—so multiple spectra can be stacked in both dimensions—now allow a few hundred objects to be observed at once (see Figure 9.2 for part of such an observation) with the GMOS instrument at the 8-m Gemini telescopes (resulting, for example, in the Gemini Deep Deep Survey or GDDS, delivering high-quality spectra of numerous massive galaxies in a redshift range which is poorly probed by photometric estimates). Upgrades of the same kind at the Keck telescopes, with the DEIMOS spectrograph, are being made to support the DEEP 2 project which aims at measuring 50,000 redshifts for galaxies typically at $z = 1$, using about 85 slitlets at a time. ESO's Very Large Telescope can measure similar numbers per hour, and large faint-galaxy surveys have been carried out there as well.

For such faint targets, especially working far into the red where their redshifts place many important spectral features, the multiplex advantage of optical fibers for observing many objects at once has not been very helpful. The issue is subtraction of the glow of the night sky, not only artificial light pollution but the permanent natural light from excited molecules (especially OH) which gives the spectrum of the blank night sky a rich set of emission lines. The galaxies are much fainter than the surface brightness of the airglow, so precision in sky subtraction is the major observational problem. Fibers afford limited scope for measuring the sky in exactly the same way as the galaxy, since different fibers, or the same fibers at different times, must be used. Even minor changes in wavelength sampling, focus, or throughput can compromise data quality, so these systems become less practical for galaxies fainter than about magnitude 20 (in red light). Multislit systems, in contrast, measure the sky immediately adjacent to a target, on each side, and simultaneously, and have proven much more tractable for the deepest spectroscopy. Even so, such refinements as synchronized offsets of the telescope and motion of the accumulating charge on

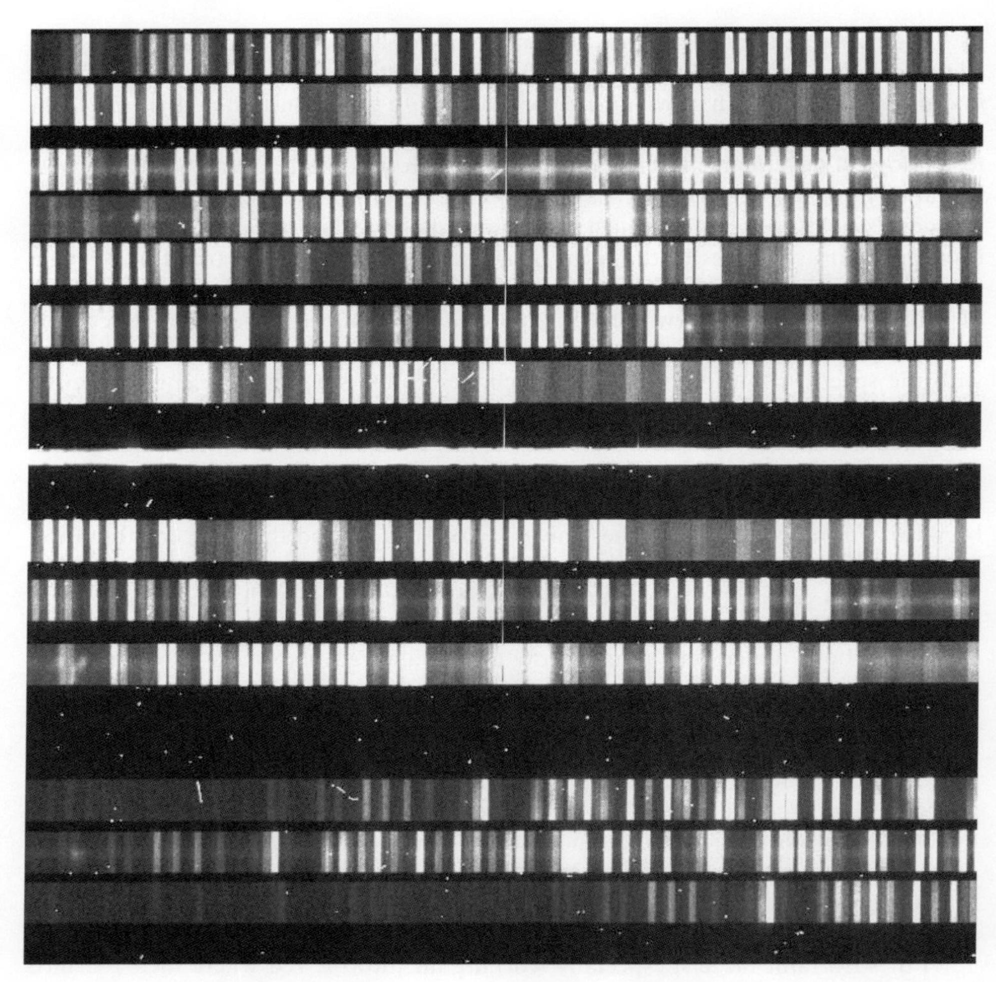

Figure 9.2. Multiobject spectroscopy of faint galaxies, illustrated by recent data obtained with the 8-meter Gemini–North telescope. Each horizontal band is the dispersed light from a short slit, placed to encompass a galaxy of interest. Thirteen spectra are shown in this section, some of which are faint enough to be invisible in the raw data. Continuum light from a galaxy appears as a horizontal streak, and emission lines are bright spots. This field has measurable galaxies at redshifts as high as $z = 3.3$, with quantitative analysis requiring careful subtraction of the night-sky atmospheric emission visible along each slit. (These data were obtained in an observing program with I. Smail, M. Ledlow, F. Owen, and G. Morrison.)

the CCD chip ("nod and shuffle") have been implemented to improve sky-subtraction limitations on faint-galaxy spectra.

The results of these surveys so far give significant evidence as to how galaxies of various kinds have evolved. The distribution of redshifts seen to a particular limiting magnitude constrains the evolution of the sample in luminosity, once the sample selection is understood. The more strongly galaxies have faded with time, the more dramatically the redshift distribution becomes weighted to high values as soon as the sample goes deep enough to see galaxies in the initial luminous phases. If we have enough information to sort galaxies by type (either morphologically or by star-formation history), this gives measures of their evolution in luminosity. Large enough samples can give more direct measures of how, for example, the star-formation rate has changed for galaxies at a fixed stellar mass (an important clue to the process of "downsizing").

These spectra also give us a nearly instantaneous measure of galaxies' evolution, through their rates of star formation as measured over several timescales. The emission lines from H II regions, which can usually be distinguished from AGN phenomena, give the current rate of star formation as modified by any dust extinction, averaged over the short lifetimes of hot ionizing stars. Other features give longer averaging times, weighted in various ways. Balmer absorption is strong for stellar populations of ages 10^8–10^9 years, with Hδ the last one to be clearly enhanced in a fading burst of star formation. The 4000-Å break gives a very long-time average of the temperature mix of the stars, essentially telling when a galaxy ceased active star formation. This has been used to advantage for several sets of galaxies, showing a decrease in break amplitude for redshifts beyond about $z = 0.9$. Estimating formation ages, which in practice means the time since the dominant stellar population formed, exploits the leverage from lookback time. When we observe populations of galaxies progressively earlier in time, any substantial episodes of star formation become clear, even if their signatures have faded beyond distinction at the current epoch. This is especially valuable for elliptical galaxies, for which the 4000-Å break amplitude is very well-behaved and systematic in our neighborhood. This tells us that the last significant star formation occurred at approximately the same cosmic epoch for most ellipticals. Measurements at redshifts closer to this epoch can clearly improve the precision with which we know when this happened, and whether the process was actually coeval for the whole class. Spectra can also show us when such galaxies have had a more recent burst of star formation, with two populations of quite different age coexisting. These have been termed "E + A" or "K + A" galaxies, from the combination of the typical spectrum of an elliptical galaxy (much like a K star) with the Balmer absorption from younger stars (as is strongest in single stars of spectral type A). These composite spectra (Figure 9.3) seem to be more common toward higher redshifts, perhaps indicating that the protracted star formation in many ellipticals was episodic.

In the data to date, we find a "red envelope": the galaxies at each redshift which have undergone the least recent star formation and thus have the strongest 4000-Å break. This envelope is closely related to the red sequence in galaxy color in the emitted frame, as smeared by redshifting of the spectrum and dimming by distance at each redshift. The amplitude of the spectral break for these reddest galaxies decreases with redshift, as we look backwards in time. The rate and pattern of this change is sensitive both to the cosmological parameters (since the age corresponding to a particular redshift depends on the Hubble constant and expansion history, while the required mix of stellar ages does not) and the star-formation history. It is

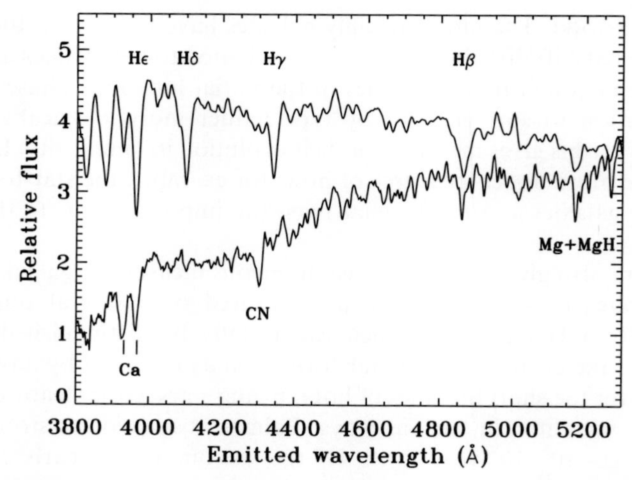

Figure 9.3. Composite galaxy spectra as tracers of old starbursts. This comparison shows an E + A or K + A spectrum on the top, represented by the local example NGC 4569 in the Virgo Cluster, compared with a normal elliptical galaxy below. The strong Balmer absorption lines of hydrogen, labeled at the top, are strongest in stellar population a few hundred million years old, which must dominate the light for these lines to be so strong. Features from an older population can also be seen, most notably in the labeled features in the lower spectrum. Such composite spectra indicate that a very strong starburst has occurred and been quenched rapidly. (Data by the author, from Mt. Lemmon Observatory.)

interesting that single-age models for the stellar populations will not fit this envelope for any simple cosmology (Figure 9.4), requiring that even elliptical galaxies still harbored significant star formation until nearly $z = 1$ (about 8 billion years ago for a typical cosmology).

This time leverage also gives a powerful way to look for the galaxies which formed most of their stars at the earliest epochs, by finding galaxies which are reddest and have the strongest spectral breaks to yet higher redshifts. In fact, two such systems attracted attention for having implied stellar ages uncomfortably close to the estimated age of the Universe, as usual depending on cosmological parameters and details of stellar evolution. The radio galaxies 3C 65 at $z = 1.1745$ and 53W091 at $z = 1.55$ have derived ages for the stellar populations of 3.5–4 Gyr, sufficient to rule out certain combinations of H_0 and q_0 by themselves. Unless there is a problem with our understanding of their stellar populations, which do not appear to be in an unusual regime, these galaxies show that some elliptical systems completed their star formation, and probably assumed essentially their observed structures, very early and rapidly. This is particularly interesting in view of the distinct models for rapid, monolithic or slower, piecemeal galaxy formation, suggesting that some systems did form much as envisaged by Eggen, Lynden-Bell, and Sandage. Perhaps there really are two distinct ways and timescales for formation of elliptical galaxies.

The distinct bimodality in galaxy colors (Chapter 4) suggests that the distinction between galaxies that are actively forming stars, and quiescent red galaxies, is a basic

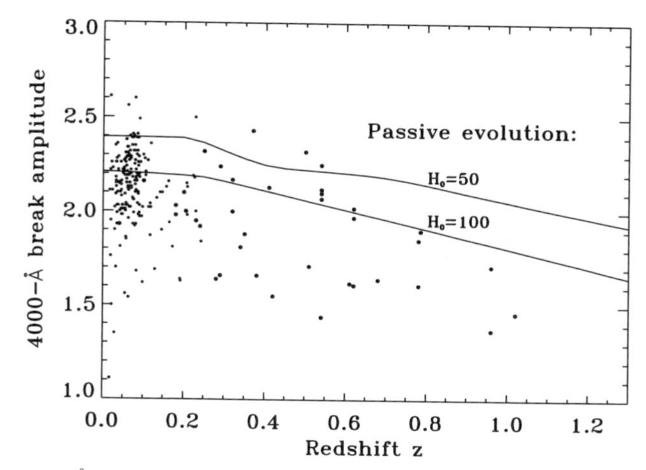

Figure 9.4. The 4000-Å break as an indicator of galaxy evolution. The points show individual elliptical galaxies, both radio sources and radio-quiet ones, from surveys by Owen, Ledlow, Keel, and Hamilton. There is a clear envelope, such that the reddest galaxies at each redshift become systematically younger with redshift. This is in the general sense predicted for galaxies whose star formation is complete. The curves show the behavior of a 13-Gyr-old population for various values of the Hubble constant (for simplicity, with a flat cosmology in both cases). The greater observed slope suggests that many ellipticals had significant star formation until as late as $z = 1$. The very existence of an upper envelope constitutes a sign of galaxy evolution.

one. The red sequence was also in place in dense environments by $z = 1.5$, albeit not quite as red as today since the stellar populations were younger. Much of the distinction between red and blue galaxies is traceable to total mass, another manifestation of the "downsizing" pattern. Massive galaxies must have formed quickly and likewise formed their stars quickly—so quickly that the classical monolithic-collapse scheme dating to Eggen, Lynden-Bell, and Sandage describes their bulk properties well. The narrowness of the red sequence also limits their development at later stages—any major mergers affecting most of the luminous red galaxies must have been "dry", lacking cool gas and resultant bursts of star formation.

Although galaxies on the red sequence do not show active evolution driven by star formation in recent epochs, it is noteworthy that their aggregate stellar mass density increases between about $z = 1.5$ and now. This may be another indication that "dry mergers" continue to be important in galaxy assembly even when starbursts are not involved. Estimates of stellar mass in galaxies are now reasonably standardized, making results from various studies at least comparable. The same kind of broadband spectral fits which yield estimates of photometric redshifts also produce models for the stellar population and luminosity, and thus imply a stellar mass. Various algorithms may incorporate reddening using empirical or theoretical prescriptions. The greatest absolute uncertainties occur due to the amount of mass in low-mass stars which are very poorly represented in the integrated light, as well as to the possibility of heavily obscured star-forming populations which would be detected only deep into the infrared.

Redshifts and spectroscopic properties of deep galaxy samples also contribute to our knowledge of the overall history of star formation, already encountered as the Madau diagram. The important quantities now are those that give some measure of recent star formation—intensity of emission lines from a galaxy's H II regions or of the ultraviolet continuum from massive stars. These must be corrected for internal extinction due to each galaxy's dust, which becomes more important at shorter wavelengths and entails unavoidable uncertainties. Even with such corrections, in fact, the average star formation rate (say, per cubic megaparsec) is a lower limit, because sufficiently dusty star-forming regions will leave no observable trace in the ultraviolet. Just such obscured regions are important in some of the most active star-forming galaxies in the local Universe, which are strong far-infrared sources with only a small fraction of the light from their stars leaking out directly.

9.4 SUBMILLIMETER GALAXIES

This bias can be addressed, at least for the most powerful systems, by selecting sources from far-infrared and radio emission. Far-infrared radiation, mostly emitted at wavelengths 50–200 μm, will be redshifted into the submillimeter range for the highest redshifts we can observe. This range has several somewhat transparent windows that can be exploited from the ground, and a handful of instruments (particularly SCUBA on the 15-meter James Clerk Maxwell telescope) have already been able to initiate deep surveys to find such galaxies. These surveys show some objects have been detected at $z = 3$–4, and that some have no detected counterpart even in Hubble Deep Field data. Until spectroscopic data, optical or infrared, is available, their redshifts can at this point be estimated only in a crude way, from the ratio of radio and far-infrared fluxes. Star-forming galaxies are significant sources of nonthermal radio emission, thought to be driven by supernova remnants as the accelerated particles interact with magnetic fields. This emission scales closely with other tracers of star formation, and is completely immune to dust absorption, making it attractive in tracing the history of star formation for that part of the galaxy population we can detect in the centimeter realm. The spectrum of the radio emission is close to a power law $F_\nu \propto \nu^{-\alpha}$ with α typically 0.7–1, rising to a longer wavelength. The spectrum of a dusty object also has a substantial bump, like a broadened blackbody spectrum, peaking at emitted wavelengths 50–100 μm. The ratio of the two fluxes can thus suggest which points on the curves we are observing at known wavelengths, and hence the redshift, if we know what the relative intensities of the two spectral components are (Figure 9.5). This is an especially rough, but useful, two-point application of photometric redshifts. The objects so far measured in detail indicate that we are seeing very dusty objects at $z > 4$. It is no surprise that galaxies could be forming stars in quantity at such epochs, from the fossil record in our neighborhood, but it is striking that the stars have already formed sufficient quantities of dust to hide themselves so thoroughly. In fact, by themselves these

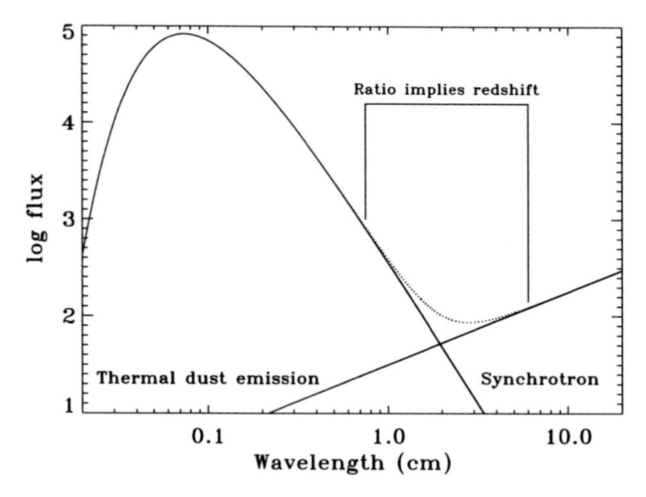

Figure 9.5. Estimating the redshift of a faint galaxy from its submillimeter and radio fluxes, based on an empirical two-component model, is illustrated here. The submillimeter radiation comes from dust grains heated by starlight (and possibly an active nucleus), producing a blackbody modified by the imperfect emissivity of the grains at long wavelengths. The radio continuum is mostly synchrotron radiation, largely from particles accelerated in supernova remnants, whose intensity scales closely with the far-IR flux in nearby galaxies. The flux ratio at known submillimeter and radio wavelengths implies a unique redshift, to the extent that the relative intensities of the thermal and synchrotron spectra agree with the model. These two components have spectral shapes so different that one dominates except in a narrow range around the 1–3 cm wavelength (where the sum is shown as a dotted curve). The thermal emission plotted is appropriate for typical galaxy dust observed at $z = 4$.

galaxies appear to make a substantial contribution to the overall star-formation rate at their most populous epoch.

There are now some spectroscopic redshifts for these "SCUBA galaxies", with typical values near $z = 2.4$. These have driven debate about their dominant energy source, as virtually all the optical spectra show emission lines characteristic of active galactic nuclei. However, the AGN do not appear capable of filling the energy budget to power the far-IR radiation, indirectly implicating enormous rates of star formation in the luminosity of these galaxies. Possibly related is a set of galaxies recently identified in Spitzer surveys, whose spectra peak near 24 μm and fall to much lower levels in the submillimeter bands. They cluster strongly enough to suggest high mass, and one suggestion has been that these are ellipticals in phases of both rapid star formation and fueling of a central AGN.

What we see of the cosmic star-formation history therefore depends critically on how accurately we can account for the effects of dust, and other detection problems such as surface-brightness selection. Various ways of operating on the data from the Hubble Deep Fields yield a broad peak in star formation near $z = 2$, a plateau for $z > 2$, or a continued increase with redshift to beyond $z = 3$. This issue will have to be resolved by deeper and more complete observations.

Spectroscopic results can give us a sharper view of galaxy evolution when compared with high-resolution imaging, from HST or adaptive-optics observations. Such a combination shows which populations of galaxies are responsible for various changes in the mix of color and star-formation rate. For example, much of the evolution for $z < 1$ happens among irregular galaxies, which dominate the population of faint blue galaxies which gave some of the first evidence for galaxy evolution. Spatially resolved imaging shows that the color and age contrast between the disks and bulges of spirals persists to at least $z = 1$, so that it still makes sense to seek different epochs and modes of star formation for these kinematic components.

We are not yet in a position to get large, complete samples of galaxies at higher redshifts, largely because only certain subclasses are easy to identify for detailed study. We can use them to learn about some of the extremes of the galaxy population at redshifts now extending to $z > 6$, keeping in mind the limitations and biases imposed by what we can and cannot yet observe. Specialized detection techniques have led to large samples of certain easily detectable kinds of galaxies at high redshifts, which we then need to integrate into our understanding of the overall galaxy population.

9.5 LYMAN α EMITTERS

Emission-line objects can be more easily detected against the sky background than continuum objects, so any population with strong emission lines is attractive simply because they can be more easily identified at particular redshifts. In the ultraviolet range that shifts into the optical band for high redshifts, the only strong emission line produced by galaxies other than AGN is Lyman α, emitted at 1216 Å. There has been a long history of searches for this line from galaxies at high redshift, punctuated by occasionally dire theoretical predictions as to how much of its radiation should escape the originating galaxy.

Early in the game, when it was suspected that young galaxies would be forming stars rapidly, there were predictions that they would be very luminous and look like nearby star-forming regions writ large, exemplified in the 1978 paper by Sunyaev, Tinsley, and Meier, entitled "Observable properties of primeval giant elliptical galaxies or ten million Orions at high redshift". Since Lyman α is by far the strongest emission line produced as ionized hydrogen recombines, searches for small sources of line radiation in specified redshift intervals (with narrowband imaging, or long-slit spectroscopy and good fortune) might be expected to yield many of these very luminous sources with strong Lyman α emission.

However, the very strength of the Lyman α transition can also be its downfall. Theoretical work showed that a photon at this wavelength travels only a short distance through the gas in a typical star-forming region before being absorbed by another hydrogen atom, to be re-emitted as another Lyman α photon or converted to another wavelength through two-photon emission, or absorbed by a dust grain, with its energy eventually emerging as infrared thermal emission. This resonant trapping made it unlikely that we would see strong Lyman α emission from star-forming galaxies, unless their gas was dust-free and had a rich velocity structure, so that even local absorbing gas would have a significant Doppler shift reducing its probability of absorbing the line radiation. There has been some tension between theoretical understanding and observational detection of Lyman α emission, with the emission line appearing more often in the real world than one might expect from straightforward theory. The first attempts to measure this emission from star-forming galaxies used the venerable *International Ultraviolet Explorer* (IUE), launched into geosynchronous Earth orbit in 1978 on a three-year mission (which in fact concluded in 1996). It was designed solely to measure ultraviolet spectra. Among the many galaxies observed during its mission, a handful were suited to detecting Lyman α emission. Not only would a galaxy need a long exposure, and have to have the requisite population of ionizing stars, but it would have to combine a large enough redshift to separate the galaxy's emission line from the residual glow of hydrogen escaping the Earth's atmosphere—the geocorona—with enough luminosity to remain a bright ultraviolet source at such distances. Nevertheless, there were several clear detections of Lyman α emission from actively star-forming galaxies in our neighborhood, to redshifts $z = 0.05$. This represented one of many remarkable feats for a 45-cm telescope. Evidently, some real galaxies manage to overcome resonant trapping, perhaps by blowing starburst winds whose acceleration provides the velocity shifts to let some Lyman α escape.

Meanwhile, searches for Lyman α at high redshift were proceeding, and bearing limited fruit. Stanislaus (George) Djorgovski, in particular, carried out a series of searches in both quasar surroundings and the general field, with none of the latter being successful. In hindsight, a little more depth and a little more luck in finding concentrations of galaxies might have done the trick. In any case the 1980s did see detection of narrow Lyman α emission sources near quasars at $z > 3$, a trend which has continued and yielded rich groups of companions to AGN at $z = 2$–4.1. The added depth available through observations with HST and the new generation of 8–10 meter telescopes on the ground has crossed the threshold of showing these objects and elucidating their properties. There is a significant population of faint objects, generally starting at blue magnitude 24 and fainter, with strong enough Lyman α emission to be picked out with narrowband filters. They occur in the Hubble Deep Fields, and have been found in clumps associated with active nuclei (radio galaxies and quasars). They may represent a very particular stage in the buildup of galaxies, looking in some ways like the protogalactic objects which enter into hierarchical schemes for galaxy growth. They are very compact, with half-light radii less than a kiloparsec (Figure 9.6). This is more like giant H II regions or complexes than galaxies, especially since their ultraviolet luminosities from these small volumes are higher than any normal galaxy in the local Universe. Many of these objects are so blue that they must have very low metallicities, either because they have just begun active star formation or because their star formation is powerful enough to initiate global winds, sweeping these (low-mass) objects free of enriched material. Their numbers in the clumps around AGN are large enough to suggest that they undergo repeated episodes of sweeping, gas accumulation, and star formation. It remains unclear what these turn into at the present epoch, since their complete

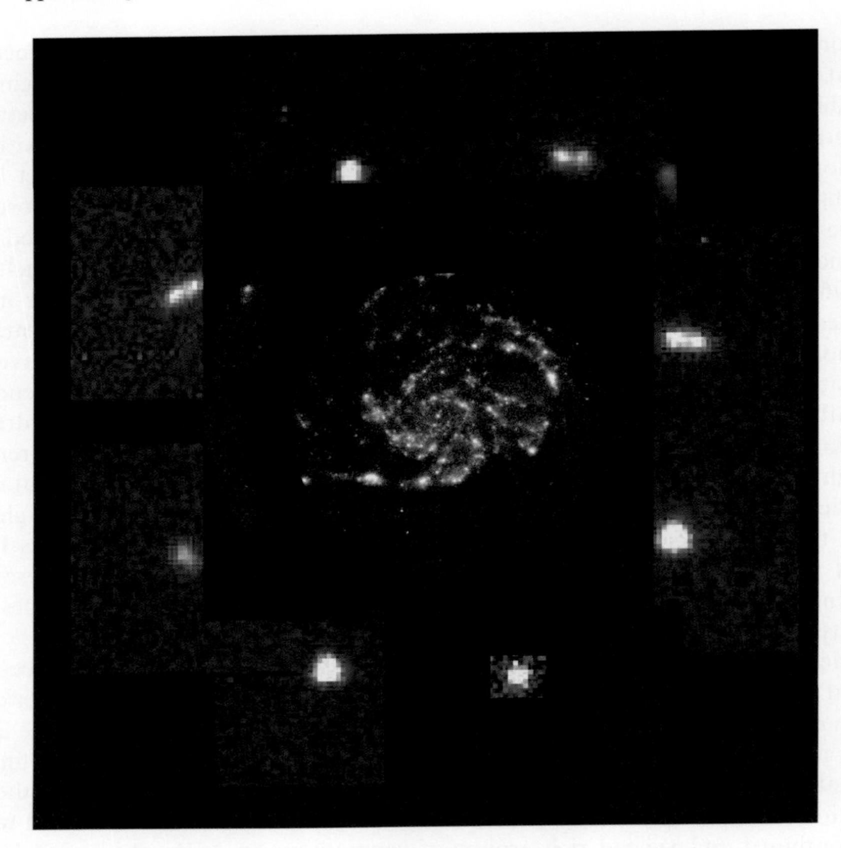

Figure 9.6. Eight Lyman α emitting galaxies extracted from a single grouping at $z = 2.4$, arranged around an ultraviolet image of the nearby spiral galaxy M101 (shown at the same linear and intensity scale). This comparison illustrates both the small size and high luminosity of many of the objects being found at high redshifts. The high-redshift objects are from Hubble Space Telescope imagery in a narrowband filter, while the M101 image was obtained using the Ultraviolet Imaging Telescope (UIT) on the space shuttle *Endeavor* during the Astro-2 mission (STS-67). (The data were retrieved from NASA's MAST archive.)

star-forming history is so uncertain. For these kinds of objects, in particular, there has been controversy as to whether we should consider them as individual proto-galactic pieces, or whether they are simply bright and compact pieces of larger systems which remain invisible because of the surface-brightness and ultraviolet selection implicit in their detection at these redshifts. So far, infrared work has failed to find any older, surrounding stellar populations.

Lyman α emitters can be found from the ground for redshifts $z > 2$, once the line and a usable bandwidth of continuum are within the atmospheric window. At somewhat higher redshifts, Lyman-break selection (Chapter 5) can find UV-bright galaxies with or without line emission, and has now produced significant numbers of galaxies at $z > 4$. Infrared spectroscopy has given estimates of the metal abundances for some of these, using the same emission lines that we understand from optical observations of nearby galaxies. They are somewhat subsolar in abundances, typically 1/3 of solar metallicity, hinting that we see them early in their history of star formation and enrichment. Here again, there is a built-in bias toward objects that are luminous in the ultraviolet and for which we can see the spectral range including the Lyman limit, so that selection of samples in other ways (particularly in the far-IR or submillimeter) may still show us quite different populations.

Searches for Lyman α emitters generally use narrowband filters tuned to some redshift of interest, usually given by some object such as a quasar or radio galaxy. A few "blind" searches have started to yield more complete statistics of their occurrence. These involve either observations with a set of stepped filters to span a range of redshifts while keeping skylight out, stepped spectrograph slits and intermediate-band blocking filters, or slitless spectroscopy (which has been effective with the lower sky background and higher resolution of the ACS camera on HST).

9.6 LYMAN α BLOBS

Some of the same data designed to detect distant galaxies via Lyman α emission revealed a population of very large, diffuse objects with strong emission, the Lyman α blobs. These have detected angular extents as large as 30 arcseconds (230 kpc at $z \sim 3$), and can be distinguished from "normal" Lyman α galaxies down to sizes near 3 arceconds. They may or may not have a detectable embedded continuum source; some are centered on active nuclei, some on faint galaxies, and some do not have an obvious ordinary galaxy associated with them (Figure 9.7). Models have been proposed involving the initial cooling of pristine gas during collapse (which is to say, what one would clearly call galaxy formation), and its opposite, outflow from an established system whether driven by a starburst or nuclear activity, as well as photoionization from embedded active nuclei. They may be related to the very extensive nebulae seen around some active nuclei selected in other ways (Figure 9.8), in which the gas often shows emission lines from such species as C IV and [O III], so it has clearly been enriched rather than remaining at primordial abundances. Spectroscopy of several of these blobs shows the kind of line-profile symmetry expected for global winds, and the brightest show weak nuclear X-ray sources. Thus, they seem to be a reaction to early energetic events in galaxies rather than a stage in the formation of galaxies as such. When an AGN is present, its observed intensity generally falls well short of that needed to photoionize the surrounding gas, so some additional process muct be acting even in the presence of the AGN.

Since Lyman α blobs can be detected only via imaging in a narrowband filter tuned to the appropriate redshift, or serendipitously through spectroscopy of otherwise blank sky, very little is known of their occurrence or its evolution with redshift. In a single overdensity at $z = 3.1$, where the first identifications of these blobs were made by Steidel, a deep survey with the Subaru telescope has revealed 35 well-resolved Lyman α blobs.

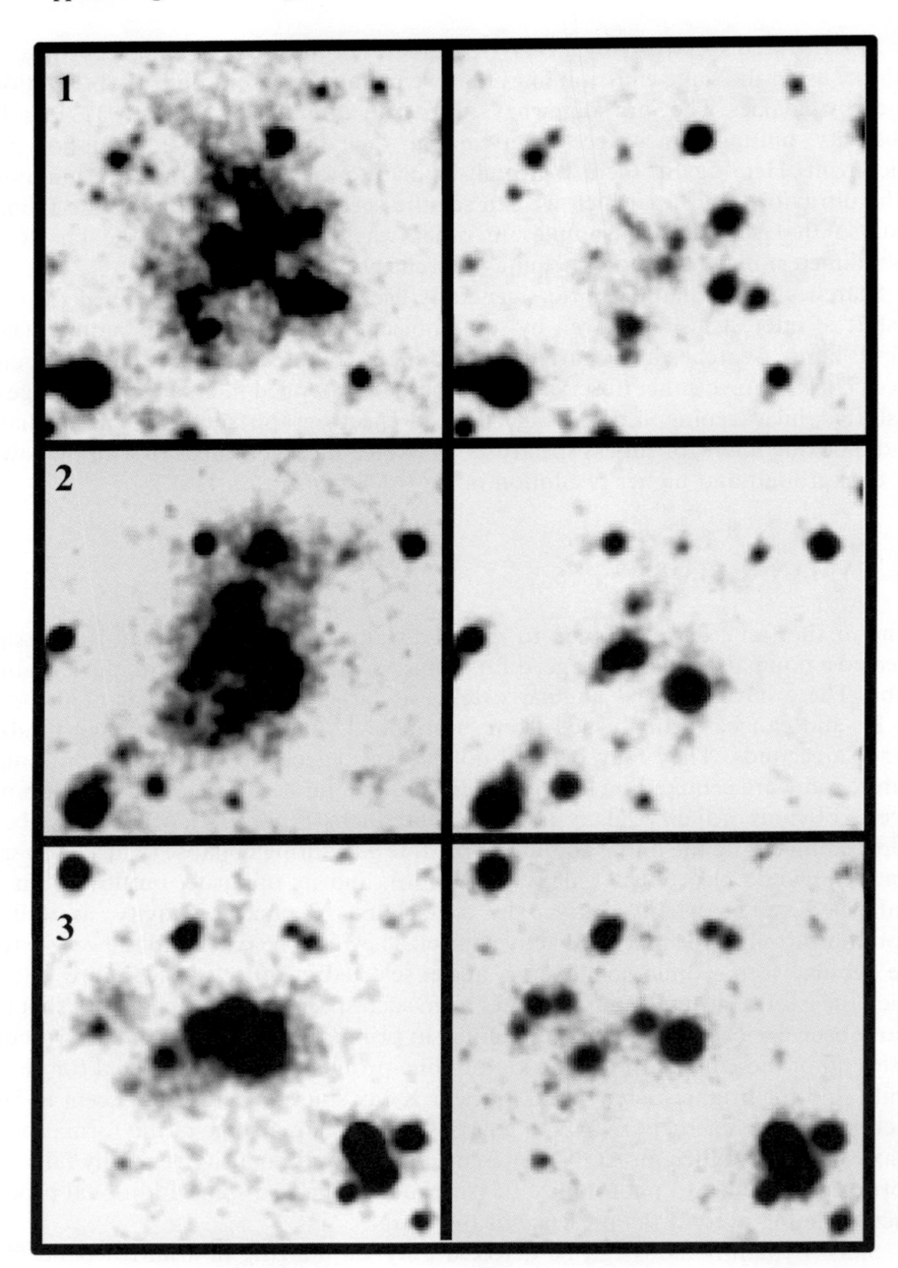

Figure 9.7. Three of the largest Lyman α blobs at $z = 3.1$ in the SSA22 field. Each horizontal pair of images shows the same region in a narrowband filter around 4970 Å (redshifted Lyman α) on the left and $B + V$ broadband continuum on the right. Each panel is 25 arcseconds square (200 kpc). (Data from the Subaru telescope of the National Astronomical Observatory of Japan, courtesy of Yuichi Matsuda, as published in *Astronomical Journal*, **128**, 569, 2004.)

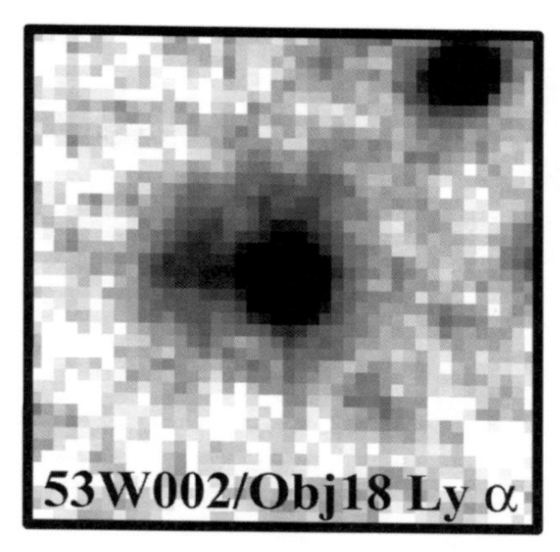

Figure 9.8. An extensive Lyman α nebula around an AGN. This is object 18 in the field of the radio galaxy 53W002 at $z = 2.39$, observed with 8 hours' exposure at the Kitt Peak 4-meter telescope. The field is 25 arcseconds on a side (205 kpc). The nucleus shows high-ionization emission lines, an X-ray source which is strongly absorbed by foreground gas, and a submillimeter source. The observed nucleus falls well below the intensity needed to photoionize this cloud, unless our line of sight is much more strongly obscured than the average (which would violate the submillimeter luminosity we see). Various sections of this nebula are detected in C IV and [O III] as well as Hα shifted into the near-infrared, so that the gas must have been processed through stars.

9.7 LYMAN-BREAK GALAXIES

Lyman-break galaxies (Chapter 5) provided the first glimpse of a wholesale galaxy population at high redshifts, and therefore the first glimpses of stellar populations, metallicities, and star-forming properties. Color selection of high-redshift galaxies is especially powerful in conjunction with HST imaging to provide structural information (Figure 9.9).

Like Lyman α emitters, the Lyman-break galaxies also show significant clustering; the growth of large-scale structure clearly began very early, and was likely part and parcel of galaxy formation to begin with. Groupings of emitters have been found associated with the radio galaxies PKS 1138−262 at $z = 2.16$ and 53W002 (and several neighboring QSOs) at $z = 2.4$, QSO absorption-line gas in another field at $z = 2.4$, associated with a QSO pair at $z = 2.56$, and the radio galaxy TN J1338−1942 at $z = 4.1$. Steidel and collaborators have similarly identified a large grouping of Lyman break galaxies at $z = 3.1$. These associations extend over scales of a few megaparsecs. Where data are available, their velocity dispersions are modest (300–400 km/s), so they are not extremely massive and relaxed. Some may be seen before they have fully decoupled from the Hubble expansion, giving additional incentive to find such early clusters as probes of the baryonic and overall mass distributions at these epochs.

Lyman-break galaxies span a significant range of luminosity and metallicity, as may be seen from average spectra of these galaxies when grouped by strength

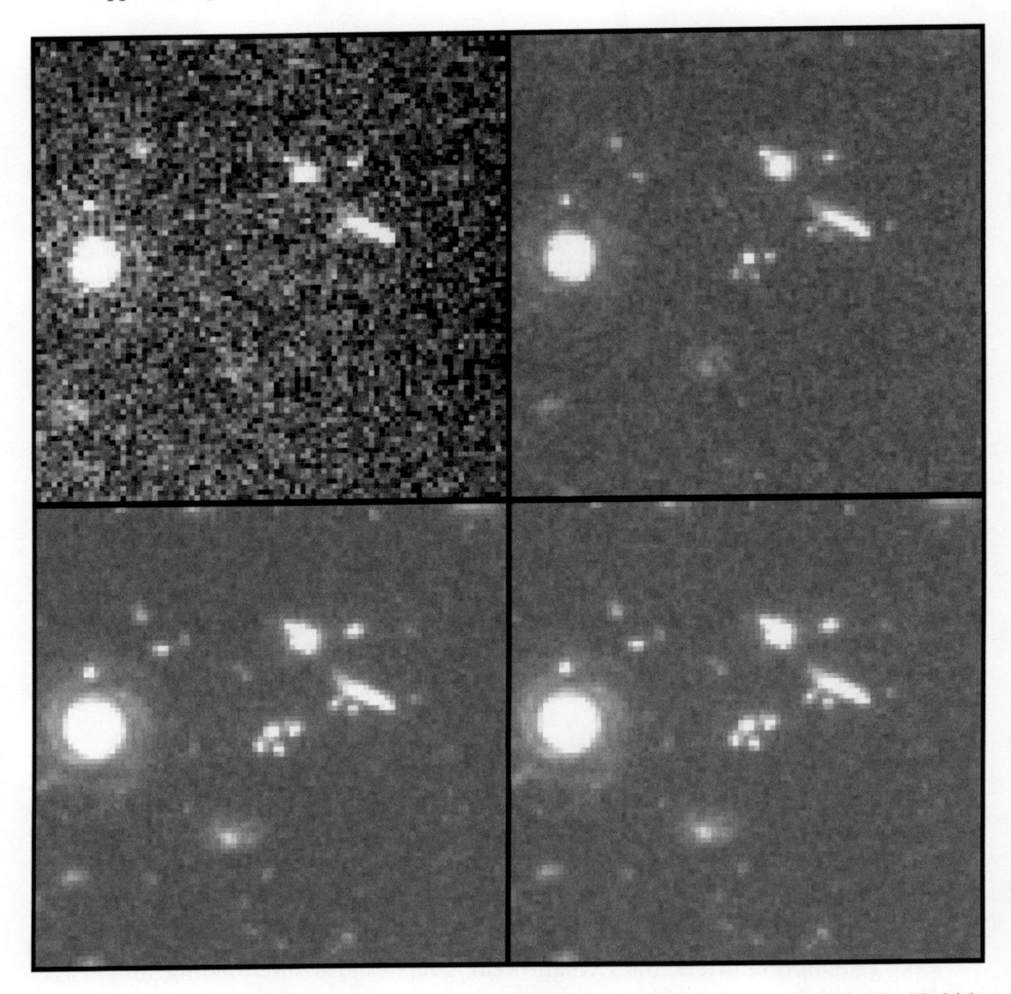

Figure 9.9. A Lyman-break galaxy. This comparison shows the brightest galaxy in the Hubble Deep Field with a known redshift $z > 3$, as the clumpy or multiple structure in the center of each frame. The entire configuration is a 24th-magnitude object in red light, spanning about 1.5 arcseconds. The four HDF bands are at 3000 and 4500 Å (*top*) and 6060 and 8140 Å (*bottom*). The near-ultraviolet band lies entirely shortward of the object's Lyman limit, making it invisible in that image, while comparably bright in the other three showing that it has a flat or blue spectral shape. This distinguishes Lyman-break systems from highly reddened, or intrinsically red, galaxies at lower redshift. (These images were taken from the publicly released data set of the Hubble Deep Field team, at *http://www.stsci.edu/ftp/science/hdf/archive/v2.html*)

of Lyman α emission (Figure 9.10). The most luminous have essentially no Lyman α emission and stronger absorption lines. These come mostly from stellar and global winds, with a few dominated by stellar photospheres, and may suggest that a luminosity-enrichment connection was already in place by $z = 3$.

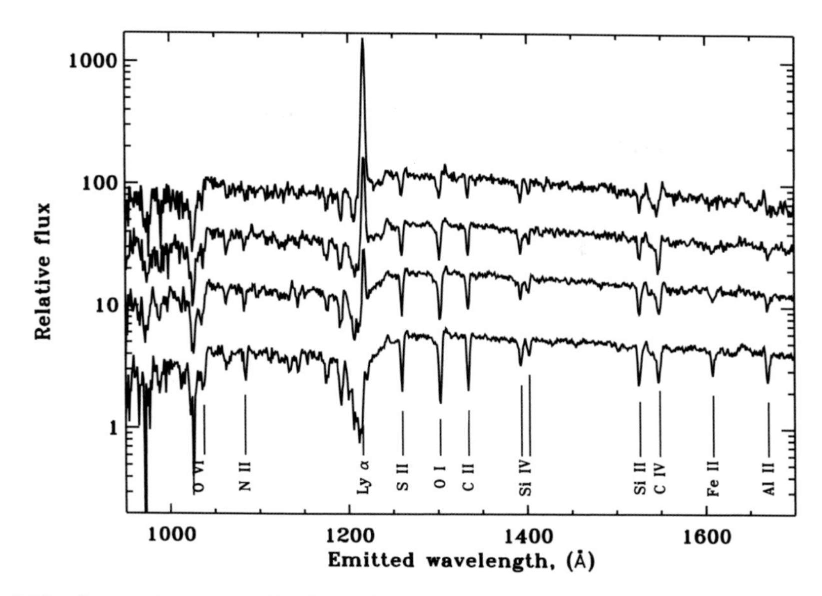

Figure 9.10. Composite spectra in the emitted ultraviolet range for groups of Lyman-break galaxies, binned according to strength of Lyman α emission. This is roughly a luminosity sequence, and the strengths of absorption lines also vary systemically along the sequence. Many of the absorption lines have important contributions from stellar winds and interstellar gas, often blueshifted indicating a global wind, so that abundance interpretations may be limited. The changing amount of Lyman α emission could, in principle, reflect different gas or grain abundances, velocity structure in the interstellar medium, or scattering geometry of the gas. The spectra are plotted on a logarithmic scale in flux per unit wavelength, with offsets to avoid overlap. (Data courtesy of Alice Shapley, as described in Shapley *et al.*, *Astrophysical Journal*, **588**, 65, 2003.)

The clear bias toward systems that are actively forming stars so as to be strong ultraviolet sources, yet not so dusty as to absorb this radiation, leads to other ways of identifying high-redshift galaxies, so that we can compare complementary kinds of objects. These means must use longer wavelengths, where cooler stars are bright and dust is less important. To date, such surveys include "blind" imaging in narrow near-infrared bands, to find highly redshifted emission in such lines as Hα, Hβ, and [O III]; surveys for optically faint submillimeter sources; and very deep radio surveys with associated optical or near-infrared followup. Such studies are in their infancy, largely for technical reasons. Infrared observations are more difficult from the ground than similarly deep optical observations, and so far the available imagers have substantially smaller formats, both factors making infrared surveys much slower than similar surveys at shorter wavelengths. Forthcoming developments in millimeter arrays (ALMA) and the James Webb Space Telescope (JWST), as described in Chapter 11, should allow enormous progress in these studies, and indeed these instruments were designed with just these goals in mind.

9.8 DEEP FIELDS: COUNTS, COLORS, AND POPULATIONS

Throughout the history of galaxy studies, there has been a powerful tradition of deep-field observations. These are selected regions observed as long as possible, to the faintest limits possible, with the aim of probing the largest possible volume of space and hence sampling the population of galaxies (and everything else in the field) to the greatest possible distance and lookback time. Even objects too faint for detailed study with spectroscopy or other detailed analysis can yield information through their numbers at various flux levels, in comparison with specific model predictions. The application dates at least to the time of Hubble, who hoped to use galaxy counts with the 200-inch Palomar telescope as constraints on cosmology. While he was eventually persuaded that such data would not be very effective for this purpose, successive deep fields later revealed distant galaxy clusters, an excess of faint blue galaxies strongly suggesting an evolving population, and have placed the start of galaxy formation at earlier and earlier epochs.

Beginning in 1996, a crucial role was played by the Hubble Deep Field and later the Hubble Ultra-Deep Field (see Box 9.1) in coordinating efforts to reach as deep as feasible with the exquisite angular resolution of the Hubble Space Telescope. Together with extensive spectroscopy from the largest ground-based instruments, exposures up to a million seconds with the *Chandra* and *XMM-Newton* X-ray facilities, and long integrations with the infrared cameras of the *Spitzer Space Telescope*, have given us the deepest and most informative windows yet on the time just after the Dark Ages. The various wavelength bands give complementary censuses—the Hubble data sample mostly light from stars, especially hot stars from galaxies at large redshifts, while the X-ray data mostly trace active galaxies powered by accretion onto massive black holes, and *Spitzer* shows us star formation so obscured by surrounding dust that the galaxies may not even be detected in the *Hubble* data.

Box 9.1 The Hubble Deep Field

A major advance both in the quality of data and in the sociology of how astronomers deal with these problems came with Hubble Deep Field projects. Impressed with the potential of HST images for problems of galaxy evolution, Robert Williams (then director of the Space Telescope Science Institute) convened an international panel to recommend the most effective way to use this capability. The outcome was a plan to use a large chunk of the director's so-called discretionary time on a single set of observations, images of a single undistinguished area of sky in four passbands, totaling an unprecedented 155 hours of exposure over nearly 150 orbits. The area to be targeted as the Hubble Deep Field was selected with some care, to make it as useful and representative of the deep Universe as possible while maximizing the efficiency of both the Hubble and other followup observations. The efficiency of the HST imaging could be nearly doubled if the region were in one of the continuous-viewing zones (CVZs), areas of the sky which at some time during the year lie near the poles of the observatory's orbit and can thus be observed without interruption due to the Earth blocking the view. In 1995 all the largest telescopes on the ground were in the northern hemisphere (more specifically, were atop Mauna Kea, Hawaii) so that the northern CVZ had more immediate prospects for supporting observations. (This is still the case for centimetric radio observations.) This confined the target to be in the northern CVZ, at a declination of $+62^\circ$. To avoid interference with the HST and followup observations, no bright stars or strong radio sources could be nearby ("bright" in this context meaning 18th magnitude or brighter). Since the objective was to study the deep Universe, no nearby clusters of galaxies should confuse the source population. Finally, it should be in a region of low gas and dust column density in our own galaxy, to make further observations in the X-ray and far-infrared domains as sensitive as possible. Further ground-based observations and test images were used to refine the selection of the Hubble Deep Field, which was finally located at epoch 2000 coordinates $\alpha = 12\,\mathrm{h}\ 36\,\mathrm{m}\ 49.4000\,\mathrm{s}$, $\delta = +62^\circ\ 12'\ 58.000''$, in Ursa Major north of the familiar Big Dipper pattern.

The observations were obtained in the period December 18–30, 1995. Almost as groundbreaking as the observations themselves was their processing and public release (Figure 9.11). The observations had been taken at a set of slightly offset telescope positions (dithering), so that they could be combined to yield further improved resolution by an algorithm known as drizzling. Perhaps most importantly, the final data products—images and object catalogs—were electronically released to the entire community at once, so that any interested researchers could have equal access. Observers at large ground-based telescopes immediately began followup work on the redshifts and infrared properties of some of the thousands of objects revealed in the HDF. Charles Steidel was able to take the initial data release and extract probable high-redshift galaxies for observations with one of the 10-m Keck telescopes within a few weeks, yielding the first large harvest of galaxies beyond $z = 3$. Since the multiobject spectrographs that could be used for such work have fields of view several times larger than the WFPC2 data in the HDF proper, additional short-exposure images had been obtained during the HDF campaign of so-called flanking fields surrounding it, to yield structural information on additional objects that would almost certainly have deep spectra obtained.

It was in the Hubble Deep Field, observed with four filters from the near-ultraviolet to the deep red, that photometric redshifts came of age, with large and uniform sets of color data plus unprecedented levels of spectroscopic "ground truth". This allowed several groups to hone algorithms for measuring the best-fitting redshifts, and their errors, from multicolor flux data alone.

Following the mission to refurbish the Hubble Space Telescope and install the new Near-Infrared Camera and Multi-Object Spectrometer (NICMOS) and Space Telescope Image Spectrograph (STIS), carried out in February 1997 by the crew of the space shuttle *Discovery* on STS-82, there were several compelling reasons to observe another deep field. While strenuous efforts had been made to observe a location for the original HDF that was as average as possible, it would be nice to

Figure 9.11. Part of the original Hubble Deep Field, seen in a color-composite spanning about one-quarter of the WFPC2 field. Spiral galaxies are distinguishable to redshifts above $z = 1$, and some of the faint red images have been identified as Lyman-break galaxies at redshifts as large as $z = 5.6$. See also color section. (Image by R. Williams of STScI, the HDF Team, and NASA.)

have our Copernican assumption of mediocrity for any piece of the sky checked somewhere else. With large telescopes being built in the southern hemisphere, especially the European Southern Observatory's VLT incorporating four individual 8-meter instruments, spectroscopic and infrared followup observations could now be carried out for targets in the southern continuous viewing zone, effectively doubling the resources available for such work. The new instruments meant that three adjacent deep fields could be observed at once—one with

WFPC2, nearly duplicating the original HDF plan; one with NICMOS in the near-infrared bands; and a field with STIS imaging, with less passband selection but going extremely deep due to its broad filters and more efficient detector. The new near-infrared bands could make important contributions, since higher-redshift objects would have much of their radiation reaching us at these longer wavelengths, and their appearance in these bands would represent the same emitted wavelengths that we are accustomed to using in classifying galaxies (so we could more easily distinguish a clumpy irregular system from a spiral with several unusually bright star-forming regions).

The one important difference in target selection for the HDF–South was designed to take advantage of the powerful spectroscopic capabilities of STIS and the added potential of comparing the distributions of galaxies and intergalactic gas. The field to be observed with STIS was centered on a quasar at $z = 2.24$, and some of the observing time was devoted to ultraviolet spectroscopy of this object. Combined with ground-based spectra, this gave an inventory of absorption-line material at redshifts up to 2.24 on that line of sight, which could be compared with the structures traced by galaxies throughout the HDF–S (Figure 9.12). Field selection thus included the same kinds of considerations as in the HDF–N, of declination and lack of interfering foreground objects, and now added searches for suitable QSOs in possible fields. The HDF–S lies in Tucana, not far from the direction of the Small Magellanic Clouds, at epoch 2000 coordinates $\alpha = 22\,\text{h}\ 32\,\text{m}\ 56.22\,\text{s}$, $\delta = -60^\circ\ 33'\ 02.7''$.

HST has been employed for one yet deeper field, the Hubble Ultra-Deep Field (HUDF). The rationale for spending so much time on a single patch of sky this time was that the Advanced Camera for Surveys (ACS), installed in 2002 by the astronauts of STS-109 (the final successful mission of the orbiter *Columbia*). Its combination of wider field of view, more sensitive detectors, and tailored filter bands gave ACS the ability to survey significant areas of sky nearly 10 times as rapdly as the earlier WFPC2. In this context, staring at a single region of the sky, ACS could detect sources 1.0–1.5 magnitudes fainter than seen in the original Hubble Deep Field, with corresponding gains in the number and typical redshifts of galaxies as well as finer spatial resolution gained from the smaller ACS pixels. The observing program entailed 412 orbits, carried out from late 2003 to early 2004. A separate program covered much of this field in the near-infrared with HST's NICMOS instrument, and simultaneously provided unusually deep total exposures in adjacent fields with the ACS. Finding a suitable field for this program proved to be an exercise in balancing constraints—visibility from both Hawaii and Chile for followup studies, lack of bright stars and strong radio sources which could compromise deep observations, desirability of having the field within Hubble's CVZ and well away from the ecliptic to reduce the brightness of zodiacal light, and lack of a bright quasar were all considered. In fact, no field could be found which satisfied all these desiderata, and the planners settled on a field (Figure 9.13) at $\delta = -27$ which had already been observed by *Chandra* in one of its deep-field programs.

Figure 9.12. Part of the Hubble Deep Field–South, in a color-composite showing a portion of the WFPC2 field. This field is somewhat richer in medium-redshift galaxies than the HDF–North, mostly from small-number statistics and the different selection requirements of the two fields as regards a nearby quasar and lack of foreground structures. The faint-galaxy counts at high redshift are consistent within the variations due to structure in the galaxy distribution. (Images by R. Williams of STScI, the HDF-S Team, and NASA.)

These fields have been the target of correspondingly deep observations with other facilities as well, to make fullest use of the great investment in the existing observations. Long exposures with the *Chandra* and *XMM–Newton* X-ray Observatories, the *Infrared Space Observatory*, and the Very Large Array at centimeter wavelengths are still under analysis. The Hubble Deep Fields will remain the benchmarks in our knowledge of the deep Universe for many years to come.

Figure 9.13. Detailed regions from the Hubble Ultra-Deep Field (HUDF or simply UDF) data, displayed as a color-composite. Some of these fields are particularly good illustrations of the prevalence of morphologically peculiar, largely interacting, systems as we look among very faint galaxies. (NASA, ESA, S. Beckwith of STScI, and the HUDF Team.)

Other kinds of extensive surveys have important comparative value in tracing the history of galaxies. For objects which are intrinsically rare, depending on the slope of their log N–log S relation, more objects and better statistics may result from spending the same exposure time on a number of (perhaps contiguous) fields in a shallower survey. This approach has been used in the GEMS, GOODS, and COSMOS surveys, all combining ground-based spectroscopy and Hubble images in at least one band, plus additional wavelenth coverage at various depths. Wider-area surveys are important in reducing the effects of cosmic variance: the fluctuations in content of various regions of space due to genuine large-scale structure rather than random Poisson statistics.

The deepest surveys we can yet undertake use not only our telescopes, but the natural amplification of gravitational lensing. When looking through massive clusters of galaxies, we can identify regions where background objects appear amplified in flux and magnified in one direction, often by factors of 30 or more. These small regions have begun to yield surveys of the deep Universe to unequaled depth.

9.9 SCENARIOS FOR GALAXY EVOLUTION

The combination of fossil evidence in the nearest galaxies and the pieces of galaxy evolution that we can see at substantial redshifts provide hints as to how galaxies might have formed, and suggest what we might be looking for as we investigate these early stages of cosmic evolution. They broadly echo the pictures derived from the ages and composition of stars in the Milky Way, taking in extreme cases a monolithic collapse of a single enormous gas cloud to form each galaxy, or the gradual accumulation of smaller units which had already begun forming stars at various rates and times before joining the growing galaxy.

There is an increasing range of data that must be explained, or at the least not contradicted, by a successful scheme for galaxy formation. Such data are as various as distribution of metallicity, age, and orbital characteristics for galactic stars, the redshift-color relation for elliptical galaxies, the overall intensity of UV radiation from galaxies, merging rate for galaxies, earliest occurrence of quasars, and the properties of high-redshift galaxies. Galaxy formation amply illustrates the principle that everything is connected to everything else.

A *monolithic formation* of the Milky Way, in a single smooth collapse from a large cloud of gas, while simultaneously forming what are today Population II stars, was posited by Eggen, Lynden-Bell, and Sandage in their classic 1962 work. If something similar happened to produce elliptical galaxies, we can understand the two stellar populations and the large ages derived for typical elliptical galaxies. The spiral/elliptical distinction would be in the efficiency of star formation during this process, with ellipticals having so little gas at its end that later supernovae could sweep the galaxy permanently clean of cool gas to give rise to later stellar generations.

In contrast, there is strong theoretical support for a picture of *hierarchical galaxy formation*. In this view the first lasting units to become bound and form stars were much smaller than today's galaxies, with the growth of galaxies occurring from these smaller objects which could already have begun star formation, bringing disparate populations of stars with differing enrichment histories into the growing galaxy. This fits with the lack of an overall gradient in abundances among galactic globular clusters (as noted by Zinn and Searle), and falls nicely in line with simulations of structure in a Universe dominated by cold dark matter (CDM). Some such growth must be happening, since we still see galaxies merge and dwarf galaxies being disrupted by encounters with larger ones. In fact, the history of globular clusters brings up the interesting question of how early there could be mergers of objects that would *not* be gaseous protogalaxies, and thus bring their own complements of stars and clusters into a system.

These extremes are echoed in ideas about how larger structures, such as clusters of galaxies, form. Both top-down notions, with galaxies forming from fragmentation of truly enormous gas clouds, and bottom-up schemes, in which galaxies form and then fall together into clusters, have been discussed. We are on firmer ground here in noting that clusters are still accreting galaxies from their surroundings, and can be caught in cluster–cluster mergers, so that a bottom-up picture explains what we see.

9.10 GLOBAL PATTERNS: DOWNSIZING

In piecing together how galaxies evolve from the observations, we are in a position like that of paleontologists. We have snapshots of the situation at various times and in different environments, variously filtered by observational limitations and what the Universe has provided for us to see. As the timespans grow and our sampling becomes sparser, it becomes less clear how we should connect the populations in these snapshots. The biological analogy uses a framework of cladistics—piecing together probably lines or trees of descent based on the least changes in those characteristics that are most robust to adaptation, and so are most diagnostic of an organism's classification. A set of rules for galaxy cladistics would include some limitations or constraints on how galaxies change with time. Generally, we do not expect the stellar mass in a galaxy to decrease with time, and the mass of heavy elements representing the metallicity of the stars should not decrease with time (although the metallicity history can be complicated by both winds and infall of pristine gas). In general, galaxies can grow in mass but not shrink. Fraix-Burnet has made progress with this approach.

Several distinct kinds of data now point to a common theme in galaxy evolution—*downsizing*. This may be expressed in several equivalent ways. The characteristic mass of star-forming galaxies has been declining monotonically since early epochs. Alternately, the star-formation history of lower-mass galaxies has continued longer. At least in the sense of the ages of the dominant stellar populations, massive galaxies formed earlier and completed the process more quickly than low-mass galaxies. These tie together the early appearance of a distinct red sequence of galaxies in clusters and the star-formation histories we deduce from the stellar makeup of nearby galaxies.

These are direct statements from multiple surveys and thus are very robust, but the physics behind the pattern remains unclear. Feedback could be important, especially if it is more likely to shut down star formation in more massive systems. Hierarchical buildup must fit this pattern, so that recent buildup either applies mostly to low-mass galaxies or involves systems that are already gas-poor ("dry mergers"). Especially for massive systems, it may be as instructive to trace the occurrence of the shutdown in star formation as the prior history of starbirth (a point particularly stressed by Dressler in the context of S0 galaxies). Slow accretion of surrounding cool gas and merger-driven star formation seem to have been important only for relatively low-mass galaxies.

The growth of massive black holes, as traced by deep surveys and particularly X-ray demographics, has paralleled this downsizing pattern. The typical mass of black holes responsible for most of the ongoing accretion has declined over cosmic time, more or less in the same way as the mass of galaxies hosting active star formation. This coincidence may reflect the fact that cool gas is important for both, or may be telling us something deeper about the connection between galaxies and the central black holes.

As seen in Chapter 5, the minimum evolution for a galaxy is the passive case, which gives a monotonic fading and reddening after a short time. Many galaxies will have a more eventful history, with active star formation and perhaps forced evolution due to gravitational interaction with another galaxy, internal dynamics such as a bar redistributing gas, or the cluster environment. In seeking the processes of galaxy formation, we need to look beyond these historical events to the initial state of the galaxies when they formed the first stars we can still see.

9.11 THROUGH THE UNIVERSE DARKLY

The Cosmos does not often give up its distant secrets gladly. There are formidable observational barriers to observation of the high-redshift Universe, even beyond those we would expect from ordinary distance. Various effects from intervening matter distort or bias our view of objects in the early Universe. Gravitational lensing changes the flux and morphology of objects seen behind galaxies and clusters. Absorption by gas or dust along the line of sight makes increasingly distant objects more and more likely to have part of their spectrum inaccessible to our observation. In each case this hindrance can also be turned to tell us things that we would not otherwise know about the intervening matter.

9.11.1 Dust extinction

The reddening and attenuation of starlight by intervening dust is familiar within our own galaxy, and we see ample evidence of dust in other galaxies from internal absorption. It may play some role on a cosmic scale as well. Wavelength selection, whereby we tend to observe high-redshift galaxies in their emitted ultraviolet for logistical reasons, means that the effects of internal dust will be enhanced compared with our experience with local galaxies.

A more insidious possibility is reddening by dust associated with galaxies whose light falls below our detection threshold, but might be so numerous that most high-redshift objects are affected by dust in at least their outer regions. This has been a concern in interpreting the evolution of quasars, since to first order it would have the observed effect: numbers increasing with redshift to a certain point, and a decrease beyond that, which would in this case be due to the accumulated absorption by dust in foreground galaxies. At this point, what we know of the nearby dust properties of galaxies and the paradigm that the dust content of galaxies is unlikely to decrease with time as star formation and production of heavy elements proceeds suggest that galactic dust is not responsible for the high-redshift cutoff. Further tests will come as identifications of deep radio and X-ray samples of active nuclei proceeds, since neither of these detection techniques is vulnerable to dust extinction.

There have been other attempts to detect dust around and between galaxies, usually by looking for a systematic (and statistical) reddening of distant background objects when seen along lines of sight close to nearby galaxies. Intergalactic space is as clear as we can measure it, though there are many cases of dust expelled from galaxies in tidal tails which might eventually dissipate into their surroundings. In clusters the hot intracluster plasma should destroy grains by sputtering within 10^7 years. Dust injected into the intracluster gas is a more efficient radiator than the hot gas, so that while such dust is present and being heated by the gas it can cool the cluster gas efficiently. In fact, the high temperature of typical clusters tells us directly that the ICM does not include significant amounts of dust.

9.11.2 The Lyman α forest

Lyman α has a special role in spectroscopy. It is the strongest spectral line arising from the ground state (i.e., a resonance transition) of the most abundant chemical element, hydrogen. At low temperatures (below about 3000 K), most hydrogen will be in this state, so this line is the most sensitive probe for cold hydrogen when suitably back-illuminated. The UV-bright continuum of quasars thus gives us the ability to probe neutral hydrogen in interstellar space with exquisite sensitivity. As soon as QSOs were found with redshifts high enough for their Lyman α emission line to come into the atmospheric window for observation from the ground (coming past the ultraviolet limit imposed by ozone absorption and aerosol scattering), the universal occurrence of the Lyman α forest was found, starting with data by Roger Lynds in 1971. The general features of this collection of narrow absorption features are very robust from object to object, depending only on redshift. Absorption features occur at redshifts up to the object's emission redshift. They are extremely narrow in velocity space, so that higher-resolution spectra generally don't resolve them but show more lines and break some into multiple components. These absorption components follow a power-law distribution in equivalent width and approximately a power-law distribution in redshift (Chapter 6).

At this point, our main concern is that for higher and higher redshifts the total opacity of the Lyman α forest increases, so that more and more of an object's spectrum is blocked, Eventually, the blocking becomes severe enough to make the whole spectrum below Lyman α inaccessible, adding to the inaccessibility between the Lyman limit and soft X-rays produced by the ionization of even a small residual of neutral hydrogen along any line of sight. If we can find objects at large enough redshifts to see before reionization was almost complete, there should be enough neutral hydrogen surrounding a young galaxy to absorb almost all emerging Lyman α radiation as well as the continuum. This means that the mere detection of Lyman α from objects at large redshifts tells us important things about not only the objects themselves but their large-scale surroundings. The current record is held by a spectroscopically confirmed detection of Lyman α emission at $z = 6.56$, described by Esther Hu and collaborators in 2002.

9.11.3 Cosmological dimming

In an expanding Universe, the inverse-square law for propagation of radiation breaks down increasingly over long spacetime intervals, because the surface area encompassed by an expanding wavefront emitted at a particular time t_0 is no longer simply the surface of a static sphere of radius $c(t - t_0)$. One consequence of this, seen in the relations from Chapter 1, is that the surface brightness of an extended source is not redshift-independent, as it would be in a static and Euclidean situation. Instead, the surface brightness scales as $(1 + z)^{-4}$, a phenomenon known as Tolman dimming or the $(1 + z)^4$ dimming. The loss in surface brightness becomes especially dramatic at redshifts $z = 4$ and above, where it exceeds a factor of 600. This gives us an unavoidable bias in favor of observing the most compact, high surface-brightness regions of galaxies at these redshifts. Using the observed ultraviolet structures of nearby galaxies and incorporating Tolman dimming, one can show that many nearby galaxies, even spirals with significant star formation, would be completely undetected in the Hubble Deep Field data for this reason. At high redshifts, we can only see galaxies which are truly exceptional by the standards of today's Universe.

9.11.4 Gravitational lensing

The central insight of general relativity—that mass curves spacetime—means that the direction of light passing a concentration of mass will be deflected, a prediction confirmed in a spectacular way by 1919 solar-eclipse observations showing the deflection of starlight upon passing near the Sun. Einstein later considered the possibility of what we would now call gravitational lensing (perhaps in this instance the French "mirage gravitationelle" is a more appropriate description), in the context of one star lensing the light from another. In a 1936 paper he calculated that the probability of our observing a sufficiently exact stellar alignment was negligibly small. Potential applications of this phenomenon in the context of galaxies were developed more completely by Zwicky, who wrote of the eventual use of natural gravitational telescopes. Much of the analytical machinery needed to understand gravitational

Figure 9.14. The "Einstein Cross" gravitational lens system 2237+030. This consists of a nearby spiral galaxy (redshift $z = 0.039$) with four images of a background quasar at $z = 1.69$ shining around the nucleus. The alignment between our position, the galaxy, and the undisturbed quasar location has to be very exact, at the level of 0.05 arcsecond, for us to see such a symmetric arrangement. (This image was constructed from NASA/ESA Hubble Space Telescope archival data originally obtained by the WFPC2 instrument team with J. Westphal as principal investigator.)

lensing was developed in advance of their actual observation, by Sjur Refsdal and by Jeno and Madeleine Barnothy. The first clear evidence for gravitational lensing came in the late 1970s, half a century after Einstein's original predictions, with the discovery of multiple QSO images, separated by 1–6 arcseconds and showing identical redshifts and spectra (Figure 9.14). In some cases the lensing galaxies can be identified, sometimes a whole group. Later, lensed images of spatially resolved galaxies were identified through rich clusters, as the background galaxy's apparent structure becomes distorted into ringlike arcs through magnification in the tangential direction. In one striking case there are five distinct images of the same ringlike background galaxy (Figure 9.15).

The splitting of images depends on the mass distribution, distances of the source and lensing mass, and the vector impact parameter of the (undisturbed) light path

Figure 9.15. A spectacular instance of strong gravitational lensing is shown in the Hubble Space Telescope image of the cluster Cl 0024+1654 at $z = 0.39$. A background galaxy at $z = 1.7$, with a distinctive ring or θ shape, is quintuply imaged into our line of sight. Each of the images is significantly magnified in a single dimension, allowing a high-resolution reconstruction of star-forming regions in this galaxy. (Image from W.N. Colley and E. Turner of Princeton University, J.A. Tyson of Bell Labs, Lucent Technologies, and NASA.)

with respect to the mass distribution. In principle, one can reconstruct the important parameters of the mass distribution given the precise relative positions of the lensed images and center of mass, and the relative brightnesses of the images. This allows a particularly elegant formulation from Fermat's principle, calculating the ray paths with extremal travel times.

Gravitational lensing also changes the intensity of an image. In split-image (or strong) lensing, a net amplification occurs, through conservation of surface brightness when angular magnification occurs even along one direction of the image. To satisfy the conservation of energy, a small deamplification occurs for lines of sight passing far from the mass in question. To satisfy conservation of surface brightness, flux amplification stems from the increase in solid angle subtended by the radiation source. Therefore, gravitational lensing will potentially alter the flux distribution of distant objects, as the likelihood of significant amplification or deamplification grows with distance (simply because the line of sight passes more mass concentrations). This

Figure 9.16. Zwicky's "gravitational telescope" is illustrated by the detection of this galaxy at $z = 4.92$, and the visibility of its internal structure. Seen in this Hubble Space Telescope image, this object has been gravitationally amplified by the foreground cluster of galaxies Cl 1358+62 at $z = 0.33$, and was identified from its strong Lyman α emission. When discovered, this was the object with the highest known redshift. The lensed object was detected as a clumpy red arc near the lower right bright galaxy. (Image courtesy Marijn Franx of the University of Groningen, The Netherlands; Garth Illingworth of the University of California, Santa Cruz; and NASA.)

has been observed for the highest-redshift infrared-bright galaxy known from IRAS, IRAS F10214+4724 at $z = 2.3$. Its enormous energy output became rather less enormous when it was found to be in the form of an arc apparently wrapped around the image of a lower-redshift galaxy, a telling signature for lensing. The expected distribution of amplification values is very asymmetric, with most values slightly less than unity and a small tail reaching very high amplifications. As a result, objects amplified by lensing will generally be overrepresented in flux-limited samples of high-redshift objects, and a stochastic amplification error will be introduced into all flux measurements at high redshift. We can use this effect to increase our sensitivity and resolution on high-redshift galaxies, by seeking lensed objects behind clusters whose colors give high photometric redshifts, and examining them closely. One notable success has been observation of two strongly lensed galaxies at $z = 4.92$, for which the cluster lensing enhances the spatial resolution of HST images by a factor of nearly 5 along one dimension (Figure 9.16).

The lumpy nature of the mass distribution on galaxy and cluster scales means that lensing becomes more and more important for higher redshifts. This can be at once a hindrance and a valuable tool. Massive clusters of galaxies can produce multiple images of background galaxies, often spread into tangential arcs which are parts of the ideal Einstein ring. These were initially observed from the ground as "blue arcs", blue because the most prominent examples were images of luminous and blue background galaxies. HST imaging has revealed marvelous instances of many background systems being distorted, so that one can almost trace the potential shape by eye (Figure 9.15).

These arcs trace the lensing geometry richly enough to enable reconstruction of the cluster potential, dark matter and all. This turns out to be in good agreement with the potential derived from X-ray mapping of the hot intracluster gas, a strong piece of evidence that the missing ingredient in cluster dynamics is in the objects rather than a lack in our knowledge of physics.

Multiply imaged quasars and lensed arcs are examples of strong lensing. Weak lensing also occurs, and is ubiquitous at substantial redshifts. When the departure from flat space is weak, the net effect is to produce systematic shear of galaxy images, which can be recognized statistically once appropriate calibration measures have been taken in wide-field images. The observed signature is a net alignment of galaxies across some region of the sky, which will be redshift-dependent because of the weighting of lensing effectiveness with path length. This gives us light-independent masses of clusters, and average masses for classes of galaxies as well. At some level, lensing and shear limit how accurately we can measure the intrinsic shapes of distant objects, and their distribution across the sky, since the Universe is lumpy and full of matter (luminous and dark).

Box 9.2 The basics of gravitational lensing

In the small-angle approximation, which applies to any gravitational-deflection situation except rays passing near neutron stars or black holes, there are analytical expressions that can predict the characteristic deflection and amplification of light from background sources. Much of this formalism was worked out by Sjur Refsdal as early as 1964. The deflection angle α is derived from the potential Φ integrated along the ray, which in this approximation is close to a straight line where the potential is significant:

$$\alpha = \frac{2}{c^2} \int \nabla\Phi \, dl$$

which yields a vector quantity on traversing an arbitary potential. For a mass distribution which is symmetric about the source–observer axis, this becomes a scalar deflection (always directed outward from that axis)

$$\alpha = \frac{4b}{c^2} \int_b^\infty (r^2 - b^2)^{1/2} \frac{d\Phi}{dr} \, dr.$$

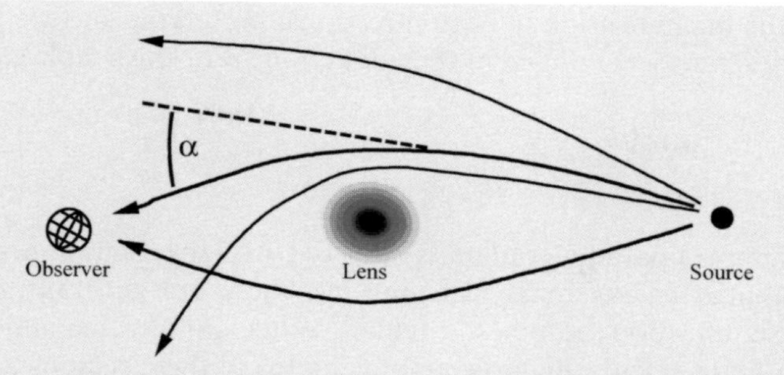

Figure 9.17. Schematic illustration of strong gravitational lensing. The deflection angle α is a function of the impact parameter of the rays (where we can speak of a simple value only because all the angles involved are small). The mass distribution and observer's location define the apparent position from which source rays can be seen. In strong lensing, there exist multiple paths (shown with heavy curves) for radiation between the source and observer. The relation between impact parameter and alpha depends on the distances and the lens mass distribution. In realistic cosmologies the geometric distances to the lens, and between lens and source, will not sum to the distance from the source to the observer.

For the simplest case of a point lensing mass, which is often appropriate for lensing by an individual star, $\alpha = 4GM/bc^2$. For small deflections, this can be generalized by letting M be the mass within the cylinder enclosed by the beam's impact parameter b. The integration is cylindrical rather than the spherical one for gravitational attraction, because the deflection depends on the inverse, rather than inverse square, of the impact parameter. Straightforward integration will yield the behavior of deflection with impact parameter for any desired density distribution (Figure 9.17). For example, the analytic King model for a galaxy's mass (without tidal truncation) gives

$$\alpha = \frac{8\pi G\rho_0 a^3}{bc^2}\ln(1 + r^2/a^2)$$

for central density ρ_0 and core radius a. This shows the generic feature of extended galaxy masses of having a maximum value (in this case $\alpha_{\max} = 45.2G\rho_0 a^3/c^2$ at $b \approx 1.8a$). On topological grounds, nonsingular mass distributions yield odd numbers of images, although one of these is often demagnified and very close to the lens center. In fact, there are no well-observed cases in which an odd number of lensed images has been seen.

To derive image locations from the lens location and (undeflected) source position, we use the requirement that the ray passes through the observer's position. This introduces the distances between observer and lens, and source and lens, both of which should be measured in angular-diameter distance since it is exactly the angular behavior that matters.

The run of deflection with impact parameter gives the image's flux amplification A for any point in the background object, simply from the areal amplification (which comes from the magnification in both directions, preserving surface brightness as the image is differently magnified in each direction). For the simple case of a point mass,

$$A = \frac{1}{2}\left[\frac{\sqrt{1+r^2}}{r} + \frac{r}{\sqrt{1+r^2}}\right] \approx (2r)^{-1} \qquad (r << 1)$$

where r is the impact parameter in units of $c/4\sqrt{GMD}$ for a lens at proper distance D. For distributed galaxy mass distributions, there are generally curves along which the magnification becomes singular. While infinite magnification never occurs over a finite extent, the large amplifications in these regions are important in modeling lenses, and in making effective use of "gravitational telescopes".

9.11.5 Galaxies

Galaxies along the line of sight to background objects naturally offer additional ways to block their light, even if it passes far out in the gaseous halo. Their signatures can be found in QSO light. Some systems of gas seen in front of QSOs have much higher column densities and metal abundances than the lines of the Lyman α forest. Depending on the redshift and amount of gas, they may be identified as *metal-line systems*, *Lyman-limit systems*, or *damped Lyman α systems*. Even the outer regions of disk galaxies have enough H I to produce very strong absorption at Lyman α. The strongest such absorption has large equivalent widths, so large that the line is in a regime where the wings reflect the intrinsic quantum properties of the hydrogen atoms rather than Doppler motions (hence the term *damped*, from the terminology of damping wings applied to these parts of a spectral-line profile). These systems have column densities above about 10^{18} atoms per square cm. Well below this, the gas will be optically thick just shortward of the ionization edge at 912 Å (which we see at the redshift of the gas, not the QSO), producing a cutoff at this point in the QSO spectrum. These kinds of absorption systems are sufficiently numerous to account for most of the neutral gas in the Universe at accessible redshifts, and have therefore been used to trace the history of gas as it accumulated into galaxies. Lyman-limit systems are common enough that they limit our ability to measure the He II Gunn–Peterson effect, and thus ionization of the intergalactic medium; we can do this measurement only along a few "lucky" lines of sight that do not pass close to any intervening gas-rich galaxy. Galaxies cover most of the sky in H I by the time we look as far as $z = 3$.

Some objects are detected only from the absorption lines of such "metals" as Al, Si, and C in QSO spectra. The hydrogen lines may be weak, or unobservable (cheaply, from the ground) because of the object's redshift. These absorbing systems bear some resemblance to the hot gas in our galaxy's halo, and are often considered to be closely linked to galaxies.

For light passing within what we would normally think of as a galaxy's inner regions (the ones bright enough to show in most images), absorption by dust in gas-rich systems will also occur. This will generally also redden the light, and can result in missing some objects from statistical samples because they are now too faint at the wavelength in question. Followup of radio surveys has led, for example, to identification of a gravitationally lensed quasar which is almost completely extinguished in visible light, and the spectra of an ensemble of gravitationally lensed quasars has been used to infer the typical reddening law in the lensing galaxies. Dust extinction has occasionally been invoked to explain the paucity of quasars at the highest redshifts, if the cumulative amount of dust extinction is large enough. This does not appear to be plausible from detailed studies of dust in nearby galaxies, so that we are left with the apparent turn-on of quasars and a space density peaking near $z = 2.2$ as real phenomena.

These observations and limitations mark the end of our ability to peer directly toward the Dark Ages. To probe earlier we must rely on theory, starting from our understanding of conditions in the early Universe and the behavior of matter under those conditions. As the next chapter outlines, we are starting to see a consensus on what happened during these times and what we should look for as new techniques allow us to penetrate some of the darkness.

9.12 BIBLIOGRAPHY

Journals

Cowie, L.L.; Songaila, A.; Hu, E.M.; and Cohen, J.G. (1996) "New Insight on Galaxy Formation from Keck Spectroscopy of the Hawaii Deep Field", *Astronomical Journal*, **237**, 3137–3145. The first clear statement of galaxy downsizing.

Einstein, A. (1936) "Lens-Like Action of a Star by the Deviation of Light in the Gravitational Field", *Science*, **84**, 506–507. The initial announcement of the possibility of gravitational lensing in the context of stars beyond the Sun. Einstein concluded that the probability of observing star–star lensing is negligible, which fits with the eventual detection of the phenomenon by monitoring millions of stars at once.

Fomalont, E.B.; Kellermann, K.I.; Partridge, R.B.; Windhorst, R.A.; and Richards, E.A. (2002) "The Microjansky Sky at 8.4 GHz", *Astronomical Journal*, **123**, 2402–2416. Examines the galaxy content of extremely faint radio sources, showing that these sources are generally star-forming, often interacting, galaxies, rather than the active nuclei which are important at higher flux levels.

Hu, E.M.; Cowie, L.L.; McMahon, R.G.; Capak, P.; Iwamuro, F.; Kneib, J.-P.; Maihara, T.; and Motohara, K. (2002) "A Redshift $z = 6.56$ Galaxy behind the Cluster Abell 370", *Astrophysical Journal Letters*, **567**, L75–L79. The current galaxy redshift record-holder, found by a combination of Lyman α emission and gravitational lensing. The strong Lyman α emission indicates that reionization was well underway by this redshift; otherwise the intergalactic medium would have absorbed the line photons close to the galaxy.

Keel, W.C.; Wu, W.; Waddington, I.; Windhorst, R.A.; and Pascarelle, S.M. (2002) "Active Nuclei and Star-Forming Objects at $z > 2$: Metallicities, Winds, and Formation Histories", *Astronomical Journal*, **123**, 3041–3054. Near-infrared observations of the compact Lyman α emitters found by Pascarelle *et al.* (1996) indicate that some have very low oxygen abundance. They are either chemically young, caught very early in their star formation history, or periodically sweep enriched gas away in starburst winds. The active nuclei in the same region have near-solar abundances, confirming that AGN are systematically associated with deeper potential wells and perhaps the earlier onset of star formation.

Kundu, A. and Whitmore, B.C. (2001) "New Insights from HST Studies of Globular Cluster Systems. I. Colors, Distances, and Specific Frequencies of 28 Elliptical Galaxies", *Astronomical Journal*, **121**, 2950–2973; and Larsen, S.S.; Brodie, J.P.; Huchra, J.P.; Forbes, D.A.; and Grillmair, C.J. (2001) "Properties of Globular Cluster Systems in Nearby Early-Type Galaxies", *Astronomical Journal*, **121**, 2974–2998. These papers correlate properties of the globular cluster systems in galaxies of early Hubble type—the best candidates for merger remnants—to the overall galaxy properties, to show that many must have completed their formation quite early. This raises the issue of how early we would speak properly of a merger rather than the acquisition of a purely gaseous cloud without constitutent stars.

Munn, J.A.; Koo, D.C.; Kron, R.G.; Majewski, S.R.; Bershady, M.A.; and Smetanka, J.J. (1997) "The Kitt Peak Galaxy Redshift Survey with Multicolor Photometry: Basic Data", *Astrophysical Journal Supplement*, **109**, 45–77. Results of one of the earliest long-term deep surveys of galaxy redshifts, from Kitt Peak. The sample includes many galaxies at $z > 0.2$ in complete subsamples, our first view of representative galaxies at these epohcs in such numbers.

Owen, F.N.; Ledlow, M.J.; and Keel, W.C. (1995) "Optical spectroscopy of radio galaxies in Abell clusters. I: Redshifts and emission-line properties", *Astronomical Journal*, **109**, 14–25

Pascarelle, S.M.; Windhorst, R.A.; Keel, W.C.; and Odewahn, S.C. (1996) "Sub-galactic clumps at a redshift of 2.39 and implications for galaxy formation", *Nature*, **383**, 45–50. Detection of compact Lyman α emitting objects in a grouping at $z = 2.4$. These are very small and have high star-formation rates, consistent with an early stage in the hierarchical buildup of luminous galaxies.

Refsdal, S. (1964) "The gravitational lens effect", *Monthly Notices of the Royal Astronomical Society*, **128**, 295–306. A detailed treatment of lensing, giving the mathematical formalism used by many investigators in interpreting observations of lensed quasars.

Spinrad, H.; Dey, A.; Stern, D.; Dunlop, J.; Peacock, J.; Jiminez, R.; and Windhorst, R. (1997) "LBDS 53W091: An Old, Red Galaxy at $z = 1.552$", *Astrophysical Journal*, **484**, 581–601. This radio galaxy at $z = 1.55$ shows a remarkably old stellar population, putting a lower limit on the time since star formation began. Similar instances are now known; systematic searches for the reddest galaxies at high redshift are one way to determine the timing of galaxy formation.

Sunyaev, R.A.; Tinsley, B.M.; and Meier, D. (1978) "Observable properties of primeval giant elliptical galaxies or ten million Orions at high redshift", *Comments on Astrophysics*, **7**, 183–195.

Teplitz, H.I.; Malkan, M.A.; Steidel, C.C.; McLean, S.; Becklin, E.E.; Figer, D.F.; Gilbert, A.M.; Graham, J.R.; Larkin, J.E.; Levenson, N.A. *et al.* (2000) "Measurement of [O III] Emission in Lyman-Break Galaxies", *Astrophysical Journal*, **542**, 18–26. Uses data on [O III] line emission, redshifted into the near-infrared bands, to examine the abundances of ionized gas in Lyman-break galaxies at redshifts up to $z = 3.4$. Their oxygen abundances are subsolar, but not extremely low as found for nearby blue compact galaxies such as I Zw 18.

Venemans, B.P.; Kurk, J.D.; Miley, G.K.; Röttgering, H.J.A.; van Breugel, W.J.M.; Carilli, C.L.; De Breuck, C.; Ford, H.; Heckman, T.; McCarthy, P. *et al.* (2002) "The most distant structure of galaxies known: A protocluster at $z = 4.1$", *Astrophysical Journal Letters*, **569**, L11–L14. Discovery of a young cluster of Lyman α emitting galaxies around a powerful radio galaxy at very high redshift.

Zwicky, F. (1937) "Nebulae as Gravitational Lenses", *Physical Review*, **51**, 290 and "On the Probability of Detecting Nebulae Which Act as Gravitational Lenses", *Physical Review*, **51**, 679. These brief, single-page outlines represent early calculations indicating that gravitational lensing may be observable in the context of galaxies as source and lens, and the use of this effect in deriving galaxy masses free of dynamical assumptions.

Internet

http://www.stsci.edu/ftp/science/hdf/hdf.html Project website for observations of the Hubble Deep Field.

http://www.stsci.edu/ftp/science/hdfsouth/hdfs.html Hubble Deep Field–South observations.

http://www.stsci.edu/hst/udf The Hubble Ultra-Deep Field.

http://cfa-www.harvard.edu/castles/ The CASTLES project (CfA-Arizona Space Telescope LEns Survey) maintains this website, with a general treatment of contemporary gravitational-lens theory and a current data compilation.

10

The processes of galaxy formation

We now come to the central problems in galaxy formation: When did galaxies form, over what timespan, and by what physical processes? What were the first self-luminous objects in the Universe? How did the oldest of today's stars come by their small, but important, allotments of heavy elements? And why do we see the Universe today dominated by galaxies, instead of individual globular clusters, single stars, or even some kind of supergalaxy spanning a billion light-years?

Understanding the process of galaxy formation must remain a theoretical endeavor, starting from the conditions in the pregalactic Universe and the known physical laws and properties of matter. The initial conditions are not known with the precision that would be needed to reconstruct the processes in detail, of course, so that a certain amount of inference and working backwards from the cosmos—as we now see it—has been inevitable.

The starting point for building galaxies is what we saw from the cosmic microwave background, just after the time of recombination: a dilute gas of hydrogen and helium, filling space almost uniformly. The "almost" is the key to our existence—the ripples in the density of this gas, of a few parts per million, were the seeds of excitement to come. These gas fluctuations need not have been the only ripples present, since dark matter might well have started to clump on its own before ordinary matter could follow, after being released from the smoothing pressure of radiation.

Various sets of fluctuations in the gas density could give rise to different outcomes, of which the two most-discussed have been the extreme cases of top-down and bottom-up formation. In the top-town scheme, the first objects to become separated from their surroundings and be gravitationally bound would correspond roughly to present-day galaxy superclusters, at the level of quadrillions (10^{15}) of solar masses. Since these vast lumps would likely be irregular or flattened, they have been known as Zel'dovich pancakes, after the work of Yakov B. Zel'dovich (1914–1987) in analyzing the process. This scenario ties in with the classical galaxy formation picture from Eggen, Lynden-Bell, and Sandage, taking this huge cloud to fragment into cluster- and finally galaxy-sized elements, at which point star formation would begin in earnest and galaxies could be said to form. A top-down sequence would be expected if the long-wavelength ripples in the pregalactic gas carried more power than the shorter ones, so that matter would clump on the largest scales first.

In contrast, if the smaller-scale fluctuations are more important, the first masses to collect and become gravitationally independent units would be small, with these objects gathering each other and their surrounding material over time. Since the typical size and mass of objects in this case would grow with time in a piecewise manner, this is known as hierarchical galaxy formation. If the growth and merging happened slowly, this would resemble galaxy mergers in today's Universe—the fragments would, beyond a certain point, already contain stars. If so, we could see the process as a buildup of the mass and luminosity of stellar systems with cosmic time.

The development of these fluctuations proceeds at a rate depending not just on their amplitude, but on their linear size. For a particular region of mass overdensity (expressed as the ratio to mean density), it will grow as its gravitational influence has time to affect elements of surrounding material at greater distances. Such a parcel of gas will begin following the Hubble expansion, being gradually decelerated by the net gravity of the nearby mass concentration. The relative velocity reaches zero at the time of *turnaround*, at which the system effectively decouples from the overall expansion. Thereafter, the gas (quite possibly itself clumpy) will fall inward to join the growing overdensity.

This process of turnaround and infall continues in clusters of galaxies today. The redshifts of galaxies within a large local volume show evidence of so-called Virgocentric flow, a net infall toward the Virgo Cluster. Telltale signs of similar flow have been seen as ripples in the redshift–distance relation for galaxies just in front of and behind additional rich clusters. The units joining clusters today are not just galaxies, but groups and smaller clusters.

As in the case of the microwave background, it is useful to consider the density field before galaxy formation through its power spectrum $P(k)$, the amount of power on scales given by inverse wavelength (or wavenumber) k, essentially the Fourier transform of the density distribution. This is of interest since theoretical scenarios predict different slopes for the $P(k)$ relation for different dominant processes, and because we can measure $P(k)$ today for comparison. The power spectrum of galaxies can be measured from redshift surveys, and a more detailed reconstruction estimating departures from a pure Hubble flow can estimate what the overall mass density field is as well.

The difficulty in reconciling all three estimates was one of the early drivers behind the notion of biased galaxy formation. The basic idea is that the distribution of galaxies is so strongly clumped (e.g., as measured by the two-point correlation function) that it's hard to concoct a process that could start from the small observed amplitude of CMBR fluctuations and end up with the observed structure in a mere 15 billion years. This led to consideration of whether galaxies fairly trace mass—that is, are the odds of a galaxy forming from a given mass of gas the same everywhere, or do they depend on the environment? This is generally quantified through a biasing parameter b, defined so that (statistically) galaxy overdensity is larger than mass overdensity by this factor:

$$dG = b\,dM$$

where each overdensity is the local mean, on some size scale of interest, divided by the cosmic mean for each value. This formulation says nothing yet about how biasing might arise physically—it could apply to a threshold density for galaxy formation, a required local peak density, or something else again.

The growth rate of perturbations at various size scales is obviously crucial in galaxy formation. On large enough scales, the growth should still be in a linear regime, in which the interpretation and history has been straightforward. On the scales responsible for galaxies, these fluctuations began nonlinear growth early on. Large fluctuations are still linear today, giving us the chance to make a connection to at least part of the early power spectrum and its growth. This was exploited by Press and Schechter (1974) in an approach which remains valuable in assessing simulations of galaxy and structure formation. They found that the larger, linear scales furnish a useful reference point for defining mass concentrations. Masses are identified using a characteristic length scale, which is important since the density field may have grown in a nonlinear way on small scales so that the structure is arbitrarily complex. This makes identification of discrete objects (such as protogalaxies) virtually impossible unless certain length scales are found to be important. The situation is analogous to some well-known examples of fractal-like behavior, such as length of a shoreline or number of peaks in a mountain range.

10.1 COOLING THE GAS

Like star formation, galaxy formation is more than a gravitational process. In order to collapse—and, in particular, in order for the baryonic material we see to collapse to a more compact configuration than dark matter—the initially uniform and warm gas must be able to cool. Virtually all mechanisms for cooling available to atomic and molecular gas are collisional, taking place only during two-body interactions. Thus, the cooling rate has a basic dependence on the collision rate, which scales as the square of particle density. On top of this dependence, the cooling rate changes dramatically with temperature as various specific processes become possible. This is expressed in the cooling function for a particular set of chemical abundances, which shows how much energy will be lost per unit mass and unit time at a particular temperature and density. There are several broad peaks in the cooling function (Figure 10.1). A gas with no energy input will drop rapidly through these regimes of efficient cooling and linger when cooling is less efficient, explaining why certain temperature ranges (e.g., around 10^5 K) are rarely encountered in cooling plasmas. At high temperatures, the material is fully ionized and cools via *bremsstrahlung* (German for "braking radiation", in which language the noun must be capitalized).

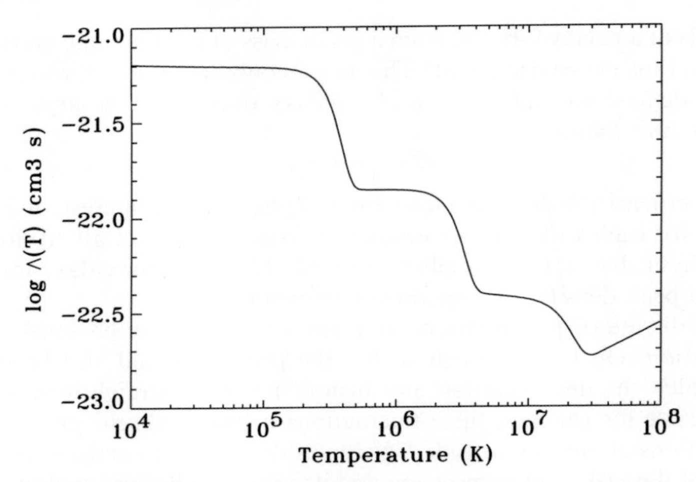

Figure 10.1. The cooling function for gas at solar abundances. The various parts of the curve correspond to distinct processes; collisionally excited emission lines (in the optical and infrared) peak near 10^4 K, with higher-energy lines in the ultraviolet and X-ray important at higher temperatures. At the highest temperatures, thermal bremsstrahlung operates; gas in many clusters of galaxies is in the slowly-cooling valley between this mechanism and the highest-temperature line cooling. At low metallicities, cooling becomes efficient only at lower temperatures and high densities, since the primary line cooling is through molecular hydrogen (including HD). The cooling rate plotted should be multiplied by the square of the particle density for simple collisional processes. (This plot uses the parametrization by Sarazin and White.)

This radiation results from electron–proton encounters in which the electron radiates because of the transient acceleration produced by Coulomb attraction. This is also known as "free–free radiation", to distinguish it from the situation in which the electron becomes bound to the proton in "free–bound radiation". This is the process by which the hot intracluster medium is detected, and its efficiency at intracluster densities is low—so low that most cluster gas has a cooling timescale exceeding a Hubble time.

The cooling processes more relevant for initial formation of galaxies are those that set in at a few times 10^3 K, after recombination. In this regime, familiar emission lines are the important cooling agents, either by recombination or through collisional excitation of various ions. The effectiveness of this cooling depends substantially on the abundances of heavy elements, since it is their emission lines that provide much of the cooling even for gas that is metal-poor by solar standards.

For cooling by collisionally excited transitions, the metallicity is obviously important. These transitions arise in excited states of ions which get their energy from collisions with electrons, whose energy depends on the gas temperature. Such ions will spontaneously decay via line radiation, in times ranging from less than 10^{-4} to several times 10^4 seconds, the high values applying to so-called forbidden lines which arise from metastable upper levels. These lines are so important for cooling because the excitation rate is a very rapid function of electron temperature, and the optical depth to internal absorption for such transitions is negligible, so the radiation escapes immediately.

In H II regions, which are being constantly heated by deep-UV starlight, we see in our own neighborhood that the equilibrium temperature correlates inversely with metallicity. The higher the abundances of heavy elements, the more cooling processes operate efficiently, and the cooler the temperature at which energy absorption balances emission. For galaxy formation and the formation of the first generation of stars, only hydrogen and helium would have mattered, which means that only recombination radiation would act as a coolant in this temperature range. In galaxy formation, in contrast to local nebulae, the gas would not be in ionization/recombination equilibrium, so that any parcel of gas would cool successively through various temperatures and cooling mechanisms.

At lower temperatures, if molecules can form (which means H_2 and HD at zero metallicity), additional cooling can occur in molecular lines. Such cooling is dramatically stronger for nonzero metallicity, in which not only can molecular lines from other species be important even at temperatures too low to excite transitions in H_2 (which, being a symmetric molecule, has zero dipole moment and therefore its low-lying transitions are extremely weak), but additional cooling can be provided by collisionally excited fine-structure lines from such neutral atoms as O and C. These atoms have very low-energy states whose decays to the ground level produce emission lines at 63 and 158 microns (for [O I] and [C II], respectively), and they are very important cooling agents for near-solar metallicities.

In addition to these atomic processes, cooling through the thermal emission from dust grains is important in today's Universe, especially during late stages of cloud collapse leading to star formation. Particle collisions and absorption of ultraviolet and visible radiation can heat the grains, which then radiate at 10–40 K. The grains' near-blackbody radiation peaks at such long wavelengths (40–160 μm) that it can escape unabsorbed. Once again, though, this avenue for cooling was not available for galaxy formation from pristine material, even with the salting of heavy elements expected from the first stars.

Cooling of the pregalactic gas allowed baryonic matter to decouple from the potential wells defined by dark matter and thus collapse to configurations dense enough to make stars. There is an interesting coincidence between the sizes of objects that would have relatively short cooling timescales and the masses of present-day galaxies, originally pointed out by George Blumenthal *et al.* (1984). This is just what we might expect as the conditions for objects to be able to collapse under their own gravity, and suggests that indeed these basic physical processes can have very large results.

10.2 MODELING GALAXY FORMATION

Numerical simulation of galaxy formation, incorporating enough physics to be realistic, is a formidable computational problem. It must be done in a cosmological context, ideally so as to have self-consistent connection to the larger-scale structures forming at the same time. This then falls out of the large-scale structure problem, one of the major "Grand Challenge" computational topics which have been important in recent years (others being turbulence and weather forecasting). A range of algorithms has been applied, leveraging various approximations to improve both the speed and accuracy of the codes. Fortunately for our collective sanity, a test of about a dozen such routines from the same initial conditions gives reassuringly similar outcomes. Some of the most successful models for galaxy growth have been semi-analytical, using approximate expressions for the impact of such small-scale processes and star formation and energy input from supernovae, while purely numerical techniques track the larger-scale structures. Comparison of semianalytical and purely numerical approaches confirms that they give similar results on large scales, with divergence on scales of individual galaxies, as might be expected from the resolution of these techniques.

Contributing to the difficulty of modeling galaxy formation is that the simulations just dealt with really track the mass (usually both baryonic and dark matter, in some assumed ratio). What we see is further modulated by biasing and by the history of star formation, gas heating, and mass loss of individual systems. This differs from the problem of star formation (already difficult enough) in that the dark matter and baryonic matter have separated dynamically, and the material we observe does not trace the mass distribution directly.

Ideally, a credible model for galaxy formation would reproduce the statistical observables of galaxies today as well as the populations we can observe at high redshifts. It should reproduce the luminosity function and mass-to-light ratios of present galaxies, and their observable guises in the Tully–Fisher and fundamental-plane relations. It should also predict the observed correlation function of galaxy clustering in both scale and amplitude. Matching the metallicity and color distributions of observed galaxies would also be a good sign. With all these desiderata, it is not surprising that we don't yet have a truly comprehensive model for galaxy formation and growth, but the state of the art does match the actual Universe in some important respects (Figure 10.2).

Simulations with mass density dominated by cold dark matter, in a form which interacts only gravitationally with itself and ordinary matter, have proven successful in many respects. For plausible prescriptions for how to go from mass to light and mass distribution to galaxy type, they give the right range of masses, Tully–Fisher relation, clustering statistics, and probably chemical-abundance distributions. However, there are some predictions that these simulations generically miss. They show a few hundred dwarf companions for each Milky Way-like galaxy, far beyond the numbers we can find. An obvious solution would be that these companions have long since merged with massive galaxies. Indeed, we see some evidence of this in the Milky Way's own halo, but it is not clear that the data can accommodate hundreds (rather than tens) of former companions in this way. Deep galaxy counts to lower surface-brightness limits would be powerful in testing this idea, since we would expect to see a clear distinction in the immediate environments of galaxies with redshift if companions are being absorbed so rapidly. At the other extreme, non-interacting

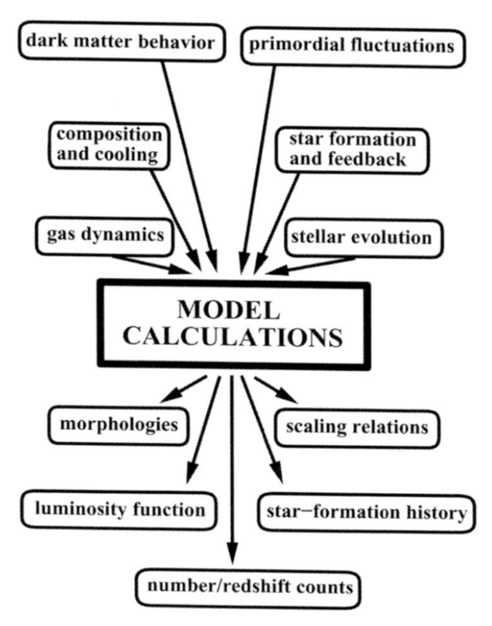

Figure 10.2. A schematic flow chart of the inputs and checks for calculations of galaxy formation. The inputs come from cosmology, via the spectrum of fluctuations at recombination and the distribution and properties of dark matter, and astrophysics, in the forms of gas dynamics, cooling behavior, and stellar lifecycles for given mass and composition. The outputs can be, in principle, any observables for galaxies at various redshifts that can be connected to the simulations. Because of the huge range in physical scales that are involved, various approximations have to be made in these connections, so the comparison with real galaxies cannot yet quite "close the loop". (Adapted from an original by Carlton Baugh.)

CDM models predict that the central mass distribution of galaxies peaks more strongly than we find in spirals. One solution being explored is that dark-matter particles may be able to interact with each other beyond gravity, which would decrease their central concentration. In each of these areas some astronomers feel that the basic physical processes are in hand, but that properly dealing with such dissipative "gastrophysics" will require a new generation of computational acumen. Others remain unconvinced that we're not missing something.

10.3 FIRST LIGHT

What should we look for as the earliest observable signature of protogalaxies, whether large or small? Starlight is a natural choice, since one perfectly defensible way to define a galaxy includes containing stars. But, even so, what should we expect? Star formation at essentially zero metal abundance is hardly guaranteed to act like what we find here and now. Several pieces of theory suggest that such stars would be much more massive than can form today, possibly ending their lives in supernova explosions even more powerful than we see now. Calculations of stars formed from the initial complement of elements have converged to a consistent, and sometimes surprising, picture of what the first stars were like. It is this generation of stars that eased all subsequent star formation, by producing the first input of heavier elements whose presence gives us the patterns of stellar formation, structure, and evolution that we see today.

Numerical simulations show that, before the first stars formed, the cosmic web of gaseous structure extended from scales of megaparsecs down to hundreds of parsecs, being almost scale-free over this wide dynamic range. Just as seen for galaxies later, the first stars would form at the dense nodes where sheets and filaments crossed.

This first generation of stars differed in only one respect from their successors, but this one difference changes everything. They formed with no elements heavier than a tiny amount of lithium. The lack of heavier elements means that the possible cooling of protostellar gas must be completely different from what we see today. It also means that the stars would have had hotter cores, larger masses, smaller sizes, and much weaker winds than we are accustomed to.

Just as in star formation today (or galaxy formation, for that matter), these first stars could form when self-gravity of some region of enhanced gas density overcame the internal (here, thermal) pressure. To collapse far enough for fusion to set in, significant cooling of the gas is needed, preferably some form of cooling that becomes stronger at high densities so that the collapse becomes nonlinear and runs away. In today's Universe, the cooling of molecular gas in star-forming regions relies on heavy elements, in both dust and gas phases. Dust particles absorb radiation most efficiently in the ultraviolet, and can be heated by atomic impacts as well, cooling by radiation deep in the infrared where it can escape the cloud freely. Heavy elements provide gas-phase cooling through emission lines which are excited by collisions with other particles (usually electrons), and which are at wavelengths long enough to escape from dense and dusty environments. Even for abundances well below the solar level, it is this small fraction of heavy ions which dominate the cooling over hydrogen and helium processes. Some of the most important coolants are fine-structure lines of oxygen and carbon at 63 and 158 μm, respectively; the latter C^+ line often accounts for 1% of a galaxy's entire energy output.

At zero metallicity, the situation is dramatically different. None of these cooling processes, mediated by heavy atoms, is available. Cooling must happen via the H_2 and HD molecules. These molecules would not form as efficiently as they do in current molecular clouds, lacking solid grains where temporary surface adsorption catalyzes the formation of molecules. Instead, their formation was catalyzed by the tiny residual ionization at these early epochs, when perhaps 10^{-4} of electrons had not combined to form neutral H atoms. These electrons in turn produce a small abundance of the weakly-bound H^- ion by the reaction

$$H^0 + e^- \rightarrow H^- + \gamma.$$

This tiny fraction of H^- has far-reaching effects, because it allows the formation of

H_2 directly in the gaseous phase, without violating conservation of energy and momentum as an ordinary collision of neutral atoms would do, via the reaction

$$H^- + H^0 \rightarrow H_2 + e^-.$$

It is thus important that the original hydrogen is partially ionized, even if the ionized fraction is very small; neither fully ionized nor neutral gas would form H_2. This molecule can carry away energy through rotation and vibrational energy states whose transitions fall in the near-infrared. The slowest cooling occurs when the gas cools to around 200 K, such that this is where the cooling of a parcel of gas will slow, so that much of a collapsing cloud will naturally be near this temperature. Numerical simulations using these cooling processes, starting with cosmological structures for the density distribution, can follow the fate of a collapsing object as it separates from the surrounding expanding material. This happens on small enough scales that dark matter ceases to dominate events.

Even with this molecular mechanism operating, the gas could not cool to the temperatures we see in the cores of molecular clouds today. The cosmic microwave background was at that time still the cosmic far-infrared background, with a temperature at an epoch we see at redshift z higher than we now see, by a factor $(1 + z)$. For redshifts 10–20, this gives minimum temperatures for matter in equilibrium with this radiation bath in the range 30–55 K.

For genuine gravitational collapse, an object must have a cooling timescale more rapid than the freefall time of the system (a condition set out by Martin Rees and Jeremiah Ostriker). This requirement suggests clumps with masses of order 10^6 solar masses, which would have decoupled from the overall expansion at $z = 20$–30. These would have gotten a gravitational "head start" from sitting in peaks of the dark-matter distribution, but thereafter cooling and particle interactions allow the normal baryonic material to run ahead, collecting more tightly. The first smaller "protostellar" clumps of gas to become gravitationally bound in this environment are about as predicted from the classical Jeans calculation—a match that we see here precisely, because dark matter will be smooth compared with the sizes of these clouds, which would encompass typically 10^3 solar masses. Much of this mass could go into a single primordial star, which would belong to the otherwise mythical "Population III" which has yet to be seen but must have produced the first heavy-element enrichment.

These stars could grow to very high masses, thus becoming very hot and luminous compared with any stars in the contemporary Universe. The maximum mass to which a star can grow today, as it accretes surrounding gas, is limited by several instabilities, all of which are weaker or absent for a pure H/He star. For example, a star will stop accreting mass when its luminosity exceeds the Eddington limit, where the radiation pressure blowing the gas away exceeds the gravitational force inwards. The radiation pressure is proportional to the opacity of the gas, which for the relevant temperatures is largely contributed by spectral features of heavy elements. Primordial stars should mostly lie in the range 50–300 solar masses for this reason. Their accretion would be terminated by their own ultraviolet radiation; as soon as the star is hot enough to dissociate the surrounding H_2, the accretion will be effectively terminated. So would the growth of any other protostar in a typical cloud of a million solar masses, as a single star of this mass would destroy the molecules throughout its initial cloud and thereby commit infanticide on any potential siblings. To our modern eyes, these particular galaxy precursors would seem very minimalist, containing only a single star each.

These stars would be very luminous, even more luminous than their great masses suggest. During their "main sequence" phase of core hydrogen fusion, they would be smaller and have hotter cores than we now see. Hot stars today fuse hydrogen through CNO cycles, in which even a tiny original carbon abundance serves as an efficient catalyst for fusing hydrogen to helium. Lacking carbon, hydrogen fusion must proceed through the *p–p* reactions as in the Sun's core, which balances internal energy production and gravitational pressure at a higher temperature. The emerging luminosity comes out in a harder spectrum, since the stars lack the metal atoms which have strong opacity deep in the ultraviolet, and redistribute the radiation to longer wavelengths in metal-rich stars. One of these stars could ionize the gas in most of its surrounding cloud, producing immense and spectacular emission nebulae (albeit nebulae that we may find extraordinarily difficult to observe). We already know on very direct grounds that these stars were gone by about $z = 5$, since we don't see their enormous H II regions as Lyman α sources. However, this observation is not very informative at much larger redshifts, since a neutral intergalactic medium (before reionization) could soak up Lyman α photons very effectively so that the observability of Lyman α depends critically on the density and velocity field around an ionizing source.

Just as massive stars do today, these objects (variously termed very massive objects, VMOs, or very massive stars, VMSs) would have short lifetimes and explode as supernovae. After a few million years, one of these objects would undergo one of several runaway mechanisms leading to either its complete disruption in a supernova or formation of a central black hole, with or without a brilliant explosion. Here again, the important processes are somewhat different than in today's metal-rich stars. First, these stars would keep most of their initial mass, rather than losing much of it to winds, as a direct consequence of most wind-producing processes acting through radiation on heavy elements. Next, the hot cores of stars beyond 100 solar masses become vulnerable to pair-production instability. The temperature rises so high that gamma rays can form electron–positron pairs, each of which removes slightly more than 1 MeV of energy from the surroundings. The onset of pair production can drive the collapse of the core and initiate a supernova. In stars of other mass ranges, the temperature can become high enough to reverse the ladder of fusion reactions, breaking heavy nuclei back into helium nuclei (α particles) and those in turn to protons and neutrons. These processes absorb the same amount of energy released during the original fusion processes, again leading to collapse of the stellar core. The explosion of one of these massive stars would release about 100 times the typical kinetic energy of a typical supernova, up to 10^{53} ergs. These supernovae would also produce much larger amounts of enriched material than their successors, with some models showing 100 solar masses of enriched materials being ejected, compared with the 1–5 solar mass range computed for today's type II supernovae.

This first generation of stars was important not only energetically, but above all chemically. If a few times 10^{-5} of the baryons formed such stars, their debris accounts for the base level of metallicity found in galactic stars and intergalactic gas, about 10^{-4} of solar. Alternatively, about 1% of baryons were part of the protostellar clouds that gave rise to VMOs. The ubiquity of this base metallicity, and the distinct pattern to the relative abundances of heavy elements in the oldest stars, indicates that these objects scattered their ejecta widely, almost uniformly, through material which would then form the galaxies we see now. Even though these stars did not involve as much material as the later generations that we can trace at redshifts $z = 1$–4, their importance in setting the stage for everything that was to come must be recognized. Without these massive primal stars, none of the rest would have happened.

The disruption of these initial clouds, when their embedded massive stars exploded, means that there is a sense in which galaxy formation had to begin twice. The densest baryonic clumps collapsed first and hosted the initial, very massive, stars. The deaths of these stars disrupted their parent clouds, while salting the whole Universe with its initial complement of heavy elements. Only after this did the densest remaining structures collapse to form the seeds of today's galaxies.

It was not guaranteed that stars would be the first luminous objects to form. For several years after the first refurbished HST images of QSOs appeared, there was discussion of whether active nuclei could provide the "first light", perhaps from some kind of primordial seed black hole around which galaxies grew. This has been rendered unlikely by two realizations. First, the number of "bare" quasars, those without a detectable surrounding galaxy, is small (and may be zero). Second, the Magorrian relation (Chapter 8) shows that, on average, the central dark object contains about 0.5% of the mass of the surrounding spheroid's stars. It is much easier to see how the formation of a stellar system could lead to a constant fraction falling into a central black hole than how an original black hole could always manage to cloak itself in a constant mass of stars, whatever its original environment. Thus, it appears that stars (or their dynamical precursors in gas) came first. As the central mass grows, it eventually becomes an important contributor to the overall gravitational potential, influencing stellar orbits. Some simulations have shown that the central black hole in an isolated galaxy can stop its own growth in this way, by changing the distribution of stellar orbits and reducing the number of stars in the "death cone" of orbits from which stars can be accreted.

Massive stars, of the ordinary kind we still find today, certainly appear to have been around when the earliest quasars appeared, for chemical reasons. The emission-line spectra of quasars are remarkably consistent across the whole observed range of redshift. In particular, the emission features from metals—such as carbon, silicon, nickel, nitrogen—are just as strong (if anything, some are stronger) in quasars at high redshifts $z > 4$ as they are in the nearest examples. This suggests that massive stars had already formed and evolved so as to enrich the gas around the core with these heavy elements by the early epochs at which we see the highest-redshift quasars. Enrichment is most pronounced in those elements that are released by type II (core-collapse) supernovae, as we would see from a young population of stars.

This probably means that quasars appear only where the density has been high enough for massive star formation.

Even if they didn't start the process of galaxy formation, it is striking how soon massive central objects (observed as they power luminous quasars) appear on the scene. If some of them have masses like the ones dynamically measured in nearby galaxies, characteristic growth rates of a solar mass per year were needed. This rate would give a high, though not unique, QSO luminosity, but the starting point of this accretion is important. Growth of a stellar-mass black hole would be limited by radiation pressure (as the system reached the Eddington luminosity), so the accretion rate would be proportional to the mass as long as sufficient nearby material was available. In the idealized situation of a black hole accreting from a dense surrounding cloud of gas, its mass would grow exponentially with time. However, when we consider the existence and dynamics of stars in a galactic nucleus, this regime quickly ceases to apply, and there are significant thresholds in the processes that are efficient in delivering mass for accretion. The first marks the mass at which the black hole is able to tidally disrupt passing stars with useful frequency, since some or all of the gas liberated in this way would generally be accreted.

It is possible that the collapsed stellar remnants of a massive starburst in a growing galactic nucleus could congregate closely enough to become a single massive black hole, speeding the growth of this object to the point at which tidal disruption of passing stars would be important. This idea was considered in the context of Seyfert nuclei by Weedman in a 1983 paper. The remnants of massive stars would preferentially sink to the middle of the potential, following the principle of equipartition in celestial mechanics, under which a system of interacting particles will evolve toward equal distribution of total energy among them. Thus, the more massive ones will have smaller velocities and sink to the center. This could act on 10^7-year timescales for plausible starbursts, collecting as much as 10^7 solar masses with a parsec-sized region. If much of this mass could be quickly incorporated into a single massive black hole, this sequence would give a plausible way to start the growth of the supermassive objects inferred not only today, but already at the highest redshifts we can see.

The evolution and stability of a relativistic star cluster, one whose mass is sufficiently compact for the internal motions to be relativistic, are long-standing problems which have posed significant challenges, both analytic and numerical, for the last three decades. There are general reasons to expect a rapid collapse, but the details and whether a stable state can be reached seem to depend on the details of the mass and velocity distributions. *If* such a collapse occurs to an ensemble of stellar remnants, it would yield a black hole then capable of growing substantially on its own. This is far from certain—not only do we need to understand the fate of relativistic star clusters, but we need to know whether these stellar remnants would congregate quickly and tightly if they formed, but were not otherwise embedded in, the nucleus of a young galaxy.

A more exotic mechanism for the early growth of black holes has been suggested in connection with the possibility that the nonbaryonic dark matter consists of a species of sterile neutrino. "Sterile" in this context refers to a neutrino which mixes so weakly with the flavors we know of (electron, μ, and τ) as not to partake in a yet-observable way in the oscillations seen in reactor and solar neutrinos. Such particles would decay over long timescales into whatever neutrino flavors have lower mass, and would preserve some interesting pieces of physics first explored when initial reports suggested a large mass for the electron neutrino. In particular, such neutrinos would form degenerate configurations in mass concentrations (protogalaxies) comprising 10^4–10^6 solar masses within radii of a few parsecs. Such a configuration would collapse very quickly—once a stellar-mass black hole appeared within it—from normal processes of stellar evolution and not requiring a supermassive star. There is a mass range, rougly 5–10 keV, in which existing data allow such neutrinos to exist.

Whatever the initial mechanism, such a rapid growth to 10^6–10^7 solar masses would make the central black hole able to tidally disrupt passing stars often enough to grow at a substantial fraction of a solar mass per year, at least in dense galaxy centers. Giant stars would be the most frequent victims, with envelopes much less tightly bound than their main-sequence counterparts. As long as the velocities of stars are distributed isotropically, the black hole can continue to grow, with the accretion rate increasing with its mass and tidal influence. We might expect that this epoch of rapid growth and accretion is related to the strong evolution of quasar luminosity to the peak near $z = 2.2$. Its termination would then be traced by the precipitous drop in the density of quasars after that.

Within an undisturbed galaxy, a black hole can starve itself once it is massive enough to alter the distribution of stellar orbits, which happens when it approaches 1% of the bulge mass. The potential shape then becomes spherical near the black hole, whatever its form farther out, so that stellar orbits become systematically more circular and few stars are scattered by gravitational encounters into the plunging orbits that could bring them close enough for tidal disruption. At this point, the growth of the black hole is effectively halted unless an interaction with another galaxy changes the overall potential and introduces stars and gas into such near-radial orbits. This cutoff may account for the Magorrian relation, since it predicts roughly the correct mass ratio between the stars and central black hole. In this view, much of the QSO activity seen at low redshifts would be associated with some disturbance of a galaxy's dynamics, which is plausible for many (although not all) of the QSO host galaxies imaged in detail, including a substantial number with close, compact companion galaxies or evidence of tidal disturbance.

The entire black-hole mass we see today need not have formed so early. Since merger models suggest that central black holes and stellar bulges would grow at approximately proportional rates as the systems merge, for both gas-rich and gas-poor systems, the growth of the black holes could have continued in this way to the present day, subject to such constraints as X-ray source counts.

10.4 REIONIZATION

The timing and nature of the earliest sources of radiation, as galaxies are formed, controls another major cosmic event—reionization. The intergalactic medium, which had been all of the ordinary matter at the time of recombination, became neutral at that epoch, so we can see the microwave background through it. However, the intergalactic gas that we see more recently is ionized—and very highly ionized at that (Chapter 6). How did it get that way? Stars and active nuclei might both be plausible contributors of the amount of ionizing radiation that would be required. Young stars include the massive, hot ones that ionize surrounding gas within galaxies, and a sufficient rate of star formation could ionize all the gas within and around galaxies. Much less ionizing radiation is needed to maintain the ionization of the intergalactic medium than to ionize a neutral medium in the first place, since the density of gas expanding with the Universe scales at $(1+z)^{-3}$ and the rate of ionization needed to balance recombinations in the ionized gas scales as the square of this density, giving an extremely rapid drop in the overall energy requirements with cosmic time.

At the present epoch, quasars are easily capable of keeping the whole IGM ionized. Their continua can be observed below the Lyman limit, so we know the level of ionizing radiation that they typically produce (and that the IGM even to redshifts above $z = 5$ is ionized, since we can make this measurement). If the background flux of photons at frequency ν is J_ν, the rate of photoionization is

$$\Gamma = 4\pi \int \frac{J_\nu}{h\nu} \sigma_\nu \, d\nu$$

where σ_ν is the cross-section for photoionization at the given frequency and h is Planck's constant. Photons just above the ionization energy are most effective at ionization, since the cross-section varies approximately as ν^{-3} near the ionization edge. This factor, plus the fact that the energy per photon varies with frequency, introduces a dependence of ionization rate on spectral shape in addition to the overall intensity of ionizing radiation. Comparing quasar continua with models for hot, massive stars, their spectral shapes make a difference of about a factor of 5/3 in ionization rate for a given ionizing flux. The ratio is less favorable for starlight, since their spectra are steeper up to the peak of their emission deep in the ultraviolet.

Existing measurements can help narrow down just when reionization happened. It cannot have begun before $z = 35$ or the CMBR spectrum would be distorted from its observed Planck curve. The recent Gunn–Peterson detection in a QSO at $z = 6.3$ tells us that reionization was almost complete (neutral hydrogen fraction 10^{-3} at that epoch). These limits place reionization still in the observational Dark Ages, although it clearly overlapped with the occurrence of stars or active nuclei. For several reasons, stars (of the normal, nonzero metallicity variety) are the favored agents for this energy input. They would have been widespread, had high space density, and the observed galaxies at $z = 3$ provide a plausibly extrapolated density of ionizing radiation sufficient to ionize the whole intergalactic medium. Quasars, on the other hand, were much rarer at these redshifts, even accounting for their high luminosities. The kinds of quasars that we might still be missing systematically at high redshift—those which are strongly reddened by surrounding dust—are the ones that contribute minimally to ionizing radiation, since only their X-rays would escape to ionize the surrounding gas. The final phases of reionization, seen in absorption against QSOs at $z > 6$, overlap with times when galaxies are already known to have been forming stars. This is too late for the initial massive (but rare) stars to have contributed, since they were long gone, and too few to ionize the IGM. In fact, the total flux from starlight during the early, intense epochs of star formation may have been high enough for radiation pressure to become important in shaping the history of mass infall into galaxies. Recent deep measurements of the luminosity function of star-forming galaxies, using photometric redshifts in several deep fields, indicate that most of the reionizing radiation came from low-luminosity star-forming galaxies, unlike the luminous galaxies that we can study in detail at these epochs. Dwarf galaxies may once have shaped our present Universe.

The situation in the present Universe reverses the roles of galaxies and quasars. The energy density required to maintain the high ionization of the IGM drops rapidly as its mean density decreases from the Hubble expansion, so that much less ionizing flux is needed to maintain its ionization over time. The contribution of star-forming galaxies to the overall ionizing continuum today seems to be small, leaving the field to quasars. The importance to galaxies hinges on what fraction of the Lyman continuum (ionizing) radiation escapes individual galaxies, to be available elsewhere. There are low limits to this fraction based on far-ultraviolet observations of a few starburst systems, seen at redshifts large enough that their Lyman limit is significantly red-shifted compared with the absorption in our own galaxy.

The ionizing flux from quasars has observable effects on galaxies, as well as on the intergalactic medium. At the low densities typical of H I in the outer parts of galaxy disks, hydrogen will be ionized to the depth where a column density of a few times 10^{19} atoms per square cm has been reached. Compared with the column densities of clouds in our part of the Milky Way, this is small, but in the disks well beyond typical optical radii, this value becomes comparable with the total column density through the disk. This is close to the column density at which extended H I disks are observed to be truncated, which suggests that the hydrogen extends still farther out, but is ionized by the integrated quasar flux. At these low densities, the recombination time becomes extremely long and the recombination radiation from the gas becomes very weak. Sensitive narrowband detections have shown diffuse Hα emission from the H I halo of the Andromeda galaxy and the Magellanic Stream of tidally stripped H I, giving independent and consistent estimates of the ionizing flux coming from outside the galaxy. Andromeda is a better test case for this process than the halo of our own galaxy, both because the viewing geometry is better for the outer disk and because its lower rate of star formation reduces the role of ionization by stars within the galaxy.

With discrete sources of ionizing radiation, reionization would have proceeded starting from their locations, to gradually encompass more of the Universe. Nick Gnedin (2000) has recognized three phases (Figure 10.3). In the first, there are scattered bubbles of ionized gas around star-forming regions (at this stage, probably to be identified with protogalaxies). The surrounding neutral medium would limit not only the growth rate of these bubbles, but the transfer and observability of any escaping Lyman line emission. This is followed by a stage of consolidation in which the bubbles have grown large enough to overlap and merge, so that most of the intergalactic volume becomes ionized. Ultimately, the densest

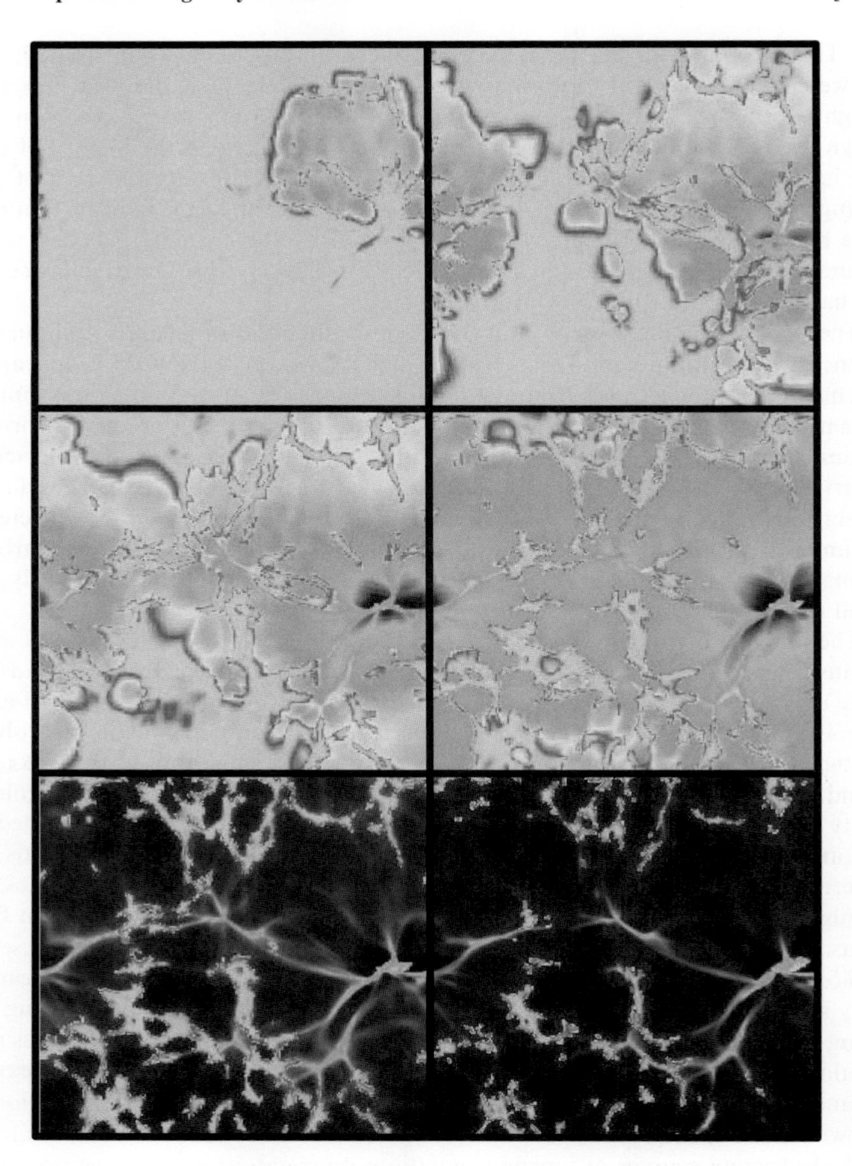

Figure 10.3. The history of cosmic reionization by individual sources. This simulation traces the ionized fraction of the intergalactic medium once star formation is widespread enough to begin ionizing the surrounding gas. The ionized regions are small, expanding around the young galaxies, and eventually merge, finally encompassing the densest regions which have yet to begin star formation. The color scheme runs from black for fully ionized material, through blue, green, yellow, and red for neutral gas. These snapshot slices cover a region 4 Mpc on a side in comoving coordinates, beginning at a redshift $z = 11.5$ and continuing to $z = 6$, where we now have our first direct glimpses of the end of reionization. See also color section. (Courtesy of Nickolay Gnedin.)

regions of the gas are ionized last, as there are finally enough ionizing photons available to reach these zones with their faster recombination timescales.

10.5 EARLY GALAXY GROWTH

Theory to this point provides us with a scheme for enriching the primordial gas to an appropriate level, concentrating it into vaguely galaxy-mass objects, beginning star formation in much the same way it happens today, and perhaps growing the central black holes that are ubiquitous in early-type galaxies. Still, there is much to do before reaching the richness of galaxy structures that we find now.

Coupling between the dynamical development and star-forming history is clearly important, since some ellipticals have stellar populations so old that there would not have been time to form them by merging disk systems. For disk systems, and the ellipticals without such tight time constraints, the simulations suggest a more gradual buildup by merging (or accretion) of smaller units. This may well have begun so rapidly as to mimic a monolithic collapse, which would fit with the small chronological age spreads for halo stars and globular clusters in our own galaxy.

While galaxies with elliptical-like morphology might result from a merger at any epoch, the thin gas-rich disks of spirals are more delicate and demanding of their own histories. A major merger, with mass ratio too close to unity, will heat the disk stars in vertical, as well as radial, motion unless the encounter geometry is unusual. The thinness and dynamically cold nature of disks have been used to constrain the merger history of disk galaxies. However, sufficiently tenuous companions could evidently be absorbed without significant effects on the disk, as we see from tracers of former companions within the stellar halo of the Milky Way. Furthermore, merging occurring early enough could still leave the gaseous disk time to settle into a thin configuration, as long as no powerful starburst swept the gas away before it could form the disk. These limits to merging history are especially important in view of the large number of (possibly merged) companions that CDM simulations predict for massive galaxies. If indeed such companions have vanished by merging, they must have done so quietly enough to leave thin, cold disks for spirals today.

Disks should become more prominent with cosmic time, as more stars have had time to form. Disks in spirals have clearly had a more extended star-forming history than ellipticals or many bulges, since they are still forming stars; in some cases their spectra are consistent with a constant star-formation rate for more than 10 Gyr. We might thus expect to find that spirals are not present beyond a certain redshift, because their disks are intrinsically dimmer than we see today, on top of the $(1+z)^4$ (Tolman) surface-brightness dimming, which would make the disks virtually undetectable at high redshifts. Unmistakable spiral galaxies are known from the Hubble Deep Field to redshift $z = 1.4$, while structures seen in galaxies at $z > 2$ are difficult to put into the Hubble system, and often include very compact star-forming regions.

We would like to know whether bulges formed and reached their present structure quickly, or whether some galaxies built their bulges over time. Some simulations show that a thin gaseous disk can slowly construct a central bulge by scattering gas clouds out of the plane. This may apply to some spirals, especially of later types such as Sc, in which the "bulge" component has an exponential light profile and therefore looks more like a product of dissipation in gas rather than violent relaxation in a stellar ensemble.

In some ways we might expect galaxy mergers today to mimic conditions during galaxy formation. We see the rapid formation of a smooth, symmetric, and bulgelike structure, collisions of gas clouds at high velocity, enormous bursts of star formation, and supernova-powered winds sweeping gas completely away from the remnant. Minor mergers, in which a small and often gas-rich galaxy is absorbed into a much more massive system, have been part of galaxy growth for a very long time, and probably do operate in the same ways today as in the early Universe. For major mergers, though, care is needed in extrapolating to the epoch of galaxy formation. The issue of disassembling two relaxed dynamical systems, with a substantial pre-existing population of stars that have undergone chemical enrichment in each galaxy, is not quite the same as taking multiple clumps of gas and merging them before much star formation or enrichment has gone on.

The hot gas in clusters gives us a window on the early history of galaxies, which may be fairly applicable to all galaxies since there had been so little time for the environment to influence events. The heavy-element enrichment found from the earliest X-ray spectroscopy of clusters tells us that the gas has been enriched by massive stars. In more detail, we find that the abundance pattern shows various ratios of processing by supernovae of types Ia and II for various clusters, depending on the cluster temperature and escape energy. Type II supernovae would have begun to appear very quickly with the first formation of normal stars in future cluster members, while type Ia supernovae would appear only $\approx 10^9$ years later as their binary progenitors evolved. The contribution of SN II appears more important for hotter, more massive clusters, while the fraction of material attributed to type Ia is fairly constant. A simple interpretation is that all clusters retain much of the SN Ia ejecta, while only the most massive can bind the SN II material. This may be as much a temporal as an escape sequence, since the SN II would have detonated earlier and perhaps when the cluster was dynamically less evolved than we now see it. In any case, only the most massive clusters seem to have retained their entire complement of enriched material; lesser potential wells have leaked some of this material into the surrounding intergalactic medium.

Winds were thus widespread in the early evolution of galaxies, so that they are seen to be coupled to their surroundings almost as stars are seen to be connected to the various phases of the interstellar medium. Especially for dwarf galaxies, strong winds could have terminated star formation, or made it episodic as various cycles of gas accretion, star formation, and wind sweeping occurred. Galaxy counts and redshift data do suggest that all this action, and most of the changes due to merging, had taken place by about $z = 0.5$, since when little actual formation of new galaxies could have taken place.

10.6 IS GALAXY FORMATION REALLY FINISHED?

The early Universe has sometimes been described as a sort of cosmological Dreamtime, taking half of the meaning of the aboriginal description of an era of mythic power. However, the Aborigines also hold that the Dreamtime is a state which can be entered today; and so it may be for galaxy formation. There is no guarantee that the process has reached some definite end point. The situation is similar to what we see in our solar system, where the planets grew first by agglomeration and then by gravitational accretion of small units, so that the infall of the pieces of Comet Shoemaker–Levy 9 in 1994 may be regarded as an unusually late event in the formation of Jupiter. We certainly see galaxies' dynamical evolution continuing, not only through the spectacular mergers and strong interactions, but through interactions between galaxy components slowly changing their internal structure. Are galaxies still forming today?

The search for genuinely young galaxies in the local Universe has concentrated on chemistry—a galaxy with very low abundances of metals has never formed large numbers of stars (unless it blew all the gas away, in which case we wonder what gas we're seeing). This explains some of the interest in blue compact dwarf galaxies, some of which have the lowest gas metallicities yet observed. (Cosmological He/H is the other major reason.) It has thus been very interesting to search for the oldest stellar populations in these systems. The flashy recent star formation that makes them visible and lights up the gas which we use to identify them as low-metallicity, also drowns out the feeble light from much older stars, so that near-infrared studies are needed to even set limits on the older stars that might be present.

Strong emission lines can be used to pick objects out of surveys for low metallicity, though extensive work may be needed to tell just how low the abundances are. There exist numerous galaxies with such active star formation that the overall spectrum looks like that of an overgrown H II region (hence the name "isolated extragalactic H II region", and its derivatives such as H II galaxies). The most metal-poor among these would be distinguished in several ways. The emission lines would be dominated by the Balmer lines from hydrogen and various helium lines. Other lines should be present, though, because the behavior of the ratios among the strongest optical lines (Balmer lines of hydrogen, [O II] and [O III], [N II]) is double-valued with abundances. Ratios among the [O III] lines can break this degeneracy, and provide the temperature estimate needed to determine abundances. At the lowest metallicities, the metal lines finally become weak simply through lack of atoms no matter how efficiently they radiate. These systems are also very blue, with little dust to affect the starlight—no metal atoms, no dust grains. As it happens, there is a fair correlation between luminosity and metallicity, today if not necessarily at early times, which may reflect two aspects of the galaxy's history. More luminous galaxies have higher metallicity perhaps because they have a longer history of star formation, so that more of the gas has been processed through stars. In the same direction, low-mass galaxies may not be able to gravitationally trap the supernova debris which contributes much of the metal content.

The most extreme low-metallicity dwarf galaxies known, from gas metallicity, are I Zw 18, Tol 1214−277, SBS 1415+437 and SBS 0335−052. Some properties of these objects are listed in Table 10.1. These have been extensively studied as perhaps the best places to measure the cosmologically interesting pre-galactic abundance ratios of light element isotopes (H, He, Li).

Table 10.1. The most metal-poor local galaxies

Object	z	O/H	12 + log(O/H)	Reference
I Zw 18	0.0026	1.6×10^{-5}	7.12	Izotov *et al.* (1999)
SDSS J0113+0052	0.0048	1.8×10^{-5}	7.17	Izotov *et al.* (2006)
SBS 0335−052	0.0134	1.8×10^{-5}	7.17	Izotov *et al.* (1999)
SDSS J2104−0035	0.0040	2.3×10^{-5}	7.26	Izotov *et al.* (2006)
Tol 1224−277	0.026	3.3×10^{-5}	7.42	Fricke *et al.* (2001)
SBS 1415+437	0.002	3.5×10^{-5}	7.50	Thuan *et al.* (1999)

Ionized gas metallicities are relatively easy to measure, compared with the detailed modeling needed to retrieve the abundances from starlight (which can only be done for fairly bright and nearby stars). None of these prizewinners lies below about 0.02 of solar metallicity. This hardly compares with the conditions when some Population II stars in our own galactic halo formed; some of these are known with derived metal abundances approaching 0.0001 of solar. This—together with calculations of the chemical history of interstellar gas once stars have begun to form, build up heavy elements, and spew them into the interstellar medium—suggests that the initial round of star formation must have happened very rapidly. A closed box of gas, once it's forming stars, will rise to about the metal abundance of the Small Magellanic Cloud in only a few million years. The only ways in which we would expect to see a longer history of star formation and such low metal abundances involved infall of more pristine gas, or very poor mixing of enriched material in to the surroundings, so that the stars ionize gas which has a chemistry different from that of the stars themselves.

Nearby metal-poor galaxies certainly fit the definition of being chemically young. It is much less clear whether they are chronologically young, which would require that they existed in gaseous form until recently and made their first generation of stars only in the recent past. Searches for older populations of stars are difficult, since the recent burst produced red supergiants and bright giants which outshine any old population, and the emission from the young stars and associated gas raises the background against which older populations have to be found. Some star-forming dwarfs with higher metallicity have been found to show redder halos, suggesting that the current activity takes place in the center of a larger, old system. The metal-poor objects do show substantial envelopes in H I, indicating that only a small fraction of the available gas has been transformed into stars; in the case of SBS 0335−052, the 10^7 solar masses of stars that we can identify is imbedded in an extensive H I cloud with a hundred times this mass.

Indirect evidence that galaxy formation might still be possible would be substantial clouds of pristine gas between galaxies, such as would eventually collapse to form (low-mass) galaxies. The Lyman α forest seen against quasar light tells us that there aren't vast numbers of such clouds, in the neutral form most congenial to galaxy formation. Searches for "disembodied" clouds of H I have turned up a few nearby examples, always associated with galaxies containing stars. The best examples are arcs of gas near interacting galaxies—one around the group in Leo containing NGC 3368, 3379, and 3384, and another around NGC 5291 and neighboring group members. Each contains several billion solar masses in a structure that probably represents an evolved tidal tail from a close encounter between group members. For metal-poor galaxies, including dwarf systems of low surface brightness that are presumed to be related because of their necessarily sparse history of star formation, H I companions occur at significant rates, and are significantly more common around the dwarfs with the most active star formation. Such H I companions to star-forming galaxies may support the idea that they are chronologically quite old, and have had some previous star formation, since otherwise we would expect to see H I from objects that have yet to undergo their initial, brief burst of star formation and thus be optically invisible.

The Lyman α forest, when examined at very high signal-to-noise ratio, turns out to have detectable amounts of elements heavier than lithium—especially carbon, at a "basement" level similar to the 10^{-4} of the solar level found for the most extreme galactic stars. As noted above, this seems to be a trace of the earliest, very massive stars, marking a chemical starting point for the kinds of star formation producing all the stars which still survive. Thus, the most metal-poor galactic stars may indeed come from its initial round of star formation, with their enrichment processes dating back before the galaxies.

These issues bring us back to the G-dwarf problem as encountered in the Milky Way fossil record, now with the added features of a nonzero starting metallicity and strong evidence that smaller units, probably containing less-processed gas, were acquired by the Galaxy over its history. Such a history is probably typical for luminous galaxies. The metallicity distribution of old stars results from the coupled histories of star formation, dynamics, gas infall, and the evolution of chemical enrichment with time. In the same way as the chemistry of intracluster gas, the detailed abundance ratios of these stars can track the chemical history of the Galaxy by telling what sources of enrichment (core-collapse supernovae, binary-merger supernovae, novae, mass loss from luminous giant stars, planetary nebulae) were important when particular stars were formed. This serves as a clock, since the timescales for these populations to appear and operate varies widely. Broadly, type II (core-collapse) supernovae produce so-called r-process isotopes in abundance. These are the nuclei which are synthesized in a high-temperature environment so rich in neutrons that new ones can be added to a heavy nucleus before it has time for a normal decay, resulting in heavier isotopes and a distinctive pattern of isotopic abundances. Here, r stands for "rapid". There has been experimental demonstration of many pieces of this process, in what one astrophysicist has described delicately as "terrestrial r-process experiments carried out almost fifty years ago". Type Ia supernovae yield mostly elements around the iron peak, while red giant stars can synthesize *s*-process isotopes, in which the neutrons are added slowly (*s*). Thus, we expect the *r*-process elements to be enriched first, followed by others whose production sites take longer to appear. This is a fair description of the patterns seen in old Galactic stars, although we can measure their chemistry in enough detail to see curious issues that hint at more complicated histories. For example, Qian and Wasserburg (2002) have suggested, from phenomenological models of nucleosynthesis, that there may have been two sites for the *r*-process, with different characteristic timescales, one weighted toward heavy elements and occurring first, followed by one working an order of magnitude more slowly and producing lighter elements with the greatest yield.

10.7 FAST-FORWARD TO THE PRESENT

At this point we can sketch a likely sequence for galaxy formation, highlighting the steps that are best and least understood. We can finally turn around and begin to survey the road to galaxy formation, in the same direction that the galaxies themselves did.

Small fluctuations in the density of gas at the epoch of recombination start the process. These fluctuations themselves are the imprint of earlier processes, such as in an inflationary phase of the very early Universe. The spectrum of these fluctuations thus incorporates the fossil record of the early Universe and furnishes the initial conditions for galaxy formation.

After recombination, gravity was the dominant force, shaping the smooth background into the cosmic web still traced by galaxies and intergalactic gas. The details were controlled by the relationship between dark matter and ordinary matter that we can trace directly. The gas remained neutral until well into galaxy formation, so that absorption of radiation by hydrogen will set a limit to our ability to study its initial phases.

As the cosmic web evolved under gravity, the first dense clumps had masses of order 10^6 solar masses. These reached densities high enough to form H_2 molecules, whose radiation allowed these clumps to cool and collapse further. This led to the appearance of the first generation of stars, the long-sought "Population III". These stars were very massive and hot, by any present standards. Unlike star formation in the present Universe, these stars were isolated from one another, as the first one to form in each region produced enough ultraviolet radiation to destroy the molecules (and the possibility of forming another such monster) in its entire parent cloud.

The supernovae produced by these massive stars were correspondingly energetic, spraying enriched material throughout the Universe. Their heritage is seen in the chemistry of the oldest extant stars, and in the intergalactic gas seen absorbing radiation from distant quasars. These explosions were powerful enough to disrupt the clouds that formed the first stars, leaving the more rarefied condensations that had not yet formed stars to begin the road to galaxies.

This was the classical "era of galaxy formation". Dense regions of gas could collapse with new efficiency, thanks to the salting of heavy elements from the first stars which made cooling more effective. The structure of galaxies today shows that this star formation began hand-in-hand with the initial collapse of galaxies. Models incorporating dark matter indicate that the first protogalaxies were small by current standards, building up through merging in a hierarchical process so that today's systems are the final result of a piecemeal buildup. However, these models also predict more small companions to massive galaxies than we see, and the oldest elliptical galaxies seen at substantial lookback times may be telling us that some galaxies manage to form almost full-grown.

With the onset of star formation, now not one to a galaxy but in clusters and vast numbers, the Dark Ages were truly over. The escaping ultraviolet radiation from this cosmic burst of star formation ionized the intergalactic gas, leaving it almost completely transparent to visible and ultraviolet light right down to the present day.

The bulges of elliptical and early-type spirals formed, at least in large part, during this period. These compact areas, dense with starbirth and lying at the bottom of the galactic potential wells, were the locations where massive black holes could begin to grow, until they became massive enough to control the dynamics of the surrounding stars and in effect terminate their own acquisition of matter. This period may be reflected in the important changes we observe in the galaxy and quasar populations with redshift.

The disks of spirals certainly formed in a much more leisurely way. Some of this process was almost certainly much as Eggen, Lynden-Bell, and Sandage proposed forty years ago: gas which did not form stars quickly would settle into parallel orbits through energy loss as clouds of gas collided. However, we know from the chemistry of local stars as well as from observations of nearby galaxies that disks continued to grow for a long time, with the infall of gas clouds which (depending on whether they had already formed some stars before that) could have qualified as independent dwarf galaxies beforehand.

The formation of galaxies then blends into their evolution, which we are now starting to observe in some detail. The grand average rate of star formation and the typical luminosity of active nuclei have been dropping for much of cosmic history. Abbé Georges Lemaître mused, "Standing on a cooled cinder, we see the slow fading of suns, and we try to recall the vanished brilliance of the origin of the worlds" in the context of the Big Bang, and it is just as appropriate when we consider the subsequent history of the galaxies. Galaxies have merged, starbursts have blown galactic winds sweeping them free of cool gas and terminating star formation, and cluster environments continue to transform galaxy structures and content. And we eventually appeared on a cooling cinder, to wonder at, and about, all these things.

10.8 BIBLIOGRAPHY

Books

Doroshkevich, A.G.; Sunyaev, R.A.; and Zel'dovich, Ya.B. (1974) "The formation of galaxies in Friedmannian universes", in *Confrontation of Cosmological Theories with Observational Data* (D. Reidel, pp. 213–225).

Journals

Abel, T.; Bryan, G.L.; and Norman, M. (2002) "The Formation of the First Star in the Universe", *Science*, **295**, 93–98. Calculations of the formation and history of supermassive stars in the early, zero-metallicity Universe.

Baugh, C. (2006) "A primer on hierarchical galaxy formation: The semi-analytical approach", *Reports on Progress in Physics*, **69**, 3101–3156. Outlines many key considerations in modeling galaxy formation, and a powerful approach to bridging the gap between the cosmological background and processes within individual galaxies. Also available as astro-ph/0610031.

Baugh, C.; Cole, S.; Frenk, C.; and Lacey, C. (1998) "The Epoch of Galaxy Formation", *Astrophysical Journal*, **498**, 504–521. Presents a semianalytic model for galaxy formation with detailed comparison with the properties of observed galaxies, and clearly sets out the excess-dwarf problem posed by straightforward schemes involving cold dark matter.

Benson, A.J.; Pearce, F.R.; Frenk, C.S.; Baugh, C.M.; and Jenkins, A. (2001), "A comparison of semi-analytic and smoothed particle hydrodynamics galaxy formation", *Monthly Notices of the Royal Astronomical Society*, **320**, 261–280. Compares results of the semianalytic and numerical-hydrodynamic schemes for simuulating galaxy formation. The differences can be understood from the properties of these two approaches.

Biermann, P.L. and Kusenko, A. (2006) "Relic keV Sterile Neutrinos and Reionization", *Physical Review Letters*, **96**, 091301. Possible impacts of sterile neutrinos as dark matter, on issues as diverse as the growth of black holes, reionization, and pulsar velocities.

Blumenthal, G.R.; Faber, S.M.; Primack, J.R.; and Rees, M.J. (1984) "Formation of galaxies and large-scale structure with cold dark matter", *Nature*, **311**, 517–525. Discusses the role of cooling timescales for various masses in galaxy formation. Their diagram of cooling time regimes versus clump mass has been widely reproduced and annotated (e.g., like fig. 17.3 of Peacock's *Cosmological Physics*).

Bromm, V.; Coppi, P.S.; and Larson, R.B. (1999), "Forming the First Stars in the Universe: The Fragmentation of Primordial Gas", *Astrophysical Journal Letters*, **527**, L5–L8. The authors use numerical simulations to trace the cooling and collapse of metal-free gas, and the likely formation path for high-mass "Population III" stars.

Corbelli, E. and Salpeter, E.E. (1993) "Sharp H I Edges in the Outskirts of Disk Galaxies", *Astrophysical Journal*, **419**, 104–110. Shows that the H I disks of spiral galaxies are generally observed to have a sharp cutoff at a column density near 10^{19} atoms per square centimeter.

Fan, X.; Narayanan, V.K.; Strauss, M.A.; White, R.L.; Becker, R.H.; Pentericci, L.; and Rix, H.-W. (2002) "Evolution of the Ionizing Background and the Epoch of Reionization from the Spectra of $z \sim 6$ Quasars", *Astronomical Journal*, **123**, 1247–1257. A recent treatment of the ionization history of the intergalactic medium, especially with regard to the contribution of quasars based on new surveys to high redshift.

Frenk, C.S.; White, S.D.M.; Bode, P.; Bond, J.R.; Bryan, G.L.; Cen, R.; Couchman, H.M.P.; Evrard, A.E.; Gnedin, N.; Jenkins, A. *et al.* (1999) "The Santa Barbara Cluster Comparison Project: A Comparison of Cosmological Hydrodynamics Solutions", *Astrophysical Journal*, **525**, 554–582. A detailed comparison of various techniques for tracing the development of cosmic structure, all run from the same initial conditions. Fortunately for our confidence in astrophysics and cosmology, the results are all in reasonable agreement.

Fricke, K.J.; Izotov, Y.; Papaderos, P.; Guseva, N.G.; and Thuan, T.X. (2001) "An Imaging and Spectroscopic Study of the Very Metal-deficient Blue Compact Dwarf Galaxy

Tol 1214–277", *Astronomical Journal*, **121**, 169. This study combines imaging and high-quality spectroscopy to argue that some blue compact galaxies began star formation only in the recent cosmic past.

Gnedin, N. (2000) "Cosmological Reionization by Stellar Sources", *Astrophysical Journal*, **535**, 530–554. Simulations of the history of reionization, showing the initially isolated "bubbles" of ionized gas growing to overlap and finally encompass the highest-density regions which need the largest ionizing flux.

Gnedin, N.Y.; Norman, M.L.; and Ostriker, J.P. (2000) "Formation of Galactic Bulges", *Astrophysical Journal*, **540**, 32–38. Simulations and analysis showing that scenarios with cold dark matter can yield clumps with 0.1–1 billion solar masses by $z = 6$, appropriate to form central bulges. A key issue is that the baryonic matter can cool and collapse to densities greater than the dark matter.

Gott, J.R., III (1973) "Dynamics of Rotating Stellar Systems: Collapse and Violent Relaxation", *Astrophysical Journal*, **186**, 481–500. Describes a technique for tracing the history of a collapsing stellar system. The results suggest that a starburst might, in fact, leave remnants that are found within a 1-parsec radius after about 10 million years, within a pre-existing stellar bulge. Some such process would dramatically speed the initial growth of central black holes.

Gratton, R.G.; Carretta, E.; Matteucci, F.; and Sneden, C. (2000) "Abundances of light elements in metal-poor stars. IV. [Fe/O] and [Fe/Mg] ratios and the history of star formation in the solar neighborhood", *Astronomy and Astrophysics*, **358**, 671–681. Uses chemical abundances to trace the history of star formation in the early Milky Way. They argue that the halo took less than a billion years to collapse toward a disk since the chemical signature of type Ia supernovae is not important in extreme halo stars.

Haehnelt, M.G. and Kauffmann, G. (2000) "The correlation between black hole mass and bulge velocity dispersion in hierarchical galaxy formation models", *Monthly Notices of the Royal Astronomical Society*, **318**, L35–L38. Demonstrates that the observed relation between derived black-hole mass and stellar velocity dispersion (which parallels that for bulge stellar luminosity) can be preserved through merging, allowing ellipticals today to include many merger remnants without violating observational correlations.

Izotov, Y.; Chaffee, F.B.; Foltz, C.B.; Green, R.F.; Guseva, N.G.; and Thuan, T.X. (1999) "Helium Abundance in the Most Metal-deficient Blue Compact Galaxies: I Zw 18 and SBS 0335–052", *Astrophysical Journal*, **527**, 757–777. Heavy-element abundances in the lowest-metallicity galaxies known.

Izotov, Y.I.; Papaderos, P.; Guseva, N.G.; Fricke, K.J.; and Thuan, T.X. (2006) "Two extremely metal-poor galaxies in the Sloan Digital Sky Survey", *Astronomy and Astrophysics*, **454**, 137–141.

Madsen, G.J.; Reynolds, R.J.; Haffner, L.M.; Tufte, S.L.; and Maloney, P.R. (2001) "Observations of the Extended Distribution of Ionized Hydrogen in the Plane of M31", *Astrophysical Journal Letters*, **450**, L135–L138. Uses the interstellar medium in the Andromeda Galaxy to estimate the ionizing radiation flux coming from outside its disk, by setting limits to the amount of Hα emission in areas close to the edge of the neutral hydrogen disk.

Maloney, P.R. (1993) "Sharp edges to neutral hydrogen disks in galaxies and the extragalactic radiation field", *Astrophysical Journal*, **414**, 41–56. Demonstrates that the observed truncation of galaxy disks as seen in H I would result from ionization by the mean extragalactic UV radiation field.

Oh, S.P. (1999) "Observational Signatures of the First Luminous Objects", *Astrophysical Journal*, **527**, 16–30. Considers various ways to detect a supermassive first generation of stars, among which are the most promising techniques that detect emission from the surrounding pockets of ionized gas.

Partridge, R.B. and Peebles, P.J.E. (1967) "Are Young Galaxies Visible?", *Astrophysical Journal*, **147**, 868–886. An early estimate of the observed properties of young galaxies at high redshift, one of a series of such efforts.

Press, W.H. and Schechter, P. (1974) "Formation of Galaxies and Clusters of Galaxies by Self-Similar Gravitational Condensation", *Astrophysical Journal*, **187**, 425–438. This treatment sets out what became a standard formulation for the instability expected for collapsing gaseous objects on various size and mass scales.

Qian, Y.-Z. and Wasserburg, G.J. (2002) "Determination of Nucleosynthetic Yields of Supernovae and Very Massive Stars from Abundances in Metal-Poor Stars", *Astrophysical Journal*, **567**, 515–531. Presents calculations of how much supernovae of types Ia and II, and the hypothetical very massive first-generation stars, need to have contributed to the observed abundances of the elements.

Rees, M. and Ostriker, J.P. (1977) "Cooling, dynamics, and fragmentation of massive gas clouds—Clues to the masses and radii of galaxies and clusters", *Monthly Notices of the Royal Astronomical Society*, **179**, 541–559.

Scannapieco, E. and Broadhurst, T. (2001) "The Roles of Heating and Enrichment in Galaxy Formation", *Astrophysical Journal*, **549**, 28–45. Argues that the influence of early galactic winds shaped the subsequent history of star formation through stripping of cool gas, and that this process is important in solving the G-dwarf problem and the excess dwarf problem from dark-matter simulations.

Schneider, S.E.; Helou, G.; Salpeter, E.E.; and Terzian, Y. (1983) "Discovery of a large intergalactic H I cloud in the M96 group", *Astrophysical Journal Letters*, **273**, L1–L5. Discovery of an extended intergalactic neutral hydrogen cloud, evidently a tidal remnant in the Leo galaxy group.

Sunyaev, R.A.; Tinsley, B.M.; and Meier, D.L. (1978) "Observable properties of primeval giant elliptical galaxies or ten million Orions at high redshift", *Comments on Modern Physics, Part C—Comments on Astrophysics*, **7**, 183–195. A simple model of young galaxies, considering ways to distinguish them from quasars. The authors drew attention to UV absorption lines from hot stars, thermal infrared emission from star-heated dust, and the X-ray and radio emission from supernova remnants. They also mention (and underestimate) the Lyman break. In hindsight, they already saw most of the ways we now use to find high-redshift star-forming galaxies.

Taylor, C.L.; Thomas, D.L.; Brinks, E.; and Skillman, E.D. (1996) "A Survey of Low Surface Brightness Dwarf Galaxies to Detect H I-rich Companions", *Astrophysical Journal Supplement*, **107**, 143–174. A search for neutral hydrogen clouds near dwarf galaxies, whose existence would be relevant to the number of isolated H I clouds which have yet to form stars and thus would be candidates for delayed galaxy formation.

Thuan, T.X.; Izotov, Y.; and Foltz, C.B. (1999) "The Young Age of the Extremely Metal-deficient Blue Compact Dwarf Galaxy SBS 1415+437", *Astrophysical Journal*, **525**, 105–126. The authors use the chemical makeup and colors of this metal-poor galaxy to argue that it began star formation only within the last 100 million years or so.

Weedman, D.W. (1983) "Toward explaining Seyfert galaxies", *Astrophysical Journal*, **466**, 479–484. The fate of massive, compact stellar remnants of a starburst and their possible shrinkage toward a relativistic (unstable) configuration.

11

Forward to the past—what can new eyes expect to see?

All the foregoing discussion serves, perhaps, to highlight some of the things we don't know about galaxy formation. Some of these unknowns can be addressed only by new kinds of data, spanning a wide range of wavelength and science goals. Progress in astronomy, as in so much of science, has been driven jointly by instruments and ideas. On the hardware side, new instrumentation projects are now years-long efforts by large teams. We therefore know what facilities are to be expected for the coming decade, and can foresee some of the kinds of results that should be coming in the effort to watch galaxies form. And on the idea side, who knows what to expect?

This final chapter reviews prospects for advances in our picture of galaxy formation that might be expected from upcoming instruments. It works in order from high to low energies. These expectations make it clear why the formation of the first stars, galaxies, and quasars are areas of very active work and high hopes for pushing back the "Dark Ages". Questions of galaxy formation and evolution are prominent in plans for a new generation of extremely powerful instruments, everywhere from the gamma-ray to meter-wavelength radio bands.

11.1 GAMMA-RAY PROSPECTS

A few years hence, the multipurpose Gamma-Ray Large Area Space Telescope (GLAST) will function as a true successor to Compton. Instrumented for a wide range of studies from 10 keV to 300 GeV. In the context of gamma-ray bursts, it should yield time-resolved spectra for hundreds of events per year, constraining the degree of relativistic beaming and the possible kinds of shock interactions involved in producing the gamma rays. Onboard software will be able to autonomously point the narrower-angle instruments toward bursts detected with GLAST's all-sky burst monitor.

Additional capabilities for following the fading X-ray tail of gamma-ray bursts may be provided by the Russian Spectrum X-Gamma mission. Its detectors have better sensitivity to hard X-rays than either *Chandra* or *XMM–Newton* currently provide. The satellite is well advanced in construction, but programmatic uncertainties indicate that it is difficult to foresee the actual date.

11.2 X-RAY SATELLITES AND THE EARLY UNIVERSE

Both active nuclei and hot gas in galaxy systems are copious sources of X-rays, giving this piece of the spectrum an important role in tracing their early history. Indeed, X-rays may be the most effective way to find active nuclei in young, gas-rich, and probably dust-rich galaxies. They are obviously the most effective way to find hot gas, and can provide crucial information as to when and how this gas became chemically enriched by galactic outflows. Today's flagship X-ray missions, NASA's *Chandra* X-ray Observatory and ESA's *XMM–Newton*, have only begun to exploit their complementary capabilities in probing the early Universe. The exquisite, sub-arcsecond angular resolution of *Chandra* and the formidable collecting area of *XMM–Newton* allow different kinds of approaches to important problems. As powerful as these are, because of the long lead times and technological development needed for still more powerful observatories, planning is underway for the next generation of high-energy instruments. As with *Chandra* and *XMM–Newton*, complementary approaches may prove to be very fruitful.

NASA is planning a multi-spacecraft mission, Constellation-X, in which the solution to needing a larger collecting area is simply to orbit multiple coordinated spacecraft with identical instruments. Several proposals for the basic design, incorporating as many as four spacecraft, are still in competition. Some of these designs use a framework to be unfolded in orbit, accommodating an 8-meter focal length for the individual X-ray telescopes. They would operate in the L2 region, where Earth occultation and shadow crossings are not issues. The specifications call for spatial resolution no worse than $15''$ at low energies (trading resolution for collecting area in comparison with *Chandra*), and spectral resolutions $R = \lambda/\Delta\lambda = 300$–$3000$ at energies up to 30 keV.

ESA is pursuing a quite different, and likewise very ambitious, approach with the X-ray Evolving Universe Spectroscopy mission (XEUS). No launch date has been set; the mission has not been formally approved, although contracts for detector development were let in 2002. Working from the L2 region, XEUS would deliver images with 2-arcsecond resolution, and use detectors to deliver the extraordinary energy resolution of 6 eV up to 6 keV. XEUS would be able to measure spectra for sources an order of magnitude fainter than *Chandra* can currently detect in deep surveys. This drives the angular resolution required, to avoid overlapping and confusion of the images of faint sources.

Important goals of these missions relate to galaxy formation and evolution. They should be able to trace the appearance and history (chemical and dynamical) of hot gas in bound galaxy systems, from groups to rich clusters. The metallicity of the intergalactic medium will be followed, using X-ray absorption lines seen against high-redshift quasars. The appearance of these early active nuclei should come into view, and may allow us to follow their growth in mass and spin as these affect their X-ray spectral signatures.

11.3 LARGE GROUND-BASED TELESCOPES IN THE OPTICAL AND NEAR-IR

Increasing light grasp and range has been a driving theme in ground-based astronomy, one which shows promise of continuing as technological advances allow the construction of huge instruments with performance once thought possible only from space. The widespread collaboration in studies of the Hubble Deep Fields, demonstrating the fruitfulness of combining high-resolution spaceborne imagery with the very deep spectroscopy possible using 8–10 m telescopes, gave new impetus to the design of much larger optical telescopes. A further push came from the realization that the James Webb Space Telescope will be able to detect (and resolve structure in) objects much fainter than any planned space telescope can measure spectra for.

Serious design efforts are now underway for a new generation of huge ground-based telescopes for the optical and infrared. With apertures of 30–50 meters and resolution enhanced by adaptive optics, these are designed to measure the spectra of objects that we cannot yet detect. The partner institutions which built and operate the twin 10-m Keck telescopes, the California Institute of Technology and the University of California, have joined with a consortium of Canadian institutions to design a scaled-up 30-meter version, the Thirty-Meter Telescope or TMT (Figure 11.1). In response to the National Academy of Sciences decadal review study which identified such an instrument as a priority for the entire US community, support and management is provided by the National Science Foundation through the National Optical Astronomy Observatories. At the same time, a group of institutions including the University of Arizona and the Carnegie Observatories has embarked on the construction of the Giant Magellan Telescope (Figure 11.2), which will combine six 8.4-meter primary mirrors on a single mount to give the equivalent of a 21.4-meter aperture (at, it is hoped, notably lower cost and faster schedule). Site testing for the consortium is focusing on locations near the Las Campanas Observatory in Chile. More ambitious still, the European Southern Observatory has approved detailed design for the European Extremely Large Telescope (EELT; Figure 11.3). Currently planned for a 40-meter effective aperture, it represents something of a compromise between size and feasibility; studies of 50–100 meter designs revealed daunting technical challenges which would likely defeat attempts to be completed in the immediate future. The adaptive-optics components of the 40-meter design are impressive enough, including a 2.5-meter mirror being reshaped by at least 5000 actuators, readjusted at a rate of 1000 Hz.

Each of these mammoth instruments must rely on adaptive optics to realize large performance gains over existing telescopes; if they were restricted to 1-arcsecond angular resolution, the sensitivity gains would remain modest and hardly trivial to

Figure 11.1. A design concept for the Thirty-Meter Telescope (TMT). (Image courtesy of Thirty-Meter Telescope project, a joint effort of Caltech, the University of California, and the Association of Canadian Universities for Research in Astronomy, with support from the National Science Foundation through the National Optical Astronomy Observatories.)

exploit. Major issues include the quality of correction and angular field of the corrected image. Currently operational adaptive-optics systems degrade rapidly in image quality with distance from a reference star, even if that reference is a laser guide star. This arises from a system being built to correct wavefront disturbances at a particular height above the telescope; the system is set up so that such a height is optically conjugate to the telescope pupil. In principle, a system can be built with layers of correction that are conjugate to turbulence at multiple heights, in *multi-conjugate adaptive optics* (MCAO). Such systems are already in development and have seen their first operational demonstration in 2007, at the ESO VLT. They rely on a ring of laser guide stars around the target to yield the requisite wavefront information.

In all these designs, key goals are high-imaging and spectroscopic sensitivity well into the infrared (as far as 25 μm). This is driven in large part by the science of high-redshift objects, in which informative regions of the spectrum will be observed progressively farther into the IR. In addition, adaptive corrections of atmospheric distortion are more forgiving in speed and accuracy at longer wavelengths, and

Figure 11.2. A design rendering for the Giant Magellan Telescope (GMT). In contrast to the TMT and EELT, this design uses large individual mirrors, figured as off-axis portions of a single optical system. (Giant Magellan Telescope—Carnegie Observatories.)

indeed to date adaptively compensated imaging gives its best performance in the near-
infrared.

Both the TMT and GMT projects aim to have their facilities operating in the era of JWST and ALMA, roughly 2015–2020.

The gains to be sought from these instruments are immense. The fossil record of Galactic history would come into sharp focus, as we could measure the detailed chemical compositions of stars all across the Milky Way. They would generate detailed spectra of objects to beyond $z = 10$, with enough sensitivity to study the production of elements in the explosions of primordial massive stars. Studies of galaxies at high redshift would be possible with a large number of resolution elements across each galaxy. They might reveal the primordial stars themselves, caught in their brief and brilliant lives. In many ways, these telescopes are thought of as spectroscopic partners for imaging with the James Webb Space Telescope, but their utility will go far beyond this. If adaptive optics allows them to approach the diffraction limit, they will generate substantially higher-resolution data. However, for broadband imaging, the high thermal background from the atmosphere gives a substantial sensitivity advantage to JWST, which is an important aspect of the partnership.

The performance gains foreseen, and scientifically required, for this generation of instruments depend critically on the maturity of adaptive optics, and especially on the

Figure 11.3. A design concept for the European Extremely Large Telescope (EELT). (European Southern Observatory.)

ability to correct atmospheric effects across a substantial field of view. The best uncorrected image quality obtained for long exposures on the ground has crept upward in recent years, but seldom reaches 0.4″. This limitation, even if normally available, would be devastating for the sensitivity of giant telescopes. Adaptive optics (AO) are intended to reduce atmospheric blurring in real time, by correcting the incoming wavefront for distortion before the image is detected (or dispersed for spectroscopy). In its simplest "degenerate" form, removing the zeroth-order effects of image motion, it has been in use at least sporadically since the 1950s, when excellent planetary images were obtained from Mount Wilson by tracking the mean motion of the image. Since all parts of the image do not share the same motion, this technique is limited to very small fields and bright objects (for which off-the-shelf systems are not available to amateur astronomers). For increased performance, and apertures of more than 1 m or so, a more sophisticated approach is needed. This typically entails using a nearby bright star to measure the wavefront distortion at multiple points within the telescope pupil, and feeding these results to a small, deformable "rubber" mirror which compensates for these distortions, delivering an improved image to the focal plane. The faster the corrections are updated, the better the image compensation, so brighter reference stars (and stars closer in angle to the target object) give better results. Artificial reference stars, generated by a laser backscattering from particular atmospheric layers, can solve the problem of having a near enough reference star, though real stars still have to be used to track zeroth-order image motion. After much of the technology was boosted by the declassification of US Air Force developments, important astrophysical results have been generated with AO techniques in recent years.

The GMT design includes provision for instruments using 10–20 arcminute fields of view. These would operate using ground-layer adaptive optics, in which the correction is tuned for atmospheric irregularities just above the telescope. As envisioned, such a system would allow useful image-quality improvements over a much wider field than existing adaptive-optics systems at the expense of less complete correction. As noted in the conceptual design report for the GMT, both multi-conjugate and ground-layer adaptive optics remain substantially undemonstrated, making their development a pacing factor for the next generation of large telescopes.

11.4 AFTER HUBBLE: THE JAMES WEBB SPACE TELESCOPE (JWST)

The three "formation" questions compared at the beginning of this volume—the origins of planets, stars, and galaxies—all require a similar approach for the next step in our understanding. In each case, though for different reasons, high resolution and sensitivity are required to work farther into the infrared than the Hubble Space Telescope has been able to do. Observations with Hubble have reached several fundamental limits that can only be improved with such a new instrument. HST was not designed as a particularly optimized infrared telescope, having done enough technologically new things for its time. At wavelengths 1.7 microns and longer, the telescope's own thermal emission dominates the detection thresholds, and it is almost useless for observations of interesting faint targets in the early Universe in this wavelength regime. Furthermore, the aperture of the telescope is now very modest by the standards of the largest ground-based instruments, limiting its angular resolution and sensitivity. With our current understanding of the early history of galaxies, there are specific targets that dictate a sensitivity an order of magnitude better, low thermal background much deeper into the mid-infrared, and angular resolution comparable with HST's optical performance now into the near-infrared.

This increased infrared capability and light grasp, combined with the very low infrared sky brightness from space, should bring an obvious list of phenomena in the early Universe into view. Luminous star clusters could be identified to $z > 4$. The history of galaxies can be much more completely addressed using infrared passbands than with the emitted ultraviolet which dominates deep Hubble observations, seeing older stars and giving a census less biased to UV-bright regions. This is an important factor in testing ideas for when and how galaxies form and grow, letting us decouple the ability to trace the growth of galaxies in a dynamical sense from their history of star formation. For the same reason, infrared data will give more meaningful structural parameters, morphology, and scale length for high-redshift galaxies than we have now, telling us about the temporal origins of the Hubble sequence. Primordial stars could well be detected, and the odds are reasonable of seeing their explosions as hypernovae.

These considerations led to detailed studies for the appropriate successor to Hubble, the James Webb Space Telescope (JWST). The specifications started with a wavelength range, minimum aperture of 4 meters with stretching to 8 desirable, and a survey efficiency (to be gained by the product of light-gathering power and field of view for the camera). Most challenging in comparison with Hubble was the cost cap, originally including launch on a booster of the Atlas class. Launch is now foreseen on an Ariane V booster contributed as part of ESA's involvement in JWST, around 2013, for a 5–10 year lifetime. After evaluation of competing designs, the final configuration (Figure 11.4) has a deployable mirror with 18 hexagonal segments, for an effective aperture of 6.5 meters. The thin beryllium mirror segments are now being figured and polished. Hubble has been called a lightweight ground-based telescope carried to orbit, while JWST will be a telescope designed from its inception for the space environment. JWST will be able to dispense with the intricate system HST uses to keep the entire telescope stable to a few milliarcseconds during exposure, by reference to stars. This system placed very strong constraints on the Hubble telescope's possible slew rate between targets and gyroscopic stabilization. The JWST design replaces this by a small moving mirror just ahead of the instruments, whose motion is precisely controlled by reference to stars. This allows the entire telescope to drift in pointing by much greater amounts without affecting image quality. In hindsight, it was fortunate that such a system was not adopted for HST, since the optical components needed to correct the spherical aberration in its primary mirror are very sensitive to optical alignment so that an image wandering around the telescope's field of view would have been virtually impossible to correct.

The telescope, like several others of the next generation, will be dispatched to the region of the L2 point of the Earth/Sun system, a point of (unstable but manageable) gravitational equilibrium about 1.5 million km outward from the Earth (opposite the Sun). This location was also selected for the Microwave Anisotropy Probe (MAP) launched in August 2001, and is becoming popular with this address also intended for ESA's Planck mission to improve mapping of the microwave background and Herschel far-infrared observatory. For all of these, the thermal and communications advantages of the L2 region are compelling, even though they make launch and deployment more expensive and difficult.

Compared with the low Earth orbit used by Hubble, the L2 location offers operational advantages (though bringing the drawback that it cannot be reached by current systems for human servicing). In low orbit, nearly half the sky is obscured by the Earth, so that either rapid repointing must be done, with time spent to settle the telescope pointing, or nearly half the time will be spent looking at the Earth with the camera shutter closed (essentially the case for Hubble). A more distant location allows both more efficient overall use of telescope time, and dramatically longer uninterrupted observations. This aspect has been used to great advantage by the *Chandra* and *XMM-Newton* X-ray Observatories, occupying orbits that take several days to complete and take them as far as 140,000 km from the Earth.

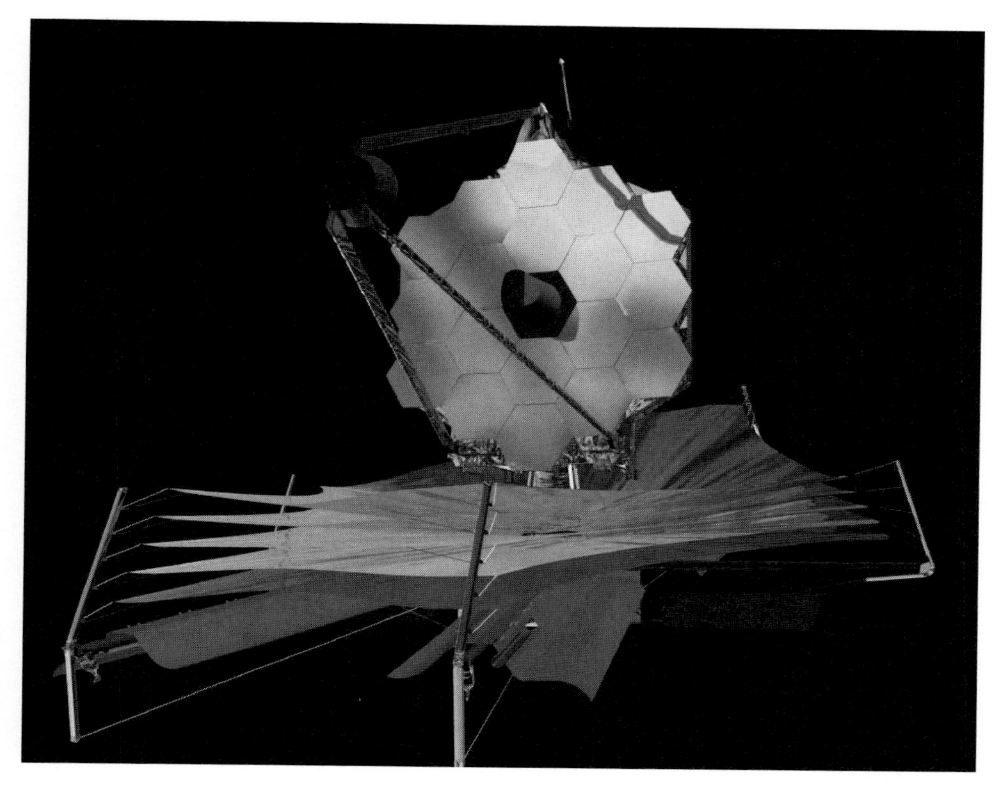

Figure 11.4. Artist's view of the final design concept for the James Webb Space Telescope. This instrument, with a segmented primary mirror with an effective aperture of 6.5 meters, will be optimized for infrared performance, and operated far from Earth in the L2 region. The segments of the primary will be deployed after launch. This instrument has, as a primary goal, exploring the "Dark Ages" immediately after galaxies formed. (Northrop Grumman Space Technology.)

Furthermore, since instruments for the deep infrared must operate close to absolute zero, the far-infrared emission from the Earth makes it a serious hindrance, and in some cases the limiting factor for how long a cryogenic instrument can last on a single load of cooling material. Paradoxically, the Earth is harder to deal with than the Sun, even though it is much fainter, since much of the Sun's light is in optical wavelengths that can readily be reflected away. Getting away from the Earth then offers considerable advantages in how cold the detectors can stay, and for how long. Infrared astronomical satellites to this point have used cryogenic systems to keep the optics and detectors cold enough to avoid their dominating the emission from faint targets. IRAS and ISO used liquid helium baths to cool the entire optical systems, limiting the mission lifetimes. IRAS stayed cold for 11 months, while ISO lasted 28 months (thanks in part of a last-minute refilling by an enterprising team at the launch pad). For HST's near-IR instrument NICMOS, a block of solid nitrogen was used for its original deployment, which lasted about 2 years, with its function recently replaced by a thermoelectric system providing long-term operation at somewhat higher temperatures.

To allow long-term operation of IR observatories beyond low Earth orbit, there has been considerable study on the use of radiative cooling as a passive way of keeping the required parts of the telescope cold enough for sensitive infrared observations. This is likely to be workable for near-infrared detectors, though it may not prove feasible for detectors at longer wavelengths, which would give JWST a limited lifetime for certain key observations.

A considerable fraction of the JWST mission, especially early on, is likely to go into a handful of very deep observations on the model of the Hubble Deep Fields. The expected performance of an IR-optimized 6.5-meter space telescope for the study of galaxy formation is stunning. It should be able to routinely find objects 50 times fainter than appeared in the deepest Hubble Deep Field data, and has wavelength-imposed redshift limits so broad that (at this point) we don't expect them to limit detection of early objects in the way that HST observations now suffer. That is, even with intergalactic absorption removing all photons below emitted Lyman α, it could see much redder and unabsorbed radiation even from objects at $z = 10$.

As with HST, scientific operation is being handled by STScI and engineering development overseen by Goddard Space Flight Center, with the European and Canadian Space Agencies involved at high levels.

11.5 FAR-INFRARED AND SUBMILLIMETER OBSERVATIONS FROM SPACE

Another occupant of the increasingly popular L2 region will be the European Space Agency's Herschel (Figure 11.5), originally known as the Far Infrared Space Observatory or FIRST. Herschel will carry a 3.5-meter passively cooled telescope, launched on the same Ariane V rocket as Planck. It will carry imaging cameras and spectrometers operating over the 60–670 micron range. This includes the range which is proving a key to understanding much of the early star-forming history of galaxies, covering the emission from dust heated by the early generations of stars, and detecting such galaxies even if the dust absorbs all the stars' direct radiation. This is a spectral range in which the Earth's atmosphere greatly restricts the sensitivity of ground-based work. The submillimeter part of the spectrum is where the dust radiation from high-redshift galaxies reaches us, once again giving us a way to trace the history of star formation even when its direct emission is blocked by the obscuring and absorbing effects of dust.

Planck is a successor to COBE and WMAP, scheduled for a 2008 launch. Using a spinning 1.5-m reflector to feed its detectors, it should generate a map of the cosmic microwave background with an accuracy of $\Delta T/T = 2 \times 10^{-6}$ and $10'$ angular resolution. Its data should extend the results from MAP to high wavenumbers in the power spectrum ($l \approx 1000$), and should recover some of the polarization structure in the CMBR. This polarized component can limit the epoch of reionization, trace

Figure 11.5. The Planck and Herschel spacecraft, depicted in their joint launch configuration atop an Ariane 5 booster. Herschel sits on top with Planck acting as a carrier during launch. Both are destined for the L2 region. (European Space Agency.)

gravitational lensing of the background radiation, and give signatures of possible directional (tensor) components to the initial fluctuations. It also allows a consistency check, since it should be possible to fit cosmological powers to the polarization power spectrum in much the same way as the intensity power spectrum. Since the polarization measurement is a differential one, agreement between the two results would add to the confidence in the overall results.

11.6 ALMA: A WORLDWIDE EFFORT IN SUBMILLIMETER ASTRONOMY

Because the technology of radio astronomy becomes more difficult with decreasing wavelength, both in receiver construction and required accuracy of the antenna dishes, millimeter radio astronomy has long lagged behind work in the centimeter waveband. Millimeter work seemed a rather narrow scientific niche for some years, being occupied largely by work on emission lines from molecules in star-forming regions and their surroundings. Much of the extragalactic community took note of the millimeter range when it was demonstrated that one could detect the coldest dust in galaxies in the submillimeter, on the long tail of their blackbody-like spectra. A few sites have been exploited for very dry conditions so that observations are often possible in several windows in the submillimeter range as well. Existing instruments (such as the James Clerk Maxwell Telescope on Mauna Kea) have been able to detect the dust components of high-redshift galaxies in this way.

As with the impetus to build JWST, it happens that questions of both star formation and galaxy formation would be handled effectively by millimeter observations of unprecedented sensitivity and angular resolution, for both spectral lines and the continuum from dust grains. It is now possible (and routine, on a small scale) to carry out interferometry at millimeter wavelengths. For a decade, there have been efforts to construct an analog of the VLA to work in the millimeter regime. Detailed studies were done by the US National Radio Astronomy Observatory, the European Southern Observatory and several partner institutions, and the Nobeyama Radio Observatory and National Astronomical Observatory in Japan. The separate projects developed close ties, and have now been amalgamated as the Atacama Large Millimeter Array (ALMA), perhaps the first genuine "World Array".

ALMA is now foreseen to include 64 12-meter antennas (Figure 11.6), with the size a compromise between the field of view that can be mapped at once and the array sensitivity. They will be spread in configurations as large as 10 km, affording angular resolution as sharp as 10 milliarcseconds—substantially better than either the VLA or HST can now deliver. The site will be on a high plain in the Andes of extreme northern Chile, 5000 meters above sea level on the Llano de Chajnantor. Site tests have shown the atmosphere in this region to be extremely dry, comparing favorably with the South Pole. Its receivers should cover bands from 0.35 to 10 millimeters in wavelength.

Observations with existing submillimeter telescopes, which have been time-consuming and at the limit of their capabilities, have given some tantalizing previews of the kinds of things ALMA can do for high-redshift objects. In the continuum, there have been detections of several high-redshift objects by the thermal emission from their dust, which redshifts from an emitted peak of 100–200 microns to peaks near 1 mm. This emission component is important in measuring the bolometric luminosity of star-forming systems; for actively star-forming galaxies, much of the luminosity emerges in reradiated far-IR radiation, and it is quite possible that many young galaxies will have produced so much dust that much of the emitted optical and UV region will be strongly obscured. The peak in thermal dust emission is so strong

Figure 11.6. An artist's depiction of the Atacama Large Millimeter Array (ALMA), a worldwide project to be erected on the Llano de Chajnantor, at an altitude of about 5000 meters in northern Chile. The large number of individual dishes is set by the need for a fairly wide instantaneous field of view, requiring small elements at millimeter wavelengths, the need for sensitivity dictating a large total collecting area, and the angular resolution to be reached by the whole array, which sets the area over which the elements must be spread. This design calls for 64 12-meter antennae. (Image courtesy of the European Southern Observatory.)

that the detectability of galaxies actually rises with redshift near $z = 4$, with the rapidly rising spectral shape more than compensating for increasing distance. ALMA should be able to detect dust emission from massive galaxies to $z = 20$.

The prominent CO emission lines, in a ladder starting at 2.6 mm and extending to multiples of this frequency (115 GHz), are an important tracer—almost the only sensitive tracer—of molecular gas, and thus of conditions for star formation. The strong lines marking transitions between molecular of low angular-momentum quantum number J have been detected from a handful of luminous galaxies and QSO hosts at $z > 2$, using the Owens Valley and Plateau de Beure interferometers. This bodes well for the systematic use of these lines to trace the history of star formation and gas content in galaxies. ALMA should be able to measure the CO lines for normal galaxies to $z = 1$, and for very luminous systems to $z = 3$, while the exceptional objects detected so far should be found to yet higher redshifts. In the more local Universe, ALMA should be able to detect CO emission and measure its velocity structure in more galaxies than any existing instrument can do in H I; in other words, for typical abundances of CO and gas content, ALMA will be more sensitive to the cold ISM than any existing instrument is, even using a minority tracer. This means that the Tully–Fisher relation can be traced to redshifts of several tenths, potentially illustrating luminosity evolution. In addition, ALMA should be able to trace the ISM by means of the strong carbon fine-structure lines which ISO has shown to be ubiquitous and strong in nearby gas-rich galaxies. For redshifts of a few these lines come through the ALMA sensitivity range (dictated by atmospheric transmission).

11.7 A REALLY, REALLY LARGE RADIO ARRAY

Radio astronomy, in the "traditional" wavelengths from a few centimeters to a few meters, has been at the forefront of research on cosmic evolution since it was realized that astronomy could trace such a history. The most powerful existing instruments—such as the giant Arecibo dish, the Very Large Array in New Mexico, the Westerbork array in the Netherlands, the MERLIN array in the UK, and the continent-spanning Very Long Baseline Array and Australia Telescope—have shown enormous capabilities in probing the distant Universe. Radio galaxies can be traced to redshifts (probably) larger than we can now identify in the optical or near-infrared, and large numbers of less powerful galaxies can be seen at high enough redshifts to examine their evolution. Polarization maps can tell us about magnetic fields around quasars and radio galaxies seen at very early times. Star-forming galaxies can be surveyed, by the radiation of ionized gas or supernova remnants, in a way independent of the emerging starlight.

Looking ahead from these kinds of results, several groups around the world had begun independent studies of substantially more powerful radio telescopes than any now existing, with work on the early Universe as an important common goal. These have melded into a worldwide project, now known as the Square Kilometer Array (SKA) since one goal is a collecting area of this order (almost fifteen times larger than the Arecibo paraboloid, and several times greater than the sum of collecting areas of all large radio telescopes now in operation). Participating communities include those which have already constructed and operated radio arrays—the USA, UK, Netherlands, Australia, India, and Canada—plus the growing Chinese astronomical community. This project is still in a somewhat informal phase, with no strict timetable for funding or construction. An international working group is in place, and a memorandum of agreement has been signed among the institutions involved. The form of the array is still under study, with a remarkable variety of proposals for the individual array elements. Cost is a major driving factor in the design, leading to interesting departures from traditional practice. Some of the ideas being considered include:

- A set of Arecibo-style fixed dishes, with moving receivers to track objects as the Earth rotates. This approach is under study by the Beijing Astronomical Observatory, which has named the effort the KARST Project, in honor of the appropriate kind of terrain (which is found in abundance in the province of Guizhou). Their proposed technology-demonstration dish would itself become the most sensitive radio telescope in existence. The ability to actively control the surface of the reflector, now routinely available for radio as well as optical telescopes, means that new designs could improve over the performance of the Arecibo spherical reflector.

- A set of flat phased-array antennae, which substitute electronic wizardry for mechanical complexity. Tracking and pointing then become software operations, an option which allows the array to observe multiple regions of the sky at once by changing the reference phases used to combine signals from various elements. Work in this area is being done by the Netherlands Foundation for Radio Astronomy and the Australia Telescope National Facility. An additional advantage to phased arrays is the ability to use recent, sophisticated software approaches to rejecting radio-frequency interference. Prototype array elements and receivers have been under development for several years. More than a million such elements would be called for in a complete SKA.
- Multiple steerable "traditional" paraboloidal dishes at each station. The option is being studied by the National Centre for Radio Astronomy in India and by the SETI Institute in California, with rather different emphases. The Indian group is building on their experience with the 45-meter dishes of the Giant Metre-Wavelength Radio Telescope (GMRT) near Pune. The SETI Institute, in contrast, is constructing an array of very small mass-produced dishes, the Allen Array, constructed from 350 6-metre commercial dishes built for satellite communication. As with ALMA, optimizing cost and performance involves tradeoffs between the size of single array elements, which makes the field accessible at once, and their number, which drives both the sensitivity of the instrument and its cost.
- Single large dishes, now fixed as Arecibo is, but made from individually flat reflecting panels, and with a focal length of half a kilometer, are under study by the Dominion Radio Astronomy Observatory in Canada. The receiver would be slung on an aerostat, a tethered balloon, allowing a telescope whose length goes beyond the usual architectural bounds. Pointing and tracking would be facilitated by precise control of the aerostat's location and attitude. This concept, the Large Adaptive Reflector, is in testing, with flights of a 1/3-scale aerostat already underway.

Whichever of these approaches proves most effective, the potential of a square-kilometer radio array is staggering. Neutral hydrogen could be detected and mapped to very large redshifts, and molecular gas traced to higher redshifts than are available to ALMA, as the redshift carries these transitions to longer wavelengths (and higher-level transitions are also available with longer emitted wavelengths, to instruments which are sensitive enough). Star formation can be traced by the thermal emission from ionized gas, and by the synchrotron radiation produced as relativistic electrons spiral through galaxies' magnetic fields. Such particles can be accelerated in supernova remnants, and account for most of the radio emission of star-forming galaxies. The SKA would see ordinary spiral galaxies (the example used in the project documents is M101) to redshifts of nearly $z = 2$. The weakest radio sources that can now be detected with the Very Large Array are often interacting galaxies at redshifts of several tenths; the SKA would give a large harvest of star-forming galaxies to redshifts beyond $z = 4$, tracing their star formation in a way completely immune to dust absorption.

This short lookforward illustrates how the study of galaxy formation, and growth, now embraces every region of the electromagnetic spectrum that we can detect. Beyond the light from stars, we can trace the signatures of star death, and the chemical enrichment left behind by these events. We are starting to see how the forms of galaxies evolve, and which galaxies have grown piecemeal or in rapid early events. We are poised to take more long steps along the road to galaxy formation.

Index

Printing: Mercedes-Druck, Berlin
Binding: Stein+Lehmann, Berlin